Optical Beam Control

Optical Beam Control

Imaging Satellites and Laser Systems

Brij N. Agrawal, Ph.D.

Naval Postgraduate School
Monterey, CA

Jae Jun Kim, Ph.D.

Naval Postgraduate School
Monterey, CA

Sachin Agrawal

WILEY

Published by John Wiley & Sons, Inc., Hoboken, New Jersey.

Library of Congress Cataloging-in-Publication Data:

Names: Agrawal, Brij N. author
Title: Optical beam control : imaging satellites and laser systems / Brij N Agrawal, Jae Jun Kim, Sachin Agrawal.
Description: Hoboken, New Jersey : Wiley, [2026] | Includes bibliographical references and index.
Identifiers: LCCN 2025021119 | ISBN 9781119830245 hardback | ISBN 9781119830405 adobe pdf | ISBN 9781119830412 epub | ISBN 9781119830429 ebook
Subjects: LCSH: Optics http://id.loc.gov/authorities/subjects/sh85095181 | Control theory http://id.loc.gov/authorities/subjects/sh85031658 | Artificial satellites in remote sensing
Classification: LCC QC355.3 .A47 2026 | DDC 621.36—dc23/eng/20250525
LC record available at https://lccn.loc.gov/2025021119

Cover Design: Wiley
Cover Image: © Brij N. Agrawal

To my wife, Shail, for her support during my professional adventures.

—Brij N. Agrawal

To my wife Sook, your love and strength have been my constant light

—Jae Jun Kim

To my parents, Brij and Shail Agrawal, whose belief in education and service to others helped shape who I am.

—Sachin Agrawal

Contents

Preface

The function of optical beam control is to meet the requirements of optical beam for pointing, pointing stability (jitter), quality of optical beam, slew maneuver, object sensing, and tracking. It is a multidisciplinary field consisting of optics, control theory, structures, thermal analyses, vibrations, atmospheric turbulence, and lasers. The wavelengths of primary interest for imaging satellites and laser systems are in the visible range and near-infrared range. Because of small wavelengths, the performance requirements are challenging. Typical telescope beam widths range from 3 to 5 microradians; to avoid most pointing loss, our residual jitter must remain below 30% of that width (that is, roughly 0.9–1.5 microradians), while absolute pointing errors must stay under one-tenth of the beam width (around 0.3–0.5 microradians). At the same time, for diffraction limited performance the primary mirror surface root-mean-square error should be less than one-thirtieth of a wavelength, which corresponds to about 30 nanometers in the visible. Achieving these numbers relies on cutting-edge optomechanical systems, advanced control laws, and a deep understanding of how optics, structures, thermal effects, vibrations, atmospheric turbulence, and lasers interact.

Because optical beam control sits at the intersection of so many disciplines, the designers of these systems should be knowledgeable not only of their own disciplines, but also of the basic principles of all the disciplines involved. In our own work at the Spacecraft Research and Design Center at the Naval Postgraduate School, our journey of understanding basic principles of optical beam control began decades ago. Initially, with our expertise in structural dynamics and control, our research focused on attitude control of flexible spacecraft, reducing settling time for slew maneuver of imaging satellites, and vibration control and isolation for optical payloads. Our focus has been on analytical simulation and validations of analytical results by experiments, resulting in the development of several testbeds. The next step was active optical beam jitter control using fast steering mirrors. For this research, we collaborated with optical scientist Dr. Segio Restaino, Dr. Ty Martinez, and Jeff Baker on developing testbeds. This work led us naturally into adaptive optics, correcting not only beam orientation but also beam wavefront error by using a deformable mirror. While participating in design reviews at Lockheed Martin of Segmented Mirror Telescope (SMT), we recognized that imaging satellite requirements are similar to those of high-energy laser beam directors, so we broadened our research scope to include laser beam control testbeds alongside our segmented mirror facility. Next, securing 3 years of funding as an Adaptive Optics Center of Excellence allowed us to assemble a truly multidisciplinary team of faculty, post-docs, and dozens of master's and doctoral students. Together, we brought our segmented mirror testbed and our laser beam control testbed into full operation. We also had workshops on optical beam control where pioneer researchers in the field, such as Dr. David Fried and Dr. Robert Fugate, participated and gave short courses.

Throughout that time, we searched for a single text that wove together basic principles of optics, control theory, structural analysis, thermal modeling, atmospheric turbulence, and laser physics of beam control into a unified whole. Individual monographs abounded, but most were narrowly theoretical or confined to a single specialty. Our students—and indeed we ourselves—learned best by collaborating with experts, by building and testing hardware, and by iterating designs in the lab. When Wiley approached us to write this book, the opportunity finally presented itself. We proposed an outline that spans the full spectrum of principles and practice, and over the past 4 years—balancing full-time teaching and research—we have distilled decades of experience into these chapters.

Chapter 1 covers the introduction of optical beam characteristics, space telescopes, ground telescopes, high-power laser systems, low-power laser systems, and optical beam control challenges.

Chapter 2 provides an introduction to optics, covering light sources, the properties of light, optical components, and optical systems. It is written from the perspective of a nonspecialist in optics, drawing on practical experience gained through work on optical beam control problems, and is intended for readers with a limited background in the field.

Chapter 3 introduces the fundamental concepts of feedback control as they relate to optical beam systems, where precision, stability, and responsiveness are important. We explore both classical and modern control techniques, including Laplace transforms, frequency-domain tools, state-space methods, and optimal control. These methods form a critical part of the engineering toolkit for understanding and shaping the dynamic behavior of high-performance optical systems.

Chapter 4 examines the sources of aberrations in optical beam control, including vibration and jitter, aberrations introduced by optical components and systems, and atmospheric turbulence. It also presents commonly used metrics for quantifying optical aberrations.

Chapter 5 covers passive vibration and jitter control, viscous devices, tuned mass dampers, active vibration and jitter control, actuators, jitter sensor, control algorithms, fast steering mirrors, reference laser beam, slew maneuvers, and active vibration isolation.

Chapter 6 provides an overview of adaptive optics, covering key components such as wavefront sensors and deformable mirrors, various system configurations, and control methods used in adaptive optics.

Chapter 7 covers telescope design, telescope performance, optical components, focal plane, operation concepts, image aberration, space telescopes, segmented mirror telescope, analytical model, and telescope design example.

Chapter 8 introduces the fundamentals of lasers, including their key characteristics, beam profiles, propagation behavior, and associated aberrations. It also presents laser beam control components not discussed in earlier chapters and explores applications of laser beam control in areas such as free-space laser communication and high-power laser systems.

Acknowledgments

We wish to express our profound gratitude to all those who have contributed to our research endeavors in optical beam control, laboratory development, and resource acquisition. The progress achieved in this work would not have been possible without the support, guidance, and expertise of many distinguished individuals.

We are especially indebted to those who shared their knowledge and played a pivotal role in the development of advanced laboratory capabilities: Dr. Sergio Restaino, Dr. Ty Martinez, Mr. Jeff Baker, the late Dr. David Fried, Dr. Robert Fugate, Dr. Jim Riker, Dr. Alan Duncan, Dr. Don Gavel, Dr. Claire Max, Dr. David Redding, Dr. Steven Griffin, Dr. David Schaechter, Dr. Glenn Tyler, Dr. Allen Bronowicki, Dr. Chris Wilcox, Dr. Lewis DeSandre, Mr. John Blackburn, Mr. Terry Brennan, Dr. Paul Merritt, Dr. Bryn Davis, and Dr. Robert Laskin.

We gratefully acknowledge the instrumental roles of Lt. Col. Andrew Adams, Dr. Phillip Howerton, Mr. Thomas Acquilina, and Dr. Leonard Ferrari in facilitating the establishment of the Segmented Mirror Telescope (SMT) at the Naval Postgraduate School. We further recognize Dr. Steven Griffin, Mr. Steve Baugh, and Mr. David DeYoung for their critical contributions to the full operationalization of the HBCRT testbed.

The realization of the Adaptive Optics Center of Excellence at the Naval Postgraduate School was made possible through the vision and support of General Ellen Pawlikowski, General Bruce Carlson, RADM Nevin P. Carr, Jr., Vice Admiral Dan Oliver, Dr. Jim Scrofani, Dr. Karl A. van Bibber, and Dr. Knox Millsaps. We are also thankful to Mr. Quentin Saulter and Ms. Sarwat Chappel for their sustained support of our research initiatives.

We extend our appreciation to the talented graduate students and research fellows whose dedication and efforts have significantly advanced the mission of the Spacecraft Research Center. Notably, Dr. Adam Yingling and Dr. Matt Allen made substantial contributions to the SMT research. Dr. Melissa Corley and Dr. Joe Watkins also played key roles in advancing our work. Our research fellows, Dr. John Bagnasco and Dr. Bautista Fernandez, were instrumental in establishing operational testbeds for experimental research.

We also acknowledge the professional collaboration of Ms. Aileen Storry, Executive Publisher, Mr. Akhil Ajikumar, Managing Editor, and Poornimaa Balakannan, Content Refinement Specialist at Wiley, whose coordination and patience were invaluable in the preparation of this book.

We express sincere appreciation to the leadership of the Naval Postgraduate School for providing the infrastructure, resources, and institutional support that enabled the conduct of this research and for permission to write this book.

Lastly, we offer our heartfelt thanks to our families, whose encouragement, patience, and unwavering support sustained us throughout the undertaking of this challenging and rewarding endeavor.

1

Introduction

1.1 Optical Beam

The function of optical beam control is to meet the requirements of optical beams for pointing, pointing stability (jitter), quality of optical beams, slew maneuvers, object sensing, and tracking. It is a multidisciplinary field consisting of optics, control theory, structures, thermal analyses, vibrations, atmospheric turbulence, and lasers. This is a broad field, and this book will cover the fundamentals of each area with its applications to imaging satellites and laser systems.

Figure 1.1 shows the range of electromagnetic wave frequencies and wavelengths.

The wavelengths of our interest are primarily in the visible range, 400–700 nm, and the near infrared range, 700–1100 nm. Because of small wavelengths, the requirements are challenging to meet, for example, typical telescope beam widths are 3–5 μrad. To mitigate most of the pointing loss, beam jitter should be less than 30% beam width, resulting in 0.9–1.5 μrad, pointing accuracy should be 0.1 times beam width, resulting in 0.3–0.5 μrad, and optical beam wave quality for diffraction limited performance, primary mirror surface RMS error, should be less than wavelength/30, 30 nm for the visible range. These requirements are very challenging and require

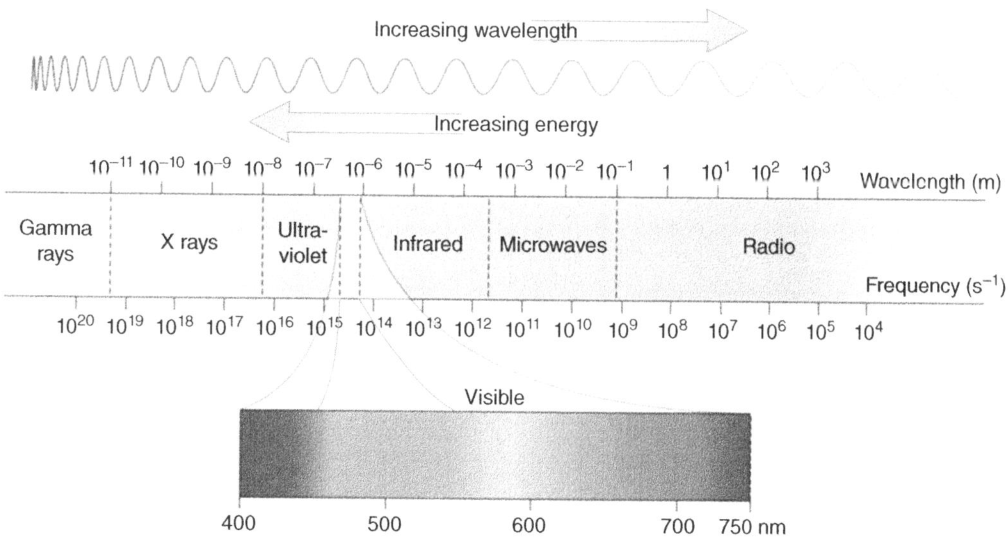

Figure 1.1 Electromagnetic wave frequencies and wavelengths.

Optical Beam Control: Imaging Satellites and Laser Systems, First Edition.
Brij N. Agrawal, Jae Jun Kim and Sachin Agrawal.

state-of-the-art technologies to meet these performance requirements. For electromagnetic waves with larger wavelengths, such as radio waves with several orders of magnitude larger wavelengths, pointing, pointing stability, and beam quality requirements will be much easier to meet in comparison with optical beams.

Laser is an acronym for light amplification by stimulated emission of radiation. Ordinary visible light, say from a household light bulb or a flashlight, comprises multiple wavelengths, or colors, and is incoherent, meaning the crests and troughs of the light waves are moving at different wavelengths and in different directions. In a laser beam, the light waves are "coherent," meaning the beam of photons is moving in the same direction at the same wavelength.

Laser light can travel hundreds of meters without being scattered. Since ordinary light is diffused, it cannot focus on a sharp point. Laser light can focus on a point with high intensity thanks to its directional structure. The intensity of ordinary light decreases rapidly with distance. For this reason, you can look at the ordinary light source without harming your eyes. In contrast, laser light is emitted in a narrow beam. Since the energy is concentrated in a very narrow area, looking at the laser light with the naked eye can damage the eye.

1.2 Telescopes

Most telescopes used by astronomers are on Earth. We call these ground-based telescopes. It is much easier and cheaper to build a telescope on Earth than in space. It is also much easier to fix if things go wrong. However, there are downsides as well. A telescope on the ground must look through Earth's atmosphere to see into space. This is a problem because the atmosphere can blur our images.

The air, as shown in Figure 1.2, also blocks out light from parts of the electromagnetic spectrum like X-rays, gamma rays, infrared, and long radio waves. This means even if we have the right kind of telescope, it cannot see this type of light from Earth. The air gets in the way. That is why some telescopes are in space.

We call the parts of the light spectrum that can get through the air atmospheric windows. These are the parts of the electromagnetic spectrum where the opacity (how much light is blocked) is

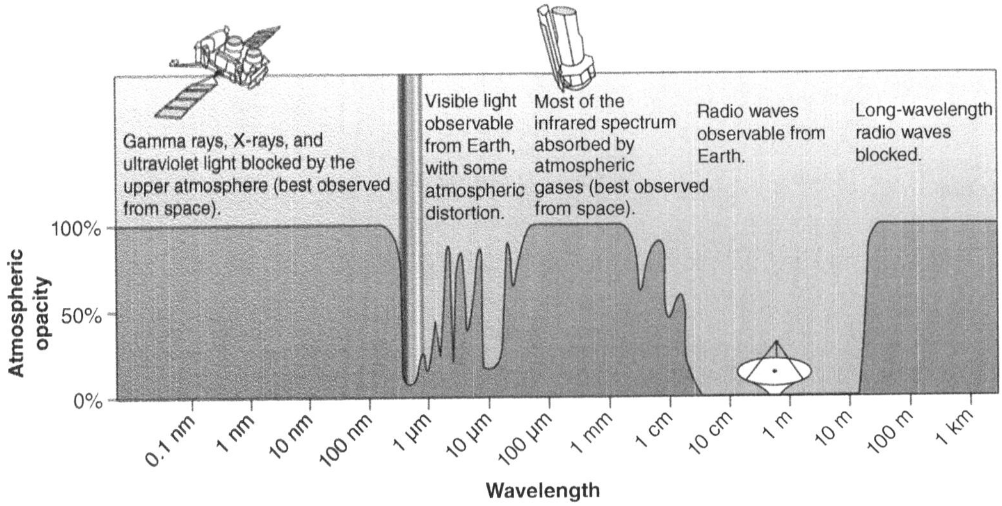

Figure 1.2 Atmospheric opacity of electromagnetic waves. *Source*: [1]/wikimedia Commons/public Domain.

close to 0%. If the opacity is 100%, then no light with that wavelength can get through the air to reach the ground.

Building and using a space telescope is a huge challenge. It also costs a lot of money. It has only been possible since the 1980s. The first space telescope was the Hubble Space Telescope (HST). It began observing in 1990. It has taken over one million images so far.

Since 1990, there have been lots of other space telescopes. Some collect light which Earth's atmosphere blocks out. Like Chandra, which observes in X-rays, and Fermi, which looks at gamma rays. Others can see microwaves or infrared. This has given us a new view of our Universe. The largest, the James Webb Space Telescope (JWST) was launched on December 24, 2021.

1.2.1 Space Telescope

Space telescopes have made many fantastic discoveries. Kepler found thousands of exoplanets. Spitzer was the first to image light from an exoplanet. Gaia observed a supernova outside the Milky Way. In the future, Laser Interferometer Space Antenna will try to detect gravitational waves in space.

Imaging satellites have been used for many civilian, scientific, and military applications, looking toward Earth and space. Here, we will briefly discuss the HST and the JWST. Both were developed by NASA for scientific research of planets.

1.2.1.1 Hubble Space Telescope

Figure 1.3 shows a picture and Figure 1.4 configuration of HST.

Since its launch in April 1990, NASA's HST [2] has continued this historic quest, providing scientific data and photographs of unprecedented resolution from which many new and exciting discoveries have been made.

This unique observatory operates around the clock above Earth's atmosphere to gather information for teams of scientists studying virtually all the constituents of our universe, including planets, stars, star-forming regions of the Milky Way galaxy, distant galaxies and quasars, and the tenuous hydrogen gas lying between the galaxies.

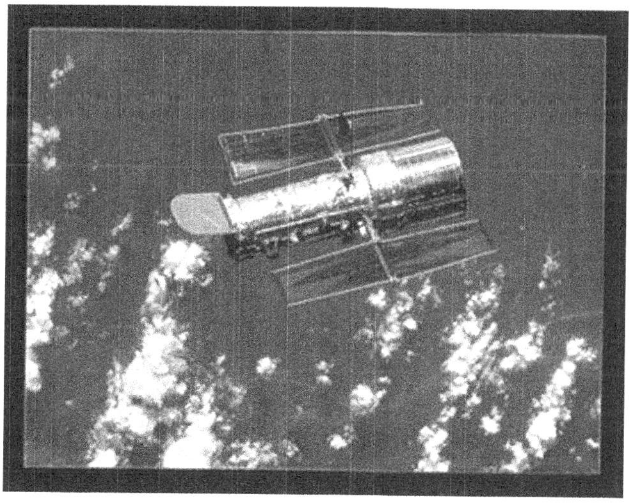

Figure 1.3 Hubble Space Telescope.

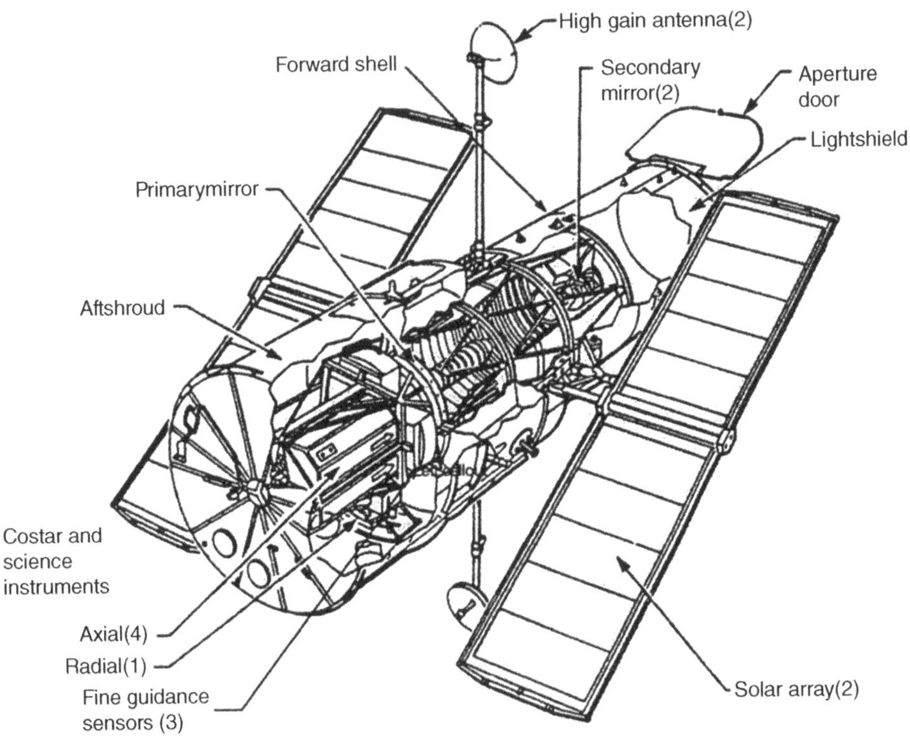

Figure 1.4 Hubble Space Telescope configuration.

The major elements of HST are the Optical Telescope Assembly, science instruments, Support System Module, and solar arrays. HST sensor operates in ultraviolet light, visible light, and infrared light with wavelengths 100 nm–1.8 μ. The main parameters are:

Launch date	April 1990
Launch vehicle	Space Shuttle Discovery
Mission duration	15 years
Total payload mass	11,110 kg
Optical system	Ritchey–Chretien design Cassegrain telescope Primary mirror 94.5 in. (2.4 m) in diameter Secondary mirror 12.2 in. (0.3 m) in diameter
Optical resolution	0.043 arcseconds (0.00001°)
Orbit	320-mile (593 km) altitude, inclined 28.5° from the equator
Orbit time	Orbit time– 97 minutes per orbit
Pointing accuracy	0.012 arcseconds (0.000003°), three-axis stabilized, zero momentum biased control system using reaction wheels.
Pointing stability	0.007 arcseconds
Cost	US$2 billion

1.2.1.1.1 Mission

The HST is the first observatory designed for extensive maintenance and refurbishment in orbit. The Discovery cargo bay is equipped with several devices to help the astronauts: the Flight Support System to berth and rotate the telescope; large, specially designed equipment containers to house the Orbital Replacement Units; and a remote manipulator arm from which astronauts can work and be maneuvered as needed.

After Hubble's deployment in 1990, scientists realized that the telescope's primary mirror had a flaw called spherical aberration (thickness of a human hair). The outer edge of the mirror was ground too flat by a depth of 2.2 μ. This aberration resulted in images that were fuzzy because some of the light from the objects being studied was being scattered. COSTAR (the Corrective Optics Space Telescope Axial Replacement) was developed as an effective means of countering the effects of the flawed shape of the mirror. COSTAR was a telephone booth-sized instrument that placed five pairs of corrective mirrors, some as small as a nickel coin, in front of the Faint Object Camera, the Faint Object Spectrograph, and the Goddard High Resolution Spectrograph. Servicing Mission 1, launched in December 1993, was the first opportunity to conduct planned maintenance on the telescope. In addition, new instruments were installed, and the optics of the flaw in Hubble's primary mirror were corrected.

On November 13, 1999, the HST was placed into safe mode after the failure of a fourth gyroscope. In safe mode, Hubble could not observe targets, but its safety was preserved. This protective mode allows ground control of the telescope, but with only two gyros working, Hubble cannot be aimed with the precision necessary for scientific observations of the sky. Controllers closed the aperture door to protect the optics and aligned the spacecraft to ensure that Hubble's solar panels would receive adequate power from the Sun. In servicing mission 3A, December 19–27, 1999, astronauts replaced all six gyroscopes.

In servicing mission 3B, March 1–12, 2022, astronauts replaced flexible, 8-year-old, solar array panels with smaller, rigid ones that produced 30% more power and improved pointing accuracy to eliminate control and structures interaction. The pointing stability requirements of 0.07 arcseconds were achieved by changing the flexible solar array to a rigid solar array to avoid structural and control interactions and changing the control design.

The earlier discussion shows how tight the requirements are to meet the mission. The surface error in the primary mirror of 2.2 μ (roughly equal to one-fiftieth the thickness of a human hair) resulted in images being fuzzy. The failure of one gyroscope failed the pointing system. To meet pointing stability requirements of 0.07 arcseconds, the solar array must be changed from flexible to rigid to avoid structural and control interactions and a new control design. These problems were resolved by satellite servicing missions by astronauts. In the absence of servicing missions, the HST, US$2B program, would have been a failure.

1.2.1.2 James Webb Space Telescope

JWST [3, 4] is the scientific successor to Hubble; its science goals were motivated by results from Hubble. JWST looks at the universe in the infrared, while Hubble studies it primarily at optical and ultraviolet wavelengths (though it has some infrared capability). Hubble's science pushed us to look to longer wavelengths to "go beyond" what Hubble has already done. In particular, more distant objects are more highly redshifted, and their light is pushed from the UV and optical into the near infrared. Thus, observations of these distant objects (like the first galaxies formed in the universe, for example) require an infrared telescope. Stars and planets that are just forming lie hidden behind cocoons of dust that absorb visible light. However, infrared light emitted by these regions can penetrate this dusty shroud and reveal what is inside. The difference between

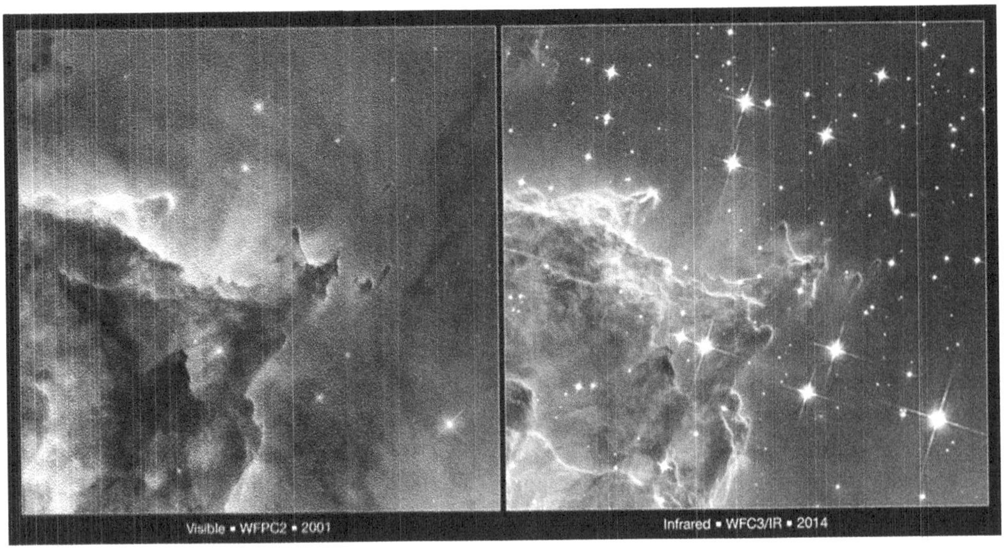

Figure 1.5 Hubble's visible and infrared views of the Monkey Head Nebula. *Source*: [2]/NASA/Public Domain.

visible and infrared images is clear from Figure 1.5 of two images of Monkey Head Nebula, a star-forming region, taken by HST. The left is visible, and the right is infrared.

A jet of material from a newly forming star is visible in one of the pillars, just above and left of center in the right-hand image. Several galaxies are seen in the infrared view, much more distinct than the columns of dust and gas.

The main parameters of JWST, as shown in Figure 1.6, are as follows:

Launch date:	December 24, 21
Launch vehicle:	- Ariane5
Mission duration:	5–10 years
Total payload mass:	6200 kg
Diameter of primary mirror:	-6.5 m
Optical resolution:	- ~3 milliarcseconds (0.0000008°)
Wavelength coverage:	-0.6 -28 μ
Size of sun shield:	-22×10 m^2
Orbit:	−1.5 million km from Earth at L2 points
Cost:	US$10 billion

JWST's primary mirror is a 6.5 m (21 ft)-diameter gold-coated beryllium reflector with a collecting area of 25.4 m^2 (273 ft^2). If it were built as a single large mirror, this would have been too large for existing launch vehicles. The mirror is therefore composed of eighteen hexagonal segments, which will unfold after the telescope is launched. Image plane wavefront sensing through phase retrieval is used to position the mirror segments in the correct location using very precise micro-motors. After this initial configuration, they will only need occasional updates every few days to retain optimal focus. The Webb telescope uses 126 small motors to occasionally adjust the optics, as there are only a few environmental disturbances for a telescope in space. JWST's optical

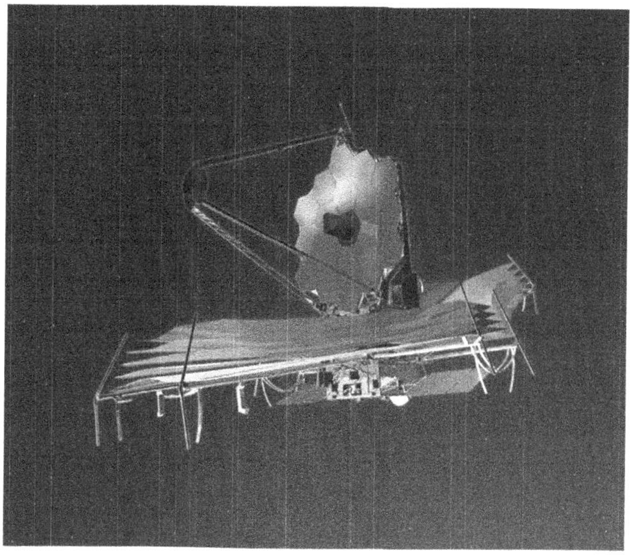

Figure 1.6 James Webb Telescope. *Source*: James Webb/Wikimedia Commons/Public Domain.

design is a three-mirror anastigmat, which makes use of curved secondary and tertiary mirrors to deliver images that are free from optical aberrations over a wide field. In addition, there is a fine steering mirror that can adjust its position many times per second to provide image stabilization.

A three-mirror anastigmat is an anastigmat telescope built with three curved mirrors, enabling it to minimize all three main optical aberrations – spherical aberration, coma, and astigmatism. This is primarily used to enable wide fields of view, much larger than possible with telescopes with just one or two curved surfaces. A telescope with only one curved mirror, such as a Newtonian telescope, will always have aberrations. If the mirror is spherical, it will suffer from spherical aberration. If the mirror is made parabolic to correct the spherical aberration, then it must necessarily suffer from coma and off-axis astigmatism. With two curved mirrors, such as the Ritchey–Chrétien telescope, coma can be minimized as well. This allows a larger useful field of view, and the remaining astigmatism is symmetrical around the distorted objects, allowing astrometry across the wide field of view. However, astigmatism can be reduced by including a third curved optical element. When this element is a mirror, the result is a three-mirror anastigmat. In practice, the design may also include any number of flat-fold mirrors, used to bend the optical path into more convenient configurations.

To make observations in the infrared spectrum, JWST must be kept under 50 K (−223.2 °C; −369.7 °F); otherwise, infrared radiation from the telescope itself would overwhelm its instruments. It therefore uses a large sunshield to block light and heat from the Sun, Earth, and Moon, and its position near the Sun–Earth L2 always keeps all three bodies on the same side of the spacecraft. Its halo orbit around the L2 point avoids the shadow of the Earth and Moon, maintaining a constant environment for the sunshield and solar arrays. The shielding maintains a stable temperature for the structures on the dark side, which is critical to maintaining precise alignment of the primary mirror segments in space. The temperature of the Mid-Infrared Instrument, operating in the range of 5–27 μm, must not exceed 6 K. It is maintained by a cryocooler.

Webb, as shown in Figure 1.7, will orbit the sun 1.5 million km (1 million miles) away from the Earth at what is called the second Lagrange point or L2. (Note that these graphics are not to scale.)

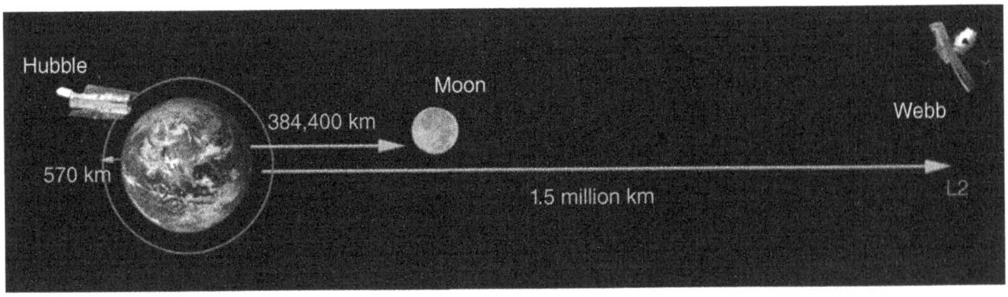

Figure 1.7 Orbit of the JWST. *Source*: [2]/NASA/Public Domain.

Figure 1.8 Keck telescope. *Source:* NASA/JPL/NASA/Public Domain.

Because Hubble is in Earth orbit, it was able to be launched into space by the space shuttle. Webb was launched on an Ariane 5 rocket, and because it would not be in Earth orbit, and it is not designed to be serviced by the space shuttle.

1.2.2 Ground Telescope

1.2.2.1 Keck Telescope

The most powerful telescope on the summit of Mauna Kea is the twin Keck observatory, as shown in Figure 1.8, which has two 10 m diameter mirrors. At the heart of each Keck Telescope is a revolutionary primary mirror made up of 36 hexagonal segments that effectively work as a single piece of reflective glass. By combining advanced optical and infrared detectors with sophisticated electronics that can combine collected light from both telescopes, the Keck observatory remains among the leading astronomical facilities in the world.

Some facts about the telescope: observatory location: Mauna Kea, Hawaii; height above sea level: 4145 m (13 790 ft); moving mass: 270 metric tons (each); mirror diameter: 10.0 m (each); distinctive feature: each mirror has 36 hexagonal segments.

Each mirror segment is kept stable by a system of active optics that adjusts its position, relative to adjacent segments, to an accuracy of 4 nm. This twice-per-second adjustment corrects any distortions due to gravity. Both telescopes are also equipped with adaptive optics (AO), which compensates for the blurring caused by turbulence in the atmosphere. This, combined with the

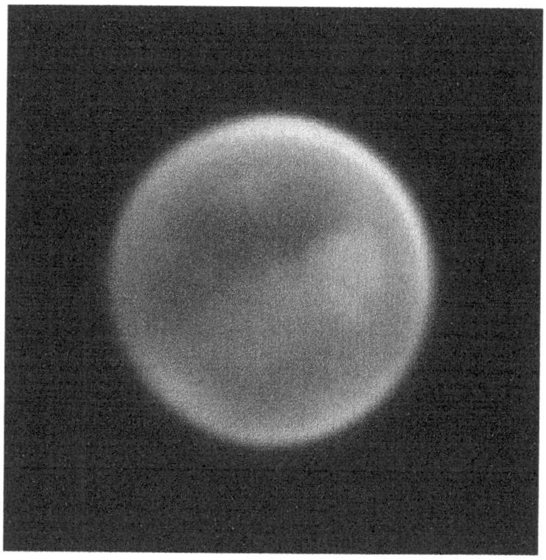

Figure 1.9 Saturn's largest moon – Titan. *Source*: Brij N. Agrawal.

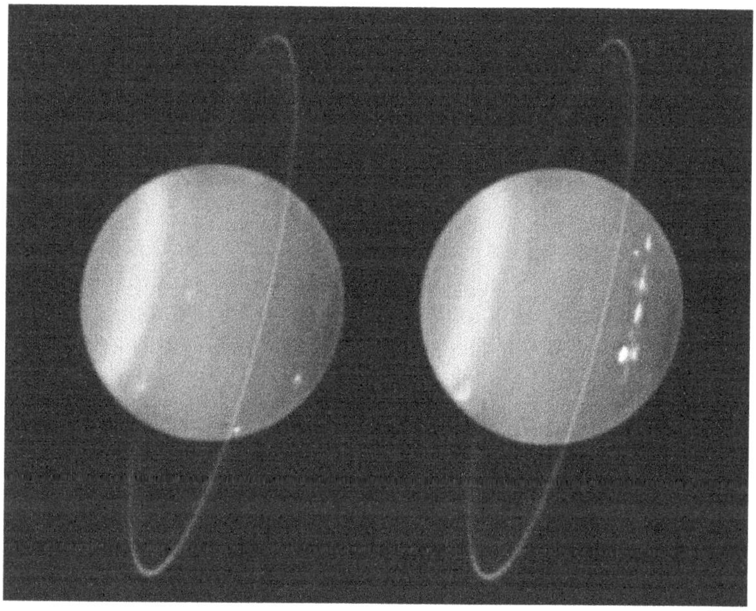

Figure 1.10 Images showing the two sides of Uranus. *Source*: Lawrence Sromovsky/NASA/Public Domain.

light-collecting power of two sizeable mirrors and some pretty sophisticated instrumentation, means that the twin Keck telescopes can produce some incredibly detailed images of distant objects in the Solar System, which are almost on a par with images from space telescopes. The following are two good examples, as shown in Figures 1.9 and 1.10, of what Keck can do.

The key advance that allowed the construction of the Keck telescopes was the use of active optics to operate smaller mirror segments, as shown in Figure 1.11, as a single, contiguous mirror. A mirror of comparable size, cast of a single piece of glass, could not be made rigid enough to hold its shape precisely; it would sag microscopically under its own weight as it was turned into

Figure 1.11 The Keck II telescope shows the segmented primary mirror. *Source*: SiOwl/Wikimedia Commons/CC BY 3.0.

various positions, causing aberrations in the optical path. In the Keck telescopes, each primary mirror is made of 36 hexagonal segments that work together as a unit. Each segment is 1.8 m wide, 7.5 cm thick, and weighs half a ton. The mirrors were made from Zerodur glass-ceramic by the German company Schott AG. On the telescope, each segment is kept stable by a system of active optics, which uses extremely rigid support structures in combination with three actuators under each segment. During observation, the computer-controlled system of sensors and actuators dynamically adjusts each segment's position relative to its neighbors, keeping a surface shape accuracy of four nanometers. As the telescope moves, this twice-per-second adjustment counters the effects of gravity and other environmental and structural effects that can affect mirror shape.

Both Keck observatory telescopes are equipped with laser guide star adaptive optics, which compensate for the blurring due to atmospheric turbulence. The first AO system operational on a large telescope, the equipment has been constantly upgraded to expand its capability. The telescopes are equipped with a suite of cameras and spectrometers that allow observations across much of the visible and near-infrared spectrum.

1.3 Laser Systems

Laser systems can be divided into two groups: low-power systems and high-power systems.

Low-power laser systems are used for commercial and civilian applications, such as laser communications. High-power laser systems are normally used for military applications.

Figure 1.12 ABL system. *Source*: [8]/Verdict Media Limited.

1.3.1 High-power Laser Systems

1.3.1.1 Airborne Laser System

The airborne laser (ABL) system, as shown in Figure 1.12, was high-power laser system [6]. The ABL was designed to detect and destroy theater ballistic missiles in the powered boost phase of flight immediately after the missile launch while the aircraft loiters at an altitude of 40,000 ft.

It was developed for the US Air Force, designated YAL-1A, and was a high-energy laser weapon system for the destruction of tactical theatre ballistic missiles, which is carried on a modified Boeing 747-400F freighter aircraft. The ABL was developed by the Air Force Research Laboratory and ABL Team, comprising Boeing, TRW (now Northrop Grumman Technologies), and Lockheed Martin. The ABL aircraft carries the chemical oxygen iodine laser (COIL) laser, which generates the killer laser beam, an infrared surveillance and high-speed target acquisition system, and a high-precision laser target tracking beam control system. The missile plume is detected by the ABL aircraft's infrared detection system at ranges up to several hundred kilometers.

The laser weapon uses three laser beam systems: the powerful killing laser beam or primary beam, a set of illuminating laser beams, and a beacon laser. The primary laser beam is generated by a megawatt COIL located at the rear of the fuselage, which lases at a 1.315 μ wavelength. The high-power laser beam travels toward the front of the aircraft through a pipe. The pipe passes through a Station 1000 bulkhead/airlock, which separates the rear fuselage from the forward cabins. The high-power beam passes through the fine beam control system mounted on a vibration-isolated optical bench. Beam pointing is achieved with fast, lightweight steering mirrors, which are tilted to follow the target missile.

A low-power, multiple beam, track-illuminating laser (TILL), being developed by Raytheon Electronic Systems, is used to determine the target's range and provides initial information on the atmosphere through which the beam is being transmitted. The TILL illuminates the target and provides aiming data for the primary beam. The beacon illuminating laser (BILL) was developed by Northrop Grumman Space Technology. The kilowatt-class BILL reflects light from the target to provide data on the rapidly changing characteristics of the atmosphere along the path of the laser beam. This data is used to control a set of deformable mirrors in the beam control system. The mirrors introduce tailored distortions into the COIL laser beam to compensate for atmospheric distortions and allow the COIL laser beam to fall on the target with full power.

A suite of infrared, wide-field telescopes installed along the length of the aircraft's fuselage detects the missile plume at ranges up to several hundred kilometers. The pointing and tracking system tracks the missile and provides launch and predicted impact locations. The turret at the nose of the aircraft swivels toward the target, and a 1.5 m telescope mirror system inside the nose focuses the laser beam onto the missile. The laser beam locks onto the missile, which is destroyed near its launch area within seconds of lock-on. Where the missile carries liquid fuel, the laser can heat a spot on the missile's fuel tank, causing an increase in internal pressure, resulting in catastrophic failure. Alternatively, the missile is heated in an arc around its circumference and crumples under atmospheric drag force or its own g-force.

On February 11, 2010, in a test at Point Mugu Naval Air Warfare Center-Weapons Division Sea Range off the central California coast, the system successfully destroyed a liquid fuel-boosting ballistic missile. Less than an hour after that first missile had been destroyed, a second missile – solid-fuel design – had, as announced by the Missile Defense Agency, been "successfully engaged," but not destroyed, and all test criteria had been met. This test was the first time that a directed-energy system destroyed a ballistic missile in any phase of flight. It was later reported that the first February 11 engagement required 50% less dwell time than expected to destroy the missile; the second engagement on the solid-fuel missile, less than an hour later, had to be cut short before it could be destroyed because of a "beam misalignment" problem. In December 2011, it was reported that the project was to ended after 16 years of development and at a cost of over US$5 billion.

Some of the performance requirements for beam control are as follows.

Parameter	Requirement
Outgoing HEL beam	1.5 m
Field of regard, azimuth	±120°
Field of regard, elevation	+45/−30°
Timeline reqts, acq to coarse track delay (plume acquisition and track)	x23 s
Timeline reqt, fine track and AO loop closure (aimpoint selection)	xx24 s
Illuminator turn-on time	xx22 s
Turret slew times	xx31, xx32,......, xx37
Inertial reference frame error	<2.5 mrad RMS (after 1 h)
TILL illumination	5 m diameter at target
TILL beamlet pointing error	<2 μrad, one sigma, one axis
HEL aimpoint bias	<200 nrad, one sigma, one axis

1.3.2 Low Power Laser Systems

Laser communications is an important application of low-power laser systems. Conventional communications systems on the ground, air, and in space use radio frequency (RF) spectrum for communications. RF spectrum has frequencies from 3 Hz to 3000 GHz (3 THz). Geosynchronous communications satellites have been heavily used for national and international communications since the 1960s. Recently, these satellites have a capacity of greater than 1 Tbps per satellite. The most used frequency bands are C band 4/6 GHz, Ku band 11/14 GHz, and Ka band 20/30 GHz. Recently, several constellations of low Earth orbit satellites have been planned for internet applications and covering areas not covered by cables. RF systems are, however, hindered by the available spectrum bandwidth limitations and regulations.

Laser communication with currently unregulated spectrum and with multi-THZ bandwidth can augment RF systems by providing order of-magnitude capacity enhancements. Laser communications systems for visible light the frequencies are 400 and 700 THz, corresponding to wavelengths of 700 and 400 nm, respectively. For infrared communications, the frequencies are lower, with a typical wavelength of 850 or 1550 nm. In comparison to RF systems, laser communication systems are an order of magnitude lighter, more compact, and have low-power requirements. There are certain advantages and disadvantages to laser communications. The beamwidth for laser communications, due to high frequency/short wavelength is three to four orders of magnitude narrower than RF communications with lower frequency. This results in some advantages, disadvantages, and challenges. The advantages include wide bandwidth and large data rate, secure communications, jam resistance, small antenna and volume, low wight, and power. There are several disadvantages in comparison to RF communications. First are the adverse effects of the atmosphere due to clouds and turbulence, resulting in deep fades and beam breakups. For high link availability requirements, such as internet, laser communications through atmospheric downlink and uplink to satellite may not meet requirements. It can be used for applications where timely delivery may not be required. For laser communications, the ground stations are on high mountains to minimize turbulence effects. To minimize turbulence, AO techniques are used that make ground stations more complex. The second challenge is that due to a very narrow laser beam, beam control for acquisition, pointing, tracking, and jitter control is difficult compared to RF communications. For a typical laser system, telescope beamwidths are 3–5 μrad, to mitigate most of the pointing loss, RMS beam jitter should be less than 30% of the beamwidth, and pointing accuracy should be at least 0.1 times beamwidth. Typical spacecraft vibrations and attitude variations are an order of magnitude higher. Therefore, passive jitter control may be required, and pointing control with very accurate sensors is required.

1.3.2.1 Laser Communications Relay Demonstration

Laser communication has been considered and developed for many applications, such as free-space optical communications from satellite to satellite. Laser communication has been used by NASA for deep space optical communication, as it requires a higher data rate than is possible with RF systems to support future science data retrieval. As an example, NASA has developed the Laser Communications Relay Demonstration (LCRD) [8] to demonstrate laser communications technologies. According to NASA, as science instruments evolve to capture high-definition data like 4K video, missions will need expedited ways to transmit information to Earth. With laser communications, NASA can significantly accelerate the data transfer process and empower more discoveries.

Laser communications will enable 10–100 times more data to be transmitted back to Earth than current RF systems. It would take nine weeks to transmit a complete map of Mars back to Earth with current RF systems. With lasers, it would take about nine days. Additionally, laser communications systems are ideal for missions because they need less volume, weight, and power. Less mass means more room for science instruments, and less power means less of a drain on spacecraft power systems. These are all critically important considerations for NASA when designing and developing mission concepts. The Figure 1.13 shows LCRD communicating data from the space station to Earth.

LCRD's modems translate digital data into laser signals, which are then transmitted via encoded beams of light, invisible to the human eye, by the relay's optical modules. LCRD can both send and receive data, creating a continuous path for flowing mission data to and from space. Together,

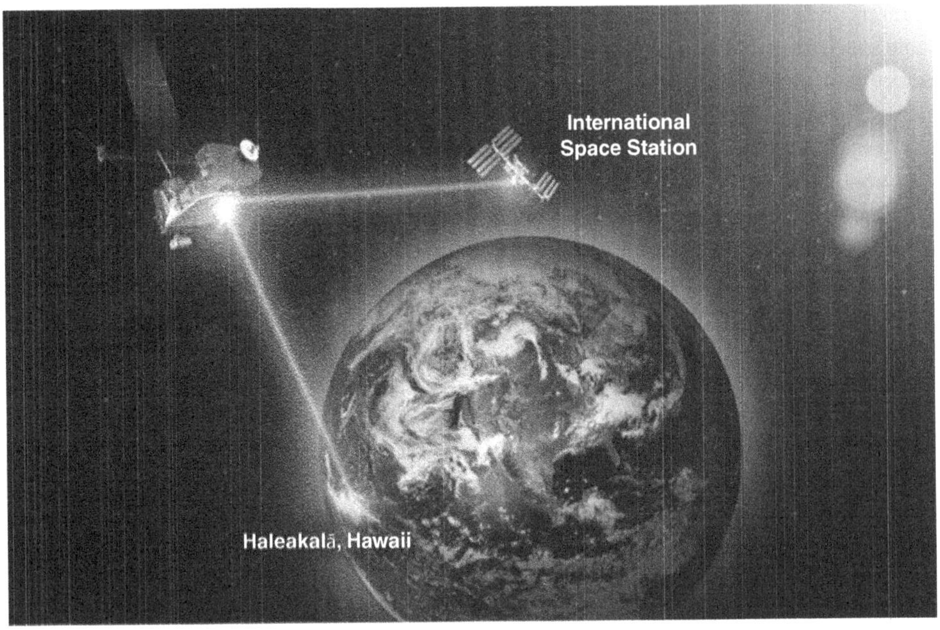

Figure 1.13 LCRD operational configuration. *Source*: NASA/Public Domain.

these capabilities make LCRD NASA's first two-way, end-to-end optical relay. Once LCRD receives information and encodes it, the payload sends the data to ground stations on Earth that are each equipped with telescopes to receive the light and modems to translate the encoded light back into digital data. LCRD uses two communication modulations, differential phase shift keying (DPSK) and pulse-position modulation (PPM).

LCRD's ground stations, as shown in Figure 1.14, are known as optical ground stations (OGS) −1 and −2 are located on Table Mountain in Southern California and on Haleakala Volcano in Maui, Hawaii respectively. While laser communications can provide increased data transfer rates, atmospheric disturbances – such as clouds and turbulence – can interfere with laser signals as they travel through Earth's atmosphere. The locations for OGS-1 and OGS-2 were chosen for their clear weather conditions and remote, high-altitude locations. Most of the weather occurring in those areas takes place below the summit of the mountains, leaving relatively clear skies perfect for laser communications.

The ground stations use AO techniques to correct aberrations in the laser beam due to turbulence.

The important parameters of the LCRD payload are as follows:

Wavelength:	1550 nm
Terminal mass:	69 kg
Optical transmit power:	130 W DC
Telescope diameter:	108 mm
Data rates:	2.880 Gbit/s using DPSK, and 622 Mbit/s using PPM.

Launch: Not yet, from Cape Canaveral Space Force Station in Florida. The LCRD payload is hosted on the US Department of Defense's STPSat-6 spacecraft, launched aboard a United Launch Alliance Atlas V 551 rocket.

Figure 1.14 LCRD's OGS. *Source*: NASA/Public Domain.

1.4 Optical Beam Control Challenges

As discussed earlier, the challenges for optical control design are primarily due to small wavelengths. The interest is primarily in the visible range, 400–700 nm, and the near infrared range 700–1100 nm. Many performance requirements are directly related to its wavelength. For example, for diffraction-limited images, the mirror RMS surface error should be less than $\lambda/30$ where λ is wavelength. For visible light at 600 nm, this will require a surface accuracy of 20 nm. As an example, as discussed earlier, the outer edge of the Hubble Telescope mirror was ground too flat by a depth of 2.2 μ (roughly one-fifth of hair's thickness). This resulted in fuzzy images due to spherical aberration. This problem was corrected by installing a correcting device by the astronauts. Similarly, for a Gaussian beam, the beamwidth is proportional to the wavelength of light (λ) and inversely proportional to the aperture size. As discussed earlier, typical telescope beam widths are 3–5 μrad. To mitigate most of the pointing loss, beam jitter should be less than 30% beam width, resulting in 0.9–1.5 μrad, and pointing accuracy should be 0.1 times the beam width, resulting in 0.3–0.5 μrad. As discussed earlier, for the HST, the pointing accuracy requirement is 0.012 arcseconds (0.000003°) and jitter 0.007 arcseconds. For laser systems, there is an additional requirement for laser beam correction due to atmospheric turbulence.

Addressing these challenges often requires a multidisciplinary approach, involving expertise in optics, control systems, structures, and thermal engineering. As technology advances, new solutions and innovations continue to emerge to tackle these design challenges in optical beam control systems. In the following chapters, the fundamentals of these areas related to optical beam control will be covered. In Chapter 2, we will focus on optics. It will cover light source, propagation of lights, lenses, and mirrors, wave interference, diffraction, polarization, and scattering. Chapter 3 will focus on beam fine control. It will cover classical control, modern control, Kalman filter. Chapter 4 will focus on sources of beam aberrations. It will cover vibration and jitters, optical aberrations, air turbulence, and the measurement of optical aberrations. Chapter 5 will focus on vibration and jitter control. It will cover vibration isolation and active jitter control. Chapter 6 will focus on AO. It will cover wavefront sensors, wavefront reconstruction, deformable mirrors, AO configurations, and AO control systems. Chapter 7 will focus on imaging satellites. It will cover telescope designs, optical train components, image aberration, and space telescopes. Chapter 8 will focus on laser systems. It will cover laser fundamentals, laser beam control hardware, laser aberration, laser performance, performance analysis, laser communications systems, and high-energy laser systems.

References

1 https://commons.wikimedia.org/wiki/File:Atmospheric_electromagnetic_opacity.svg.
2 https://en.wikipedia.org/wiki/Hubble_Space_Telescope.
3 Goddard Space Flight Center. "James Webb Space Telescope Project Mission Requirements Document," JWST-RQMT-000634, October 17, 2007.
4 https://en.wikipedia.org/wiki/James_Webb_Space_Telescope.
5 https://www.jwst.nasa.gov/content/about/comparisonWebbVsHubble.html.
6 https://en.wikipedia.org/wiki/Airborne_Laser.
7 https://www.airforce-technology.com/projects/abl/?cf-view.
8 https://en.wikipedia.org/wiki/Laser_Communications_Relay_Demonstration#.

2

Optics

Optics refers to the study of light focused on ultraviolet, visible, and infrared wavelengths. It also encompasses elements and instruments used to interact with light within those wavelengths. Optics forms the backbone of beam control systems, enabling efficient direction of light from the source to the desired target location. The primary focus of this chapter is to provide basic optics knowledge frequently utilized in beam control systems.

2.1 Light Sources

Light is electromagnetic radiation, which consists of waves of electric and magnetic fields propagating through space and a medium. Electromagnetic processes, caused by the energy loss of an atom or molecule, release a quantum of light called a photon, which carries the emitted electromagnetic radiant energy. Figure 2.1 illustrates a transverse light wave consisting of oscillating electric field (E) and magnetic field (B), which are perpendicular to each other. The amplitude of the wave (A) represents the intensity of the light, and the wavelength (λ) represents the period of oscillation that determines the color. Light waves can originate from various sources and configurations, as presented in this section.

2.1.1 Point Source and Extended Source

A point light source, or simply a point source, refers to an infinitesimally small point object that either emits the light or reflects the light to act as a light source. Ideal point source does not exist

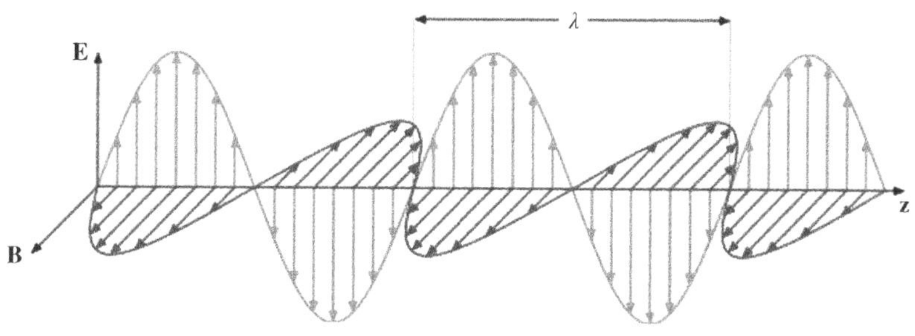

Figure 2.1 Transverse electromagnetic light wave.

Optical Beam Control: Imaging Satellites and Laser Systems, First Edition.
Brij N. Agrawal, Jae Jun Kim and Sachin Agrawal.

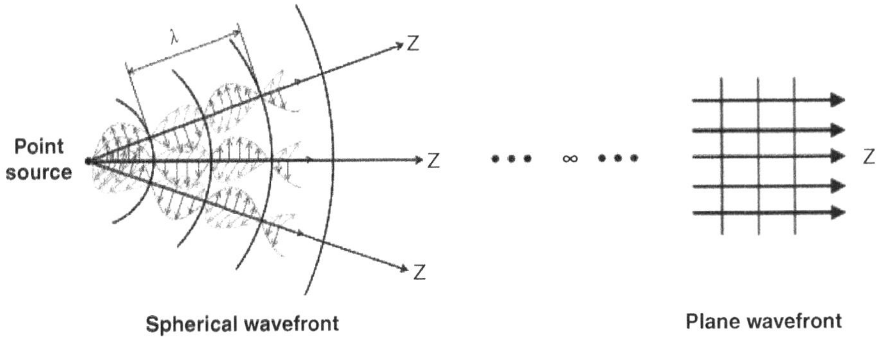

Figure 2.2 Point source and wavefront.

in real life. In practice, however, a point source also refers to a light source whose apparent size is smaller than the resolvable size of a particular optical system. For example, light from a distant star in the night sky is considered a point source for human eyes.

Figure 2.2 shows light rays originating from a point source representing the propagation directions of transverse electromagnetic waves. Assuming waves with a constant wavelength (λ), an ideal point source radiates waves with identical amplitude and phase in all directions. When we define a wavefront as a surface of points having the same phase, the shape of the wavefront from a point source in three dimensions is spherical as depicted in Figure 2.2. The light rays show the light propagation directions (z), which is always perpendicular to the wavefront. When the light waves from a point source propagate in a vacuum for a very long distance, the spherical wave from a point source becomes a plane wave. When the light wave has a plane wavefront, it is referred to as a collimated light, where the light rays are parallel to each other.

When light from a point source contains multiple wavelengths, electromagnetic waves will have multiple frequencies each with corresponding amplitude values. The shape of the wavefront and the light propagation direction are the same for each frequency component of the light.

A point source is often used as a reference source for characterization of optical systems as well as an ideal beacon for many optical beam control operations such as tracking, pointing, line-of-sight stabilization, alignment, phasing, etc. However, point source reference may not be readily available in many optical beam control applications.

An extended source refers to a light originating from an extended object whose apparent size is larger than the resolvable size of a particular optical system. For example, the Sun is considered an extended source for human eyes when it is observed from Earth. An extended source can be interpreted as a collection of discrete point sources where each point emits spherical waves with its own amplitude and phase values.

2.1.2 Coherent and Incoherent Light Sources

In light waves, there are two different aspects of coherence: temporal coherence (spectral coherence) and spatial coherence. Temporal coherence describes the correlation between light waves at one location but observed at different times, which is a measure of a light beam's ability to interfere with time-shifted samples of itself. Spatial coherence describes the correlation between light waves at different locations in space, which is a measure of a light beam's ability to interfere with spatially shifted samples of itself. In practice, a coherent light source refers to a light source that has a high degree of temporal coherence as well as a high degree of spatial coherence.

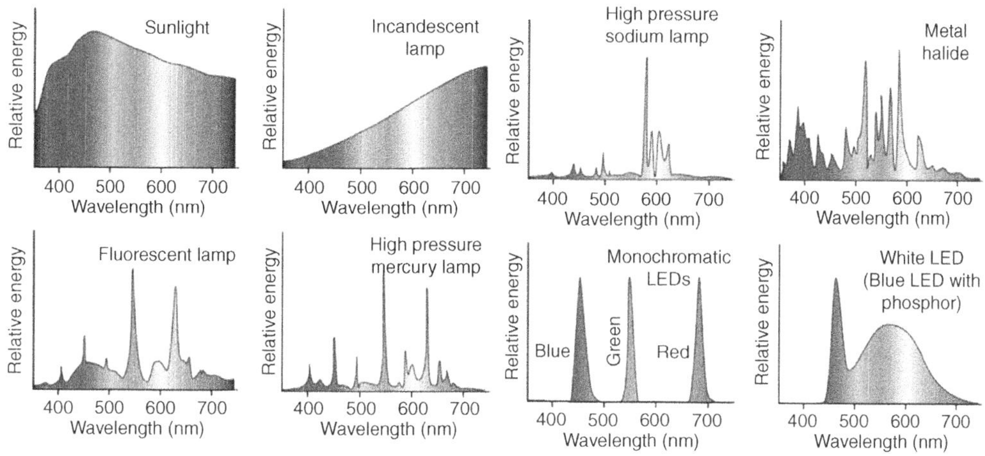

Figure 2.3 Wavelength spectrum of various light sources [1].

For a high degree of temporal coherence, the light source needs to be monochromatic light since the temporal evolution of a single wavelength light is a highly predictable sinusoidal oscillation over a long period of time. For a high degree of spatial coherence, the phase of light waves at different locations in space needs to have a fixed relationship. For example, a laser is considered a coherent light source that has a very narrow frequency spectrum (temporal coherence) and can be collimated to travel long distances over space with minimal spreading (spatial coherence).

An incoherent light source contains more than one wavelength, and waves are not in phase with one another. Sunlight is an example of incoherent light, which is a combination of ultraviolet, visible, and infrared spectrum of light when it reaches Earth's surface. The term white light is also used to represent incoherent light containing all visible frequencies of light. Figure 2.3 shows the wavelength spectrum of common incoherent light sources.

2.1.3 Polarized and Unpolarized Light Sources

Light as electromagnetic waves can have a specific direction of wave oscillations. By convention, the polarization direction of a light wave is given by its electric field vector direction. An unpolarized light source has an electric field randomly pointing in all possible directions in time. Many common light sources such as sunlight, halogen lighting, LED lights, and incandescent bulbs produce unpolarized light where the orientation of their electric fields fluctuates randomly in time. On the other hand, laser sources are often linearly polarized whose electric field vectors point along a line as shown in Figure 2.4a. By combining linearly polarized waves with different phases and oscillation directions, we can also obtain circularly polarized and elliptically polarized light waves as shown in Figure 2.4b,c. Polarization is an important consideration due to the polarization-dependent behavior of light waves in certain beam control applications.

2.1.4 Light Amplification by Stimulated Emission of Radiation

Laser (Light amplification by stimulated emission of radiation) is one of the most important light sources for many beam control applications since it is a coherent light source that can have a specified polarization state. Laser beams can be focused on very tiny spots, achieving a very high irradiance, or they can have a very low divergence to concentrate their power at a great distance.

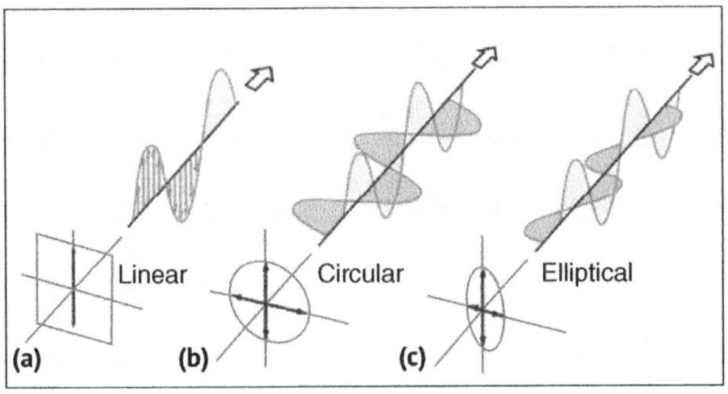

Figure 2.4 Various polarization states.

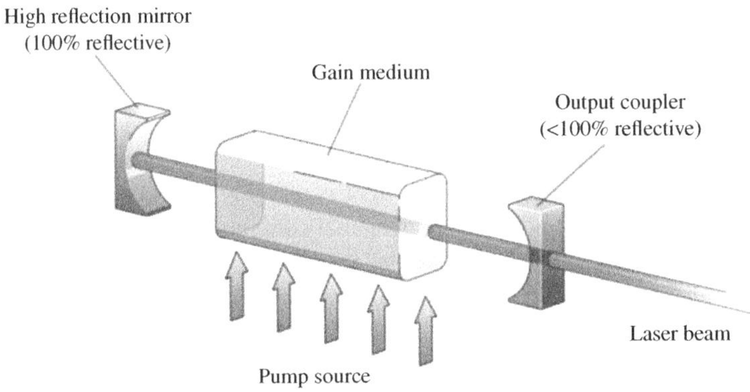

Figure 2.5 A schematic of a laser.

Stimulated emission is the process by which an incoming photon of a specific frequency can interact with an excited atomic electron, causing it to drop to a lower energy level. The liberated energy transfers to the electromagnetic field, creating a new photon with identical frequency, polarization, and direction of travel to the photons of the incident wave, giving laser light its characteristic coherence. Subsequently, both photons can trigger stimulated emissions in other atoms, creating the possibility of a chain reaction.

A generic laser, whose schematic is shown in Figure 2.5, uses an arc lamp or a flashlamp as a laser pumping source and pumps energy into a gain medium, in the form of a crystal rod, to sustain the simulated emission. To use the lamp's energy most efficiently, the lamps and gain medium are contained in a reflective optical cavity (optical resonator) that will redirect most of the lamp's energy into the lasing direction, where photons pass through the gain medium many times to receive substantial amplification. Gain medium and optical cavity design will determine the spectral coherence measured by the linewidth of a laser beam, which varies in the order of 1 kHz to 1 GHz.

If the applied pump power is too small, the gain will never be sufficient to overcome the cavity losses, and laser light will not be produced. The minimum pump power needed to begin laser action is called the lasing threshold. In a continuous wave (CW) laser, the balance of pump power against gain saturation and cavity losses produces an equilibrium value of the laser power inside the cavity; this equilibrium determines the operating point of the laser. The pulsed laser refers to any laser not classified as a continuous wave so that the optical power appears in pulses of some duration at some repetition rate.

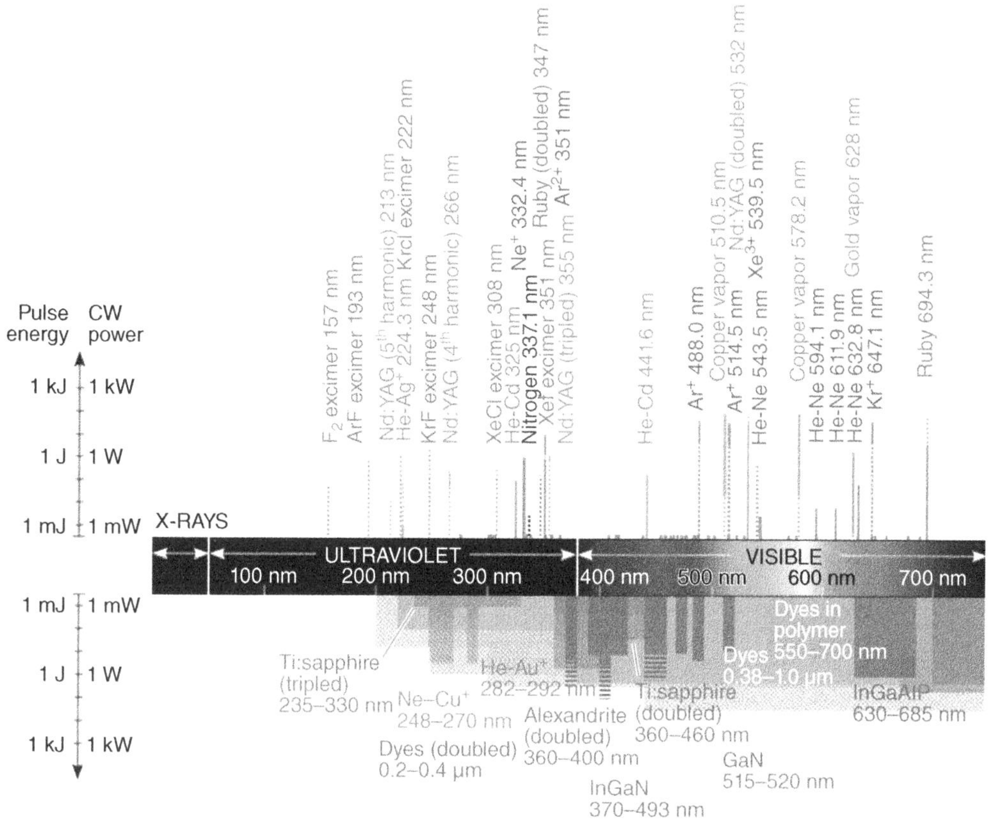

Figure 2.6 Commercial laser gain medium, wavelengths, power, and energy. *Source*: Danh/wikimedia Commons/ https://en.wikipedia.org/wiki/File:Commercial_laser_lines.svg.

A solid-state laser is a laser that uses a gain medium that is a solid, rather than a liquid as in dye lasers or a gas as in gas lasers. A fiber laser is a laser in which the active gain medium is an optical fiber doped with rare-earth elements such as erbium, ytterbium, neodymium, dysprosium, praseodymium, thulium, and holmium. Figure 2.6 summarizes various commercial lasers available with different gain medium, wavelength, power, and energy.

2.2 Properties of Light

2.2.1 Interference

Interference is a phenomenon in which light waves interact with one another in certain circumstances, causing the combined amplitude of the light wave to increase in case of constructive interference or decrease in case of destructive interference.

Assume two linearly polarized, monochromatic, plane light waves traveling in the direction depicted in Figure 2.7. Each plane wave is defined by $\boldsymbol{E}_i(r, t) = E_i \cos(\omega t - \boldsymbol{k}_i \cdot \boldsymbol{r} + \Phi_i)\hat{\boldsymbol{e}}_i$), where E_i is the amplitude, ω is the frequency, k_i is the wavevector, r is the position vector defined by the coordinates (x, y, z), Φ_i is the phase, and $\hat{e}_i$ represents the linear polarization vector, and $i = 1,2$. The interference of the two plane waves produces an interference pattern (interferogram) on a

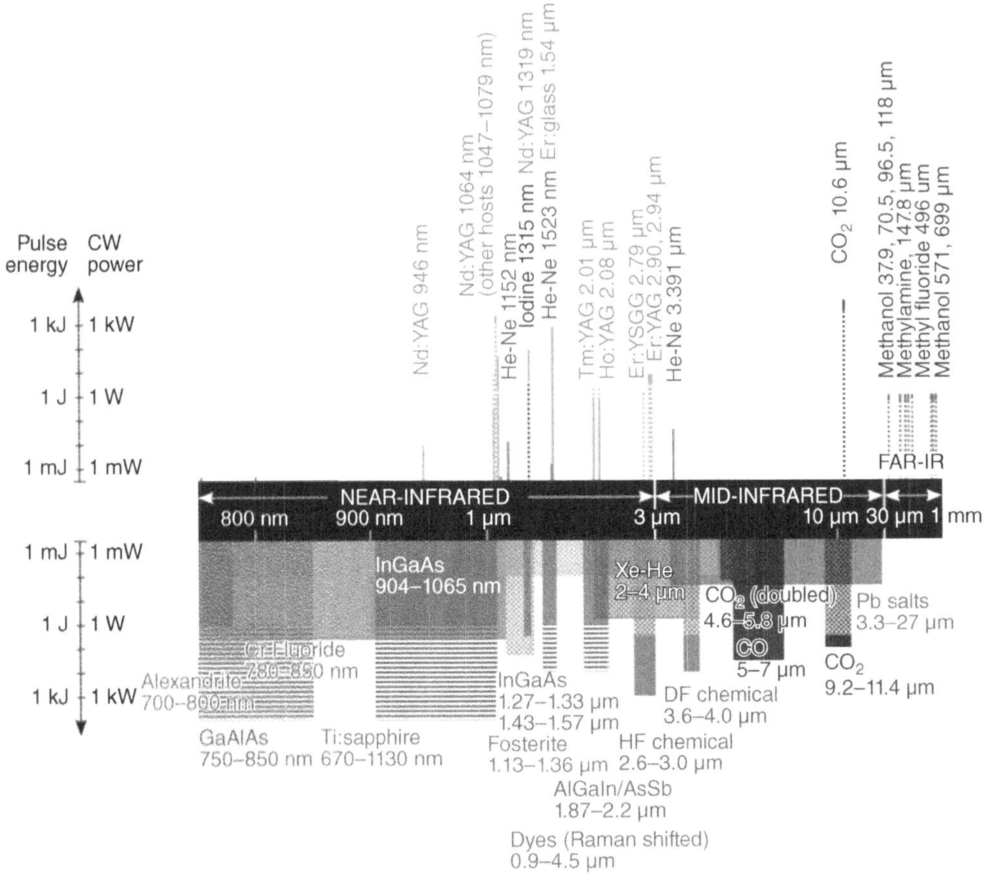

Figure 2.6 (Cont'd)

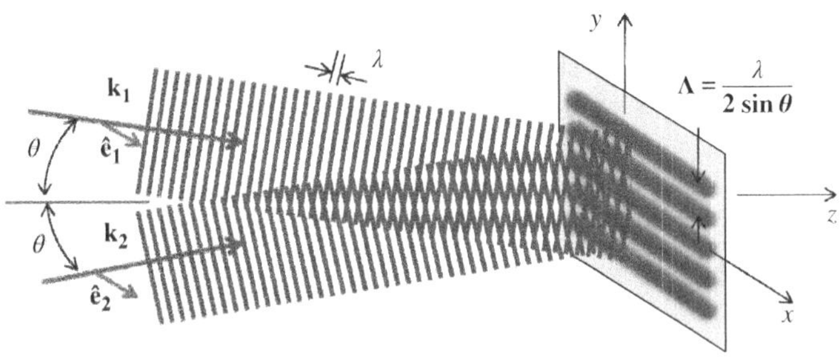

Figure 2.7 Interference pattern of two linearly polarized, monochromatic, plane light waves.

screen as one-dimensional (1D) fringe pattern with a periodicity Λ directly proportional to the wavelength λ and inversely proportional to the sine of the sum of the wavevector incidence angles θ with respect to the z axis.

Interferometry is a technique that uses the property of light wave interference to extract useful information. One common application of interferometry is the interferometer, a device that extracts information from the interference of a reference beam and a test beam. They are widely

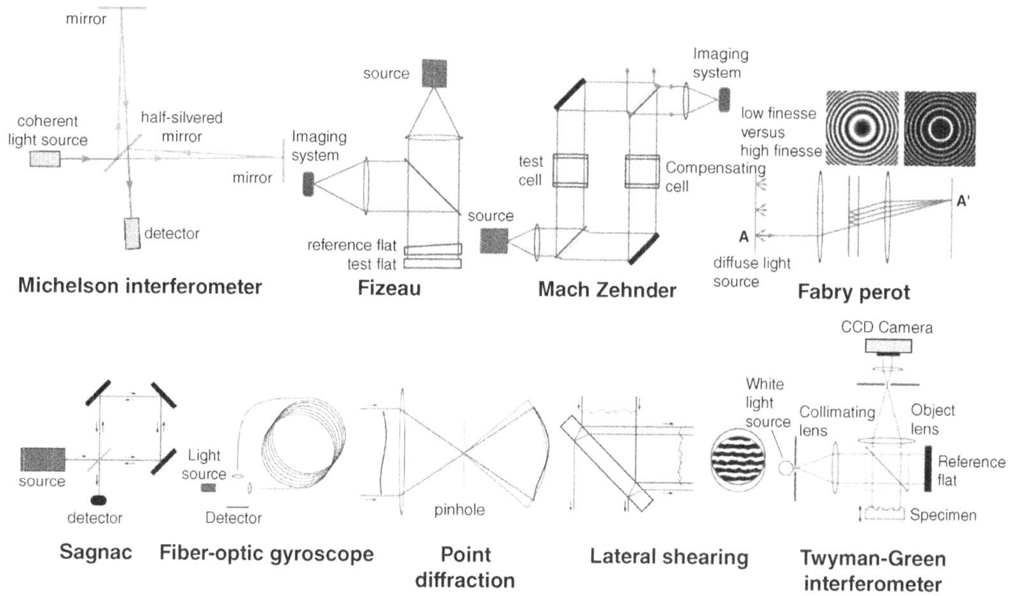

Figure 2.8 Various types of interferometers.

used in science and industry for the measurement of microscopic displacements, refractive index changes, and surface irregularities. The various types of interferometers are shown in Figure 2.8.

In homodyne interferometry, interference occurs between two beams at the same wavelength (or carrier frequency). The phase difference between the two beams results in a change in the intensity of the light on the detector. The resulting intensity of the light after mixing of these two beams is measured, or the pattern of interference fringes is viewed or recorded. The heterodyne interferometry technique is used for (1) shifting an input signal into a new frequency range as well as (2) amplifying a weak input signal (assuming the use of an active mixer). A weak input signal of frequency ω_1 is mixed with a strong reference frequency ω_2 from a local oscillator (LO). The nonlinear combination of the input signals creates two new signals, one at the sum $\omega_1 + \omega_2$ of the two frequencies, and the other at the difference $\omega_1 - \omega_2$. These new frequencies are called heterodynes. Typically, only one of the new frequencies is desired, and the other signal is filtered out of the output of the mixer. The output signal will have an intensity proportional to the product of the amplitudes of the input signals.

2.2.2 Polarization

Polarization, which represents the specific direction of wave oscillations, is a useful property of light with many beam control applications. The two orthogonal linear polarization states that are most important for the reflection and transmission of light are referred to as P- and S-polarization as shown in the polarization beam splitter cube example in Figure 2.9a. P-polarized (from the German parallel) light has an electric field polarized parallel to the plane of incidence, while S-polarized (from the German Senkrecht) light is perpendicular to this plane (Figure 2.9b).

In the case of a laser, linear polarization could be achieved using the carefully designed optical cavity and gain medium. Polarized light can be also obtained from unpolarized light by passing through a specific direction of wave oscillations while blocking other oscillation directions.

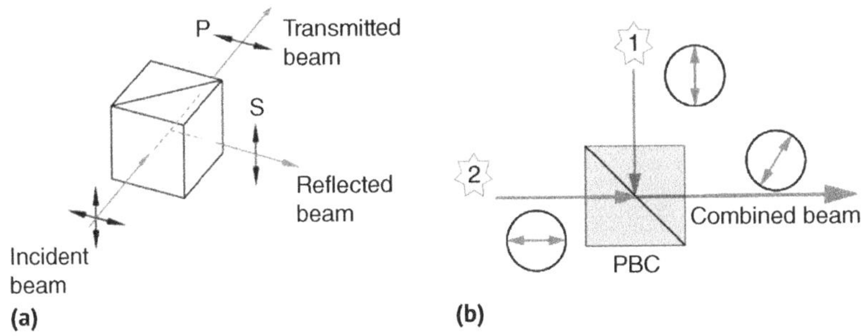

Figure 2.9 P and S linear polarization and polarization beam splitting and beam combining.

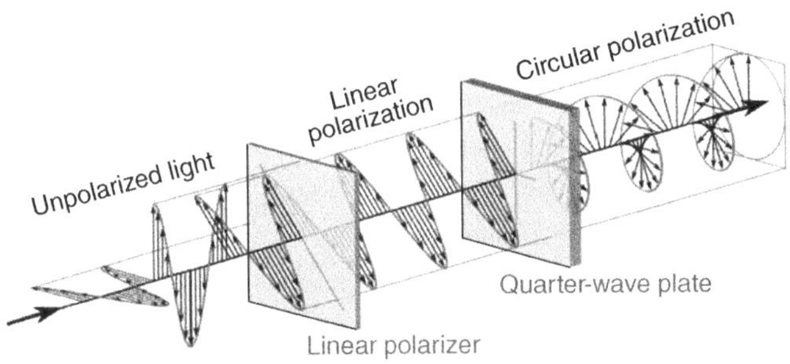

Figure 2.10 Change in polarization using a linear polarizer and quarter-wave plate.

Figure 2.10 shows an example of polarization state conversion from unpolarized to linear polarization using a linear polarizer, and linear to circular polarization using a quarter-wave plate. Polarization of light waves impacts reflectance, the focus of beams, and other optical behaviors that affect end beam control applications.

2.2.3 Doppler Effect

Similar to sound waves, the Doppler effect in light waves exhibits a change in frequency, wavelength, and amplitude of light when there is relative motion of the source and the observer. One application that uses the Doppler effect in light waves is Doppler spectroscopy for astronomy. It detects periodic shifts in radial velocity by recording variations in the color of light from the host star. When a star moves toward Earth, its spectrum is blueshifted, while it is redshifted when it moves away from Earth.

2.2.4 Light Propagation in Vacuum

Since electromagnetic waves do not require a medium for their propagation, light of all frequencies can propagate in a vacuum at the speed of light. The form of the light wave is determined by its source, and the light propagation is described by the shape of its wavefront. For example, a point source forms a spherical wavefront, and the light wave propagates in a vacuum without alteration in amplitude, frequency, and phase (Figure 2.11). For a point source with a total radiation power of P, the intensity of light (I) is inversely proportional to the square of the distance (r) from the source

$$I = \frac{P}{4\pi r^2} \tag{2.1}$$

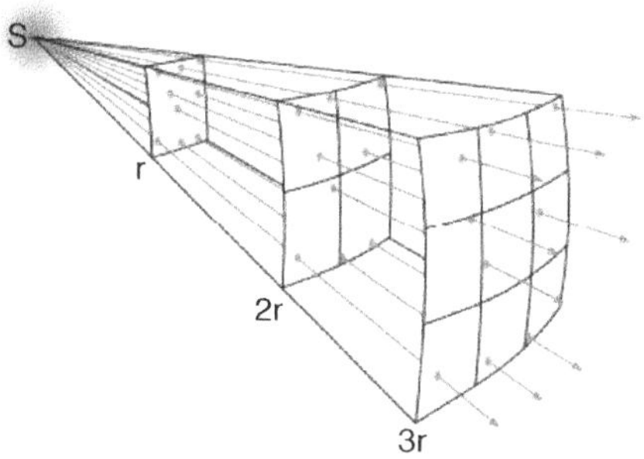

Figure 2.11 Light propagation from a point source.

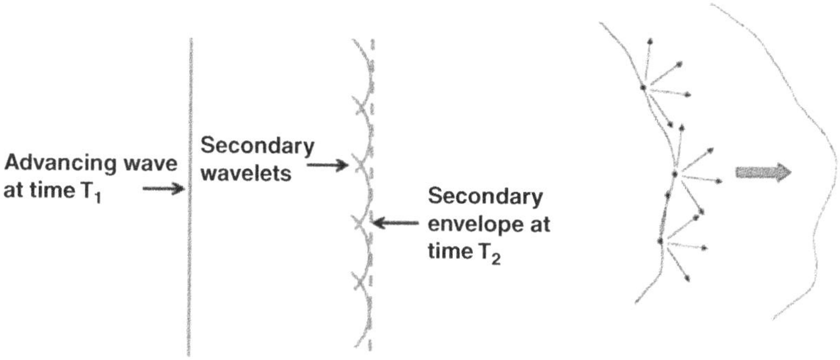

Figure 2.12 Illustration of light propagation using Huygens' principle. Plane wave (left) and arbitrary wave (right).

For light waves with arbitrary-shaped wavefront, every point on a wavefront acts as a little secondary point light source and emits a spherical wave, as shown in Figure 2.12. If these point sources have a single frequency and travel only in the forward direction, their addition at the secondary location is a result of interference of all these little spherical waves. This concept is known as the Huygens–Fresnel principle. The Huygens–Fresnel principle shows that destructive interference occurs in all but the direction of propagation perpendicular to the wavefront. Similar to the point source with a spherical wavefront, the wavefront becomes flatter as light with an arbitrary-shaped wavefront propagates a long distance in a vacuum.

2.2.5 Light Interaction with Different Media

When light waves travel through a medium other than a vacuum, the interaction of light waves with the atoms and molecules of the medium alters the light wave propagation in different ways. In general, propagating light waves will interact with the medium in the combinational form of transmission, reflection, refraction, diffraction, absorption, and scattering, as depicted in Figure 2.13.

2.2.6 Transmission and Reflection

Transmission or reflection of light waves occurs when the frequencies of light waves do not coincide with the natural vibration frequency of the electrons in the atoms of a medium. For a transparent medium, vibrations of the electrons due to the incident light waves are passed on to

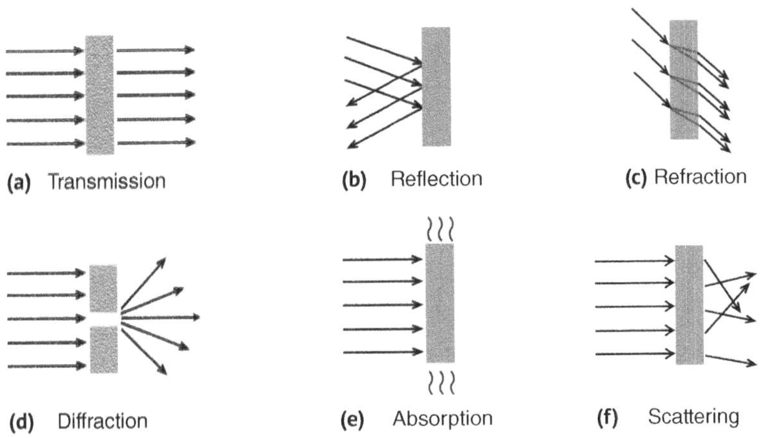

Figure 2.13 Light interaction with different media.

Table 2.1 Refractive index for common materials.

Substance	Index of refraction
Air	1.00029
Water	1.31
Fused silica (SiO_2)	1.46
Crown glass	1.52
ZnSe (10.6 μm)	2.40

neighboring atoms in the direction of light propagation, and the vibration energy is reemitted as light waves on the opposite side of the medium. For an opaque medium, vibrations of the electrons due to the incident light waves are reflected, and the reflected vibration energy is reemitted as light waves.

We call the fraction of the incident power that is reflected from the interface the reflectance (or reflectivity, or power reflection coefficient) R, and the fraction that is refracted into the second medium is called the transmittance (or transmissivity, or power transmission coefficient) T. Note that these are what would be measured right at each side of an interface and do not account for attenuation of a wave in an absorbing medium following transmission or reflection.

2.2.7 Refraction

Refraction is the bending of a light wave due to the wave's change in speed and wavelength as it passes from one medium to another. The refractive index of an optical medium (n) is a dimensionless number that relates to the speed (v) and wavelength (λ) of light traveling in a medium with respect to the speed (c) and wavelength (λ_0) of light in a vacuum.

$$v = c/n \tag{2.2}$$

$$\lambda = \lambda_0/n \tag{2.3}$$

Refractive index varies significantly even among similar materials, as shown in Table 2.1.

Light bending due to refraction is illustrated in Figure 2.14 and described by Snell's law,

$$n_1 \sin \theta_1 = n_2 \sin \theta_2 \quad (2.4)$$

where θ_1 and θ_2 are the angle of incidence and angle of refraction, respectively, of a ray crossing the interface between two media with refractive indices n_1 and n_2.

The refractive index of a medium is wavelength dependent since the velocity of a light wave depends on its frequency. This phenomenon, known as dispersion, is illustrated in Figure 2.15. For visible light, the refractive indices of most transparent materials decrease with increasing wavelength. Index of refraction could be also polarization dependent, resulting in double refraction of light in certain transparent media. This phenomenon is called birefringence.

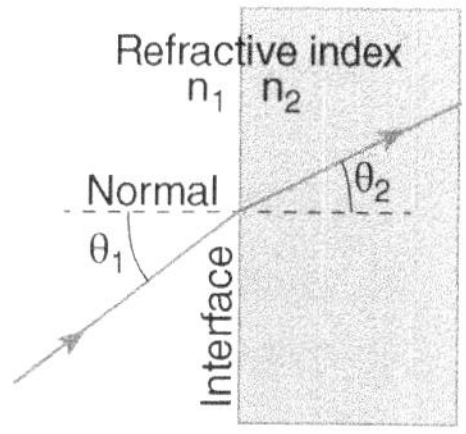

Figure 2.14 Refraction of a light ray. *Source:* Ulflund/wikimedia commons/ https://en.wikipedia.org/wiki/Refractive_index.

2.2.8 Diffraction

Diffraction refers to the bending of a propagating light wave when it encounters the edge of an object. Diffraction causes various effects in the beam propagation, and its effects are generally most pronounced for wavelengths roughly similar to the dimensions of the diffracting object ($D_0 \approx \lambda$). Since in optics we generally deal with very short wavelengths (on the order of 10^{-6} m), diffraction only becomes apparent at very small scales, e.g. in focus, compact disk, and small slit. Diffraction causes various optical behaviors in light wave propagation that affect end beam control applications.

Figure 2.16 shows the diffraction effects of the plane wave incident on a single aperture. The bending of light wave occurs by the edges of the aperture, then combined with the undisturbed

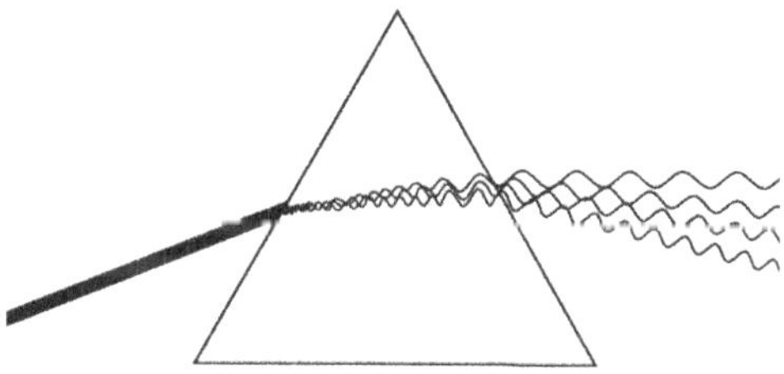

Figure 2.15 Dispersive prism. *Source:* LucasVB/Wikimedia Commons/Public Domain.

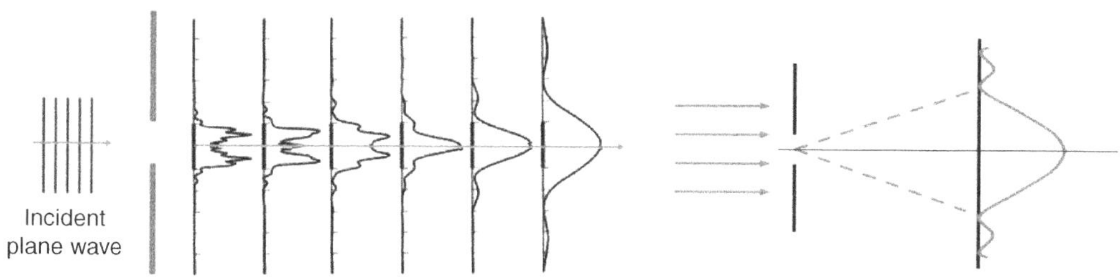

Figure 2.16 Diffraction of a plane wave due to a single aperture.

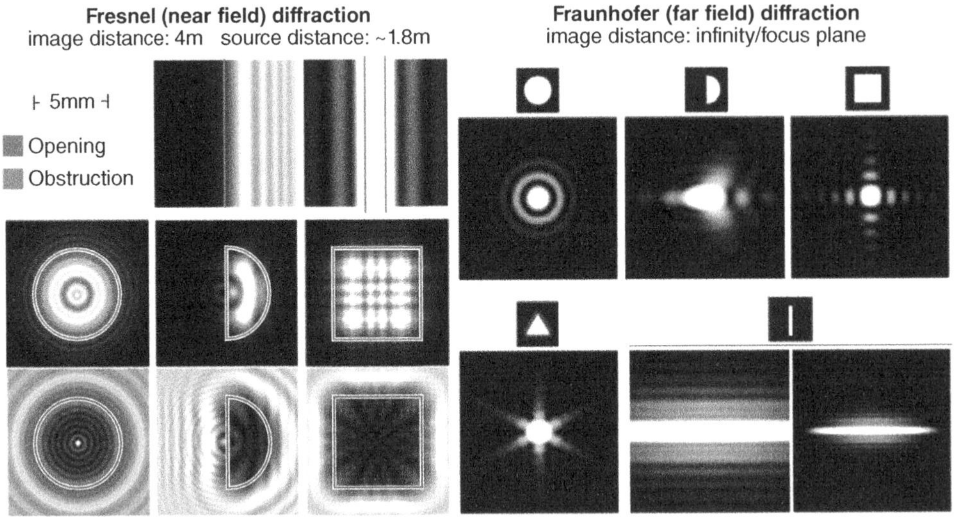

Figure 2.17 Fresnel diffraction and Fraunhofer diffraction for various aperture shapes. *Source*: Vladimir Sacek/ https://www.telescope-optics.net/diffraction.htm /last accessed on 15 March, 2025.

light wave propagating the center of the aperture. Fresnel diffraction (near-field diffraction) refers to the diffraction pattern that is observed relatively close to the aperture.

In contrast, Fraunhofer diffraction (far-field diffraction) refers to the diffraction pattern that is observed relatively far away from the aperture.

As the aperture diameter gets smaller, the transmitted light's divergence angle gets larger along with the far-field light irradiance pattern. Conversely, a larger diameter aperture yields smaller divergence angles with less diffraction effects in the far-field light irradiance pattern. Near-field and far-field diffraction patterns for various aperture shapes are shown in Figure 2.17.

For a circular aperture with a diameter D shown in Figure 2.18, the far-field diffraction pattern is called the Airy disk, where the first minimum intensity occurs at an angle

$$\theta = 1.22\frac{\lambda}{D} \tag{2.5}$$

where D is the diameter of the aperture and λ is the wavelength of the light.

2.2.9 Absorption

When a light wave has the same frequency as the natural vibration frequency of the electrons in the medium's atom, the light wave will resonate with the electron vibrational motion. During its vibration, the electrons interact with neighboring atoms to convert their vibrational energy into thermal energy. Subsequently, the light wave with that given frequency is absorbed by the medium, never again to be released in the form of light. Atmospheric absorption by the Earth's atmosphere at optical wavelengths is shown in Figure 2.19.

2.2.10 Scattering

Scattering refers to the process in which light waves deviate from a straight trajectory due to localized non-uniformities in the medium through which they propagate. When light strikes an

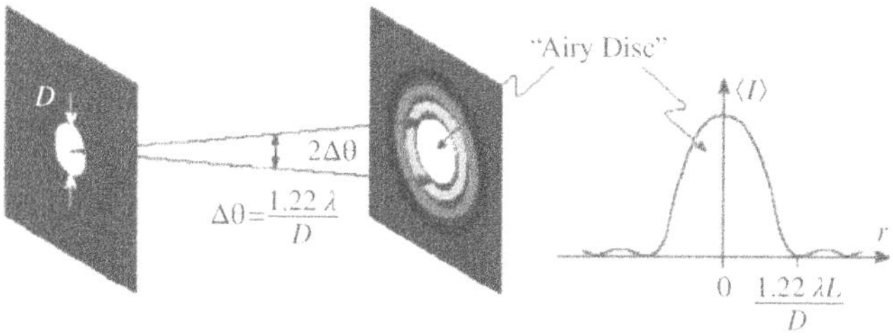

Figure 2.18 Diffraction effects of circular aperture and angular resolution limit.

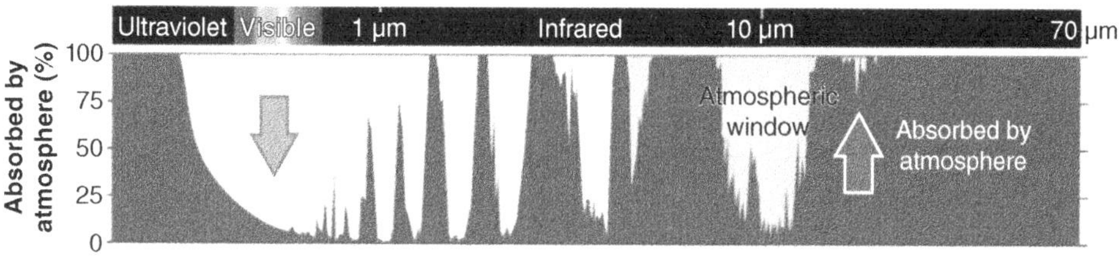

Figure 2.19 Atmospheric absorption versus Wavelength.

Table 2.2 Types of Scattering.

Type of Scattering	Condition	Example
Rayleigh scattering	Particles ≪ wavelength (λ)	Blue sky, red sunsets
Mie scattering	Particles ~ wavelength (λ)	White clouds, fog, dust
Tyndall effect	Colloidal-sized particles	Visible laser beam in mist or smoke
Raman scattering	Inelastic scattering (λ changes)	Raman spectroscopy (chemical analysis)
Brillouin scattering	Interaction with acoustic waves (λ changes)	High-resolution material studies

object, it may be absorbed and subsequently re-emitted in different directions. The various types of scattering are summarized in Table 2.2.

2.3 Geometric Optics

Geometrical optics, or ray optics, is a model of optics that describes light propagation in terms of rays. Using rays, we interpret light propagation as temporally incoherent radiation ignoring certain wave properties of light such as interference, diffraction, and polarization. Geometric optics is particularly useful for the design and analysis of the optical train (optical assembly) in the beam control system.

2.3.1 Lens

Lens is an optical element that uses the refraction property of light. For a positive lens shown in Figure 2.20a, a collimated beam of light passing through the lens converges to a focal point behind the lens. Conversely, a point source of light placed at the focal point is converted into a collimated beam by the positive lens. A negative lens makes the collimated beam diverge behind the lens with a virtual focal point in front of the lens shown in Figure 2.20b.

Focal length of a lens (f) is defined as the distance from the lens to a focal point. The focal length (f) divided by the diameter of the lens (D) is defined as f-number written in the format f/N such that $N = f/D$.

Diffraction causes the fundamental limit on an ability to focus light to an infinitely small spot even with an ideal converging lens. Due to the inevitable encounter of the aperture stop (which coincides with the edge of the lens in Figure 2.21) that blocks rays outside of the lens opening, the diffraction of rays caused at the aperture stop forces the focal spot to have a finite size with a radius of

$$r = 1.22\frac{\lambda f}{D} \tag{2.6}$$

In geometric optics, the diffraction effects of a lens are ignored. However, there are other optical aberrations that prohibit the formation of an infinitely small spot even with a perfect converging lens. Since the angle of refraction is wavelength dependent, focal point of a lens is not an infinitely small spot when the light contains multiple wavelengths. This chromatic aberration effect shown in Figure 2.22 could be reduced with an achromatic lens (achromat) that uses additional lenses bonded on the single lens to adjust focal lengths for two or more different wavelengths that are separated (typically red and blue for visible wavelengths).

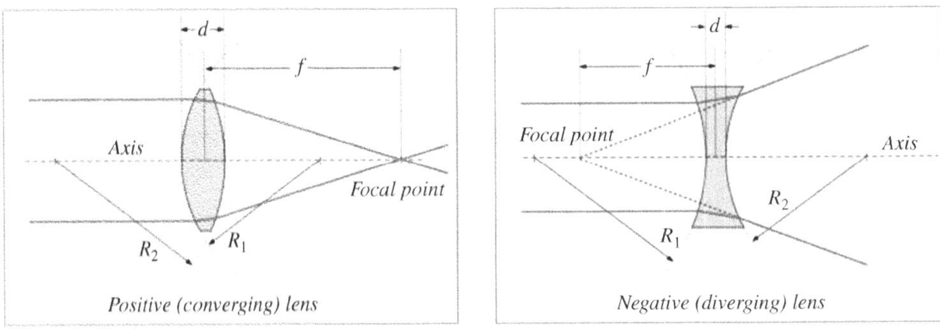

Figure 2.20 (a) Positive (converging) lens and (b) negative (diverging) lens.

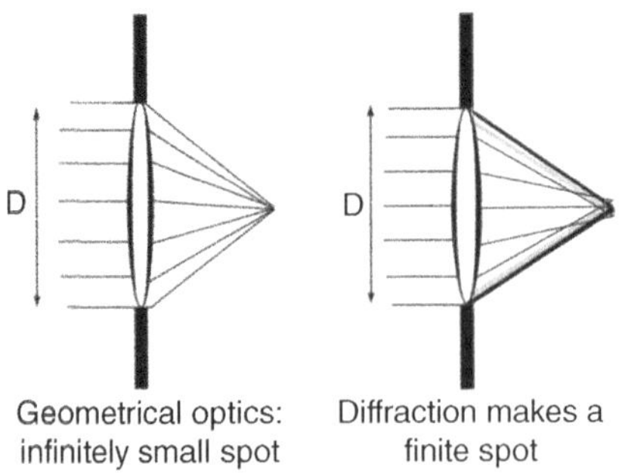

Figure 2.21 Diffraction effects of lens.

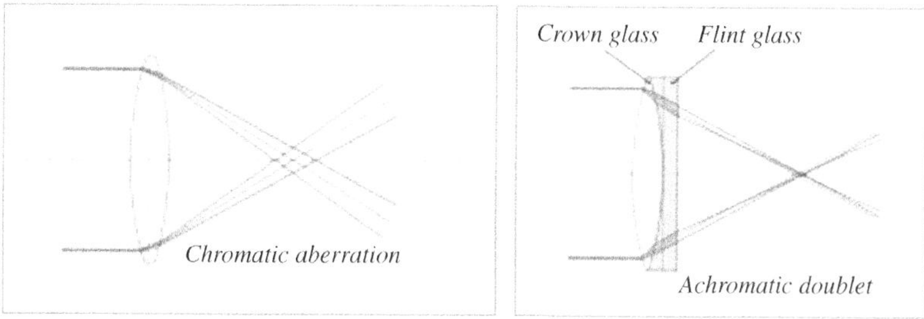

Figure 2.22 Single lens versus Achromatic lens (achromat, achromatic doublet).

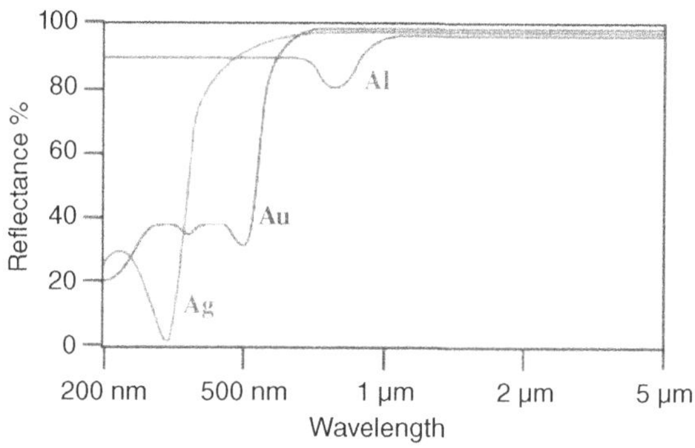

Figure 2.23 Reflectance spectrum of metals.

2.3.2 Mirrors

Mirrors are the main reflective elements. The mirror surface is typically coated with a material that can increase the reflectance of light from the surface while reducing transmission, absorption, and scatter. Two common coating materials for mirrors are dielectric coating and metallic coating.

For metallic coating, the light is reflected by loosely attached electrons in the metal that can freely oscillate with incident light waves. This mechanism works over large wavelengths and angle of incidence ranges, only breaking down when the light oscillates at a higher speed than the electrons can follow, which means that metals do not reflect well in the UV. Figure 2.23 shows the reflectance spectrum of Ag, Au, and Al, which are commonly used metals for high-reflective coating. Dielectric coating uses a thin layer of dielectric materials that are applied onto the surface of an object to improve its reflection properties and can be designed to provide a reflection on specific wavelengths of light.

Figure 2.24 shows common optical mirrors where the optical axis, a line that passes through the geometrical center of the mirror, which is also an axis of symmetry, is defined as z. For a flat mirror in Figure 2.24a, the incident angle of the ray is the same as the reflected ray about the optical axis. For a parabolic mirror shown in Figure 2.24b, all parallel rays entering the aperture are focused on a point. The focal length of a parabolic mirror (f) is defined as the distance from lens to a focal point, and the f-number is defined as $N = f/D$, where D is the size of the mirror aperture. For a spherical mirror shown in Figure 2.24c, light rays originating from the center of a sphere are reflected to its origin. Parallel rays will focus to approximately a point when the radius of the mirror is much

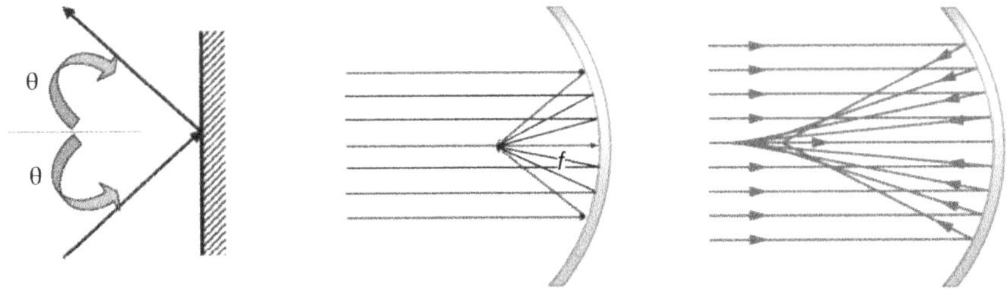

Figure 2.24 Reflection by a flat mirror (a), parabolic mirror (b), and spherical mirror (c).

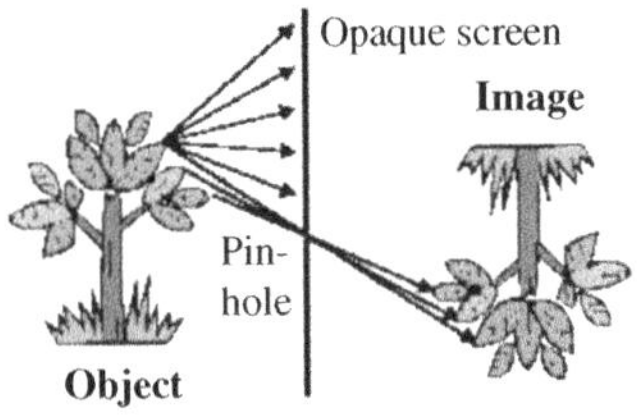

Figure 2.25 Image formation with pinhole imaging. *Source*: [5].

larger than the size of the mirror (large *f*-number). As in the case of a lens, diffraction causes the fundamental limit on an ability to focus light to an infinitely small spot even with an ideal parabolic mirror.

2.3.3 Image Formation Using Lenses and Mirrors

According to Hudgins–Fresnel principle, every infinitely small point in the object emits light as a spherical wave. The spherical waves from all the points on the object's surface get mixed together as they propagate. An imaging system redirects all the rays from a single point on the object onto another single focal point in imaging space, so you can distinguish details of the object. Therefore, the optical resolution of an imaging system is closely related to the ability to reproduce a focal point as small as possible.

Figure 2.25 shows the simplest way to form an image by one-to-one mapping of all the rays from every single point on the object onto an image point using a pinhole. Although an image is formed with a pinhole, only a small fraction of the light will pass through the pinhole and most of the light will be wasted. In addition, a small pinhole causes significant diffraction of the light that passes through it. A more efficient way to form an image is to use a lens or a mirror instead of a pinhole.

In Figure 2.26, the image point (focus) is located at the intersection of all rays passing through the lens from the corresponding object point. Assuming a thin lens, any ray that enters parallel to the axis on one side of the lens proceeds toward the focal point. Any ray that arrives at the lens after passing through the focal point on the front side comes out parallel to the axis on the other side. Any ray that passes through the center of the lens will not change its direction. This relationship is expressed in the thin lens equation

$$\frac{1}{f} = \frac{1}{d_o} + \frac{1}{d_i} \tag{2.7}$$

Similarly, Figure 2.27 shows the image formation by different types of mirrors. For a parabolic mirror, the relationship between object distance, image distance, and focal length is the same as the thin lens. For a spherical mirror, this relationship is written as

$$\frac{1}{d_o} + \frac{1}{d_i} = \frac{2}{R} \tag{2.8}$$

Paraxial approximation assumes that all rays have a small angle (typically < ~10°) with respect to the optical axis and are at a distance much less than the radius of curvature (R for a spherical mirror) from the optical axis. With small-angle approximation, the focal length of a spherical mirror is one-half the radius of curvature of the mirror, $f = R/2$, and the relationship becomes identical to the thin lens equation.

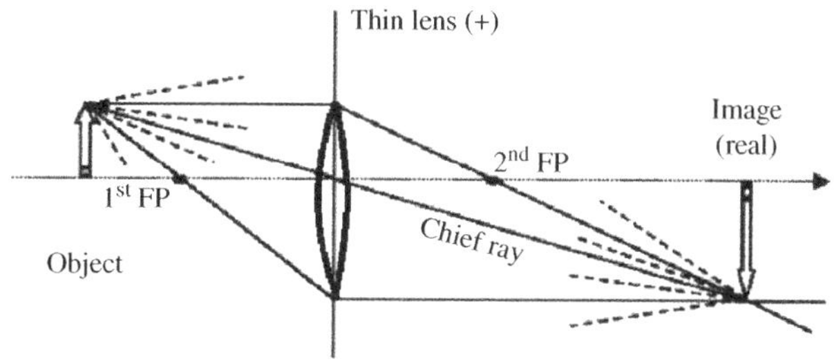

Figure 2.26 Image formation with a thin lens.

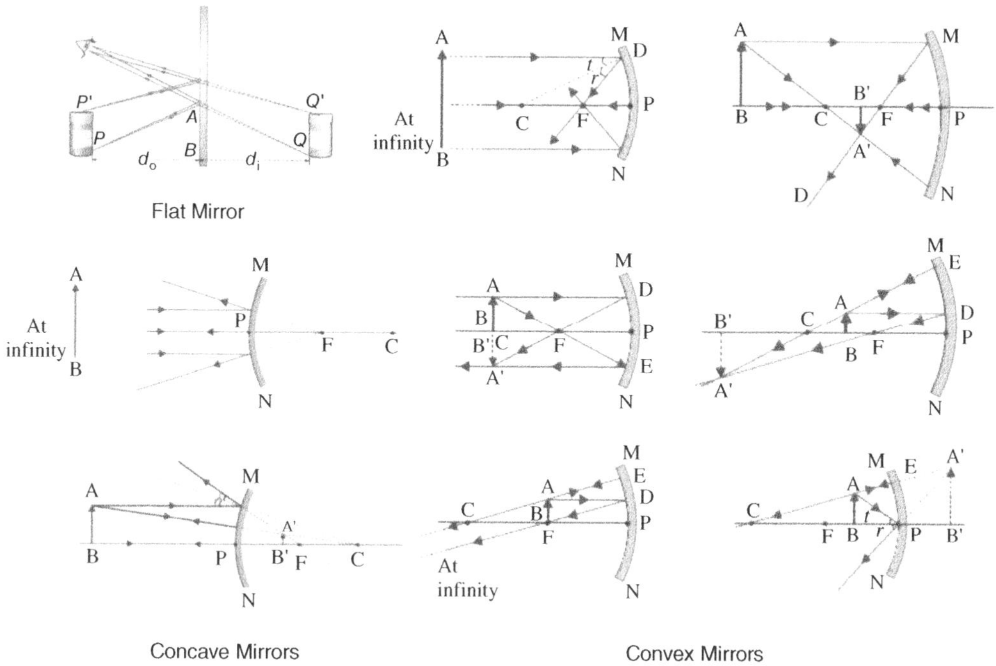

Figure 2.27 Image formation by mirrors.

Figure 2.28 Field of view.

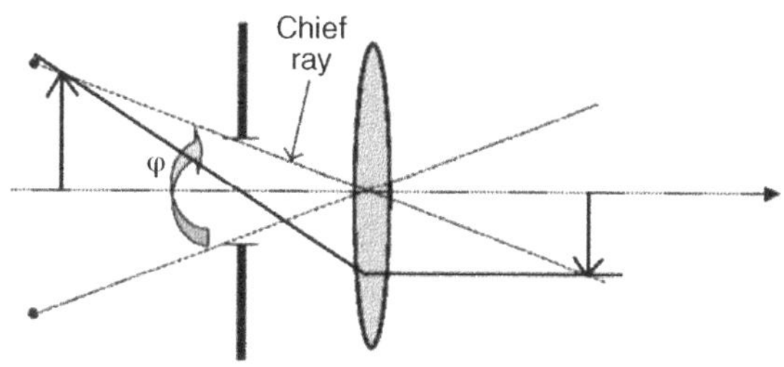

An imaging system has a finite-size entrance aperture. A chief ray is a ray that goes through the aperture stop and the center of the lens (and thus un-bent/un-deviated), as shown in Figure 2.28. Field-of-view (φ) is an angle that the chief ray from an object can subtend, given the entrance aperture of the imaging system.

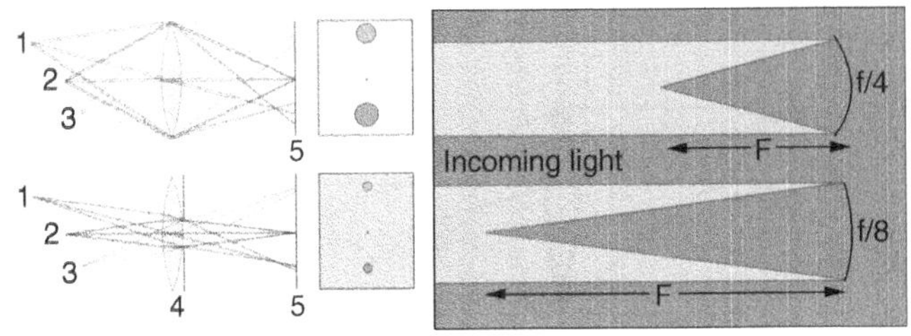

Figure 2.29 Illustration of depth of field and f-number.

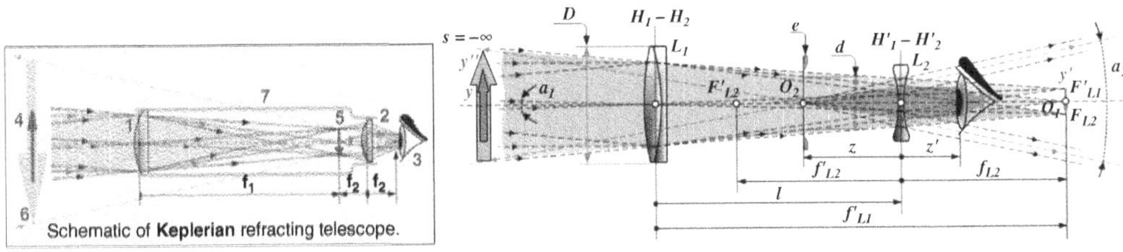

Figure 2.30 Keplerian refracting telescope and Galilean refracting telescope.

The depth of focus of an imaging system refers to its ability to maintain a small focal point if the object's position is moved closer and farther from the plane of best focus. As shown in Figure 2.29, larger f-number lenses and mirrors yield a larger depth of focus.

2.3.4 Geometric Optics for Telescopes and Microscopes

A telescope is an optical device commonly used to image distant objects. To form an image of a distant object with desired specifications in terms of optical resolution, field of view, depth of focus, light intensity, telescope dimensions, etc. telescopes often employ multiple mirrors/lenses and larger aperture to gather more light and to reduce diffraction and other undesirable effects that impair the ability to generate a small focal point.

Depending on the type of optical elements used to create a focal point, telescopes are divided into refracting telescopes and reflecting telescopes. Given a collimated light source (a point source at infinity distance), both Keplerian and Galilean refracting telescopes shown in Figure 2.30 use an objective lens with a focal length of f_1 to create a focal point and an eye lens with a focal length f_2 to re-create a collimated light output (afocal system) with a size that can fit into a human eye. Magnification of the telescope is given as

$$M = \frac{f_1}{f_2} = \frac{D_1}{D_2} \tag{2.9}$$

In geometry optics, entrance pupil is defined as an aperture stop of a system where rays within the field of view can enter the system. For telescopes shown in Figure 2.30, the objective lens is also an aperture stop, which becomes an entrance pupil, or simply a pupil, of a telescope.

In Figure 2.31, a pupil plane is defined as a two-dimensional plane that represents the location of the pupil, and a focal plane (image plane) is defined as a two-dimensional plane that represents the location of the focal point. An exit pupil is a corresponding image of an entrance pupil re-imaged by an eye lens. Since exit pupil is a re-image of an entrance pupil, the location of rays on the entrance pupil is the same as the location of rays on the exit pupil except that the re-image is

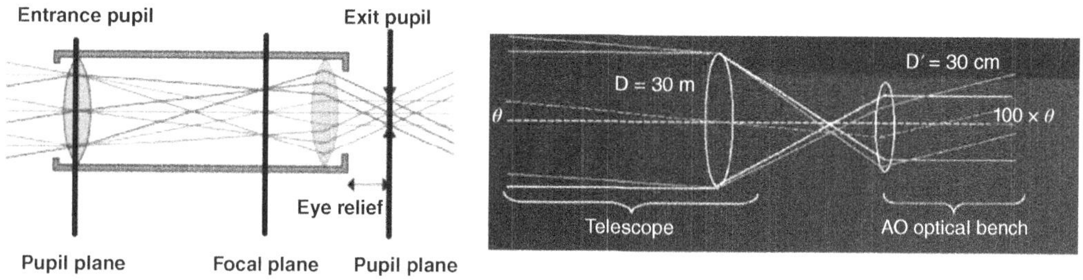

Figure 2.31 Entrance pupil and exit pupil of the refractive telescope system.

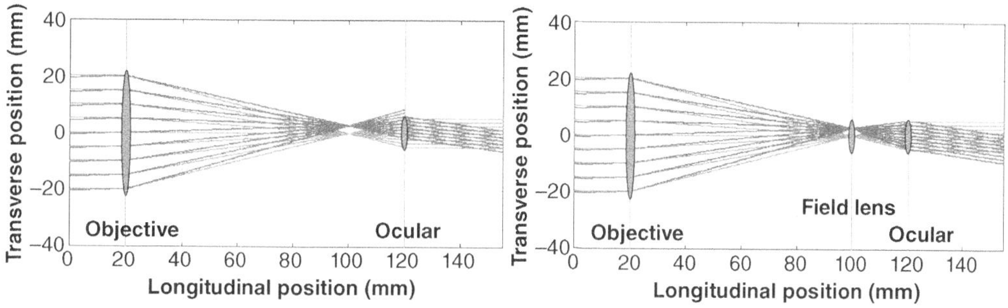

Figure 2.32 Field lens used for refracting telescope.

inverted and scaled by the magnification of a telescope. Beam control actuators are often placed at the pupil plane of the telescope because rays from different field angles overlap at this plane, allowing the actuator to influence all rays simultaneously. According to the optical invariant (Lagrange invariant), the angle of the collimated beam at the entrance pupil and exit pupil is related as

$$\Xi = \theta_1 D_1 = \theta_2 D_2 \tag{2.10}$$

$$\theta_2 = M\theta_1 \tag{2.11}$$

In geometrical optics, a field lens refers to a lens that is used to increase the field of view of the optical system. In Figure 2.32, a field lens is inserted in the focal plane of the refracting telescope in front of the eye lens to increase the field of view of the telescope. A combination of a field lens and an eye lens typically consists of a telescope eyepiece.

Reflecting telescopes replace an objective lens used in refracting a telescope with one or more mirrors to create a focal point. Reflecting telescopes are more common than refracting telescopes since reflection is wavelength independent and reflective concave mirrors are generally easier to manufacture than achromatic lenses used in refracting telescopes.

Figure 2.33 shows different types of reflecting telescopes based on the type of reflectors used to create a focal point. The primary mirror refers to a principal light-gathering concave mirror, which is also an entrance pupil for most reflecting telescopes. The reflecting telescopes also employ secondary mirror (and tertiary mirror) to further direct the light to a focal plane at a desired location. An on-axis telescope refers to a telescope with a focal point of an on-axis beam located on the optical axis of the primary mirror. On-axis telescopes simplify the telescope design with the price of light obscurations in the center of an aperture. Telescopes are also designed as off-axis systems as shown in Figure 2.34, which are usually larger than on-axis telescopes with the same primary mirror dimension.

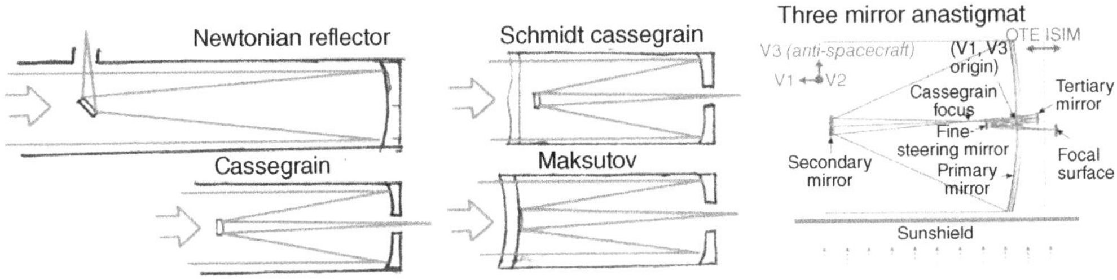

Figure 2.33 Reflecting telescope design.

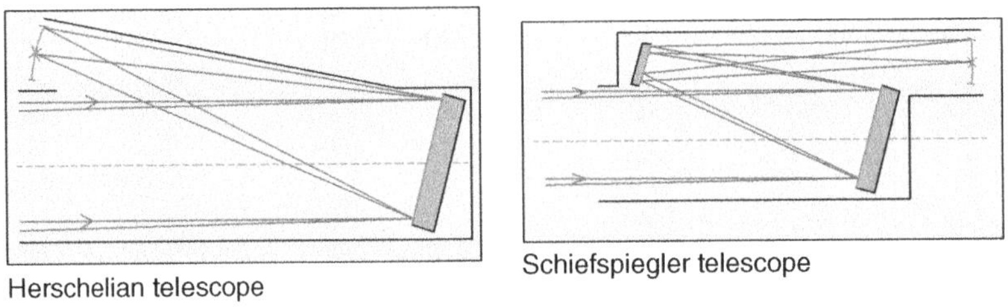

Figure 2.34 Off-axis telescope design.

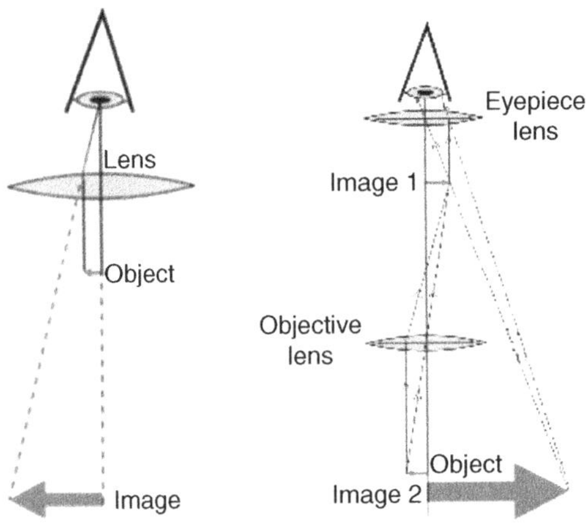

Figure 2.35 Simple microscope and compound microscope.

Optical microscopes (microscope objectives) are optical systems designed to image a close-in object in high detail. The principal design is the same as telescopes. A simple microscope shown in Figure 2.35a uses the optical power of a single objective lens to magnify the object. The compound microscope in Figure 2.35b uses an objective lens to create an image inside the microscope and uses an eyepiece to obtain a virtual image that magnifies the image created by the objective lens.

2.3.5 Geometric Optics for 4f Optical System

A 4f optical system is commonly used in beam control systems to change beam diameters and re-image optical planes (such as pupil plane and focal plane) as needed. With the system setup shown in Figure 2.36 with two lenses, a re-image of an optical plane A can be obtained in A′, and

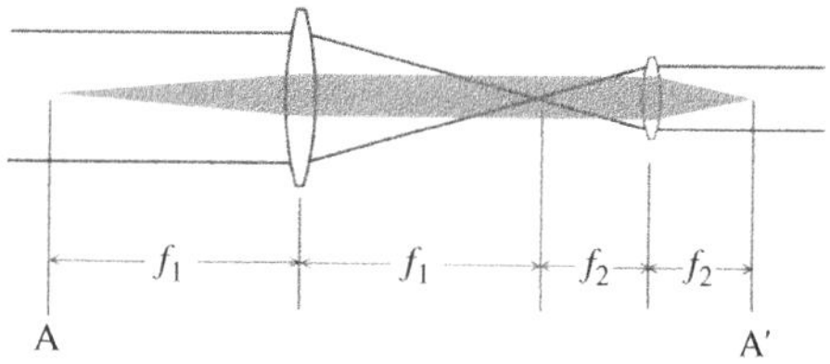

Figure 2.36 4f system with two lenses with focal lengths f_1 and f_2.

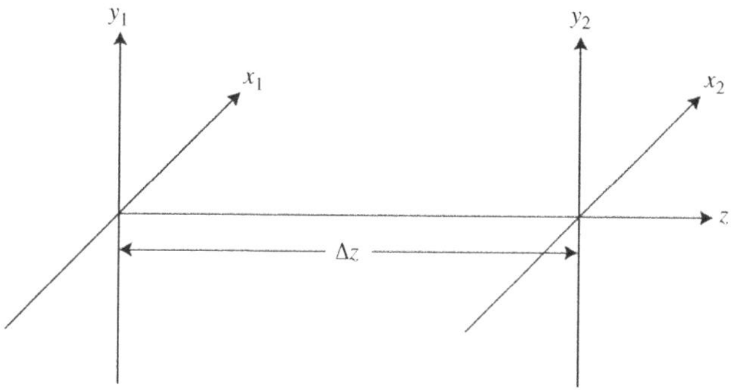

Figure 2.37 Coordinate system for optical wave propagation.

the beam diameter can be scaled by the ratio of the focal length of two lenses, f_2/f_1. Re-imaging of optical planes, also known as optical relay, is important for beam control since it enables the placement of beam control sensors and actuators at optical planes that would otherwise be physically inaccessible within the system. Certain telescope designs can be modeled as 4f optical systems, in which two lenses are used to re-image the entrance pupil plane onto the exit pupil plane, and the telescope magnification is determined by the ratio of the focal lengths of the two lenses. 4f optical systems can be also designed with off-axis parabola mirrors instead of lenses shown in Figure 2.36. A sufficiently large f-number in a 4f system ($f/\# = D_1/f_1 = D_2/f_2$), such as $f/\# > 10$, results in minimal wavefront aberrations introduced by the system. Due to the intermediate focus between the two lenses, re-imaging with a 4f system may not be feasible for certain beam control applications (e.g., high-energy lasers).

2.4 Physical Optics and Fourier Optics

Physical optics, also known as wave optics, is a branch of optics that studies interference, diffraction, polarization, and other phenomena for which the ray approximation of geometric optics is not valid. Physical optics often uses wave equations and principles like Huygens' principle to describe light behavior and involves solving differential equations to understand wave propagation and interactions. Physical optics is crucial for understanding phenomena like the design of diffraction gratings, the behavior of laser beams, and the analysis of interference patterns.

2.4.1 Scalar Diffraction Theory

When light waves encounter an aperture, diffraction occurs. Near-field diffraction (Fresnel diffraction) and far-field diffraction (Fraunhofer diffraction) refer to different regions of the diffraction pattern produced. In the paraxial regime and for apertures much larger than the wavelength, where polarization effects are negligible, the vector optical field can be approximated by a scalar field function describing the amplitude and phase of the wave along the propagation direction.

Using the coordinate system shown in Figure 2.37, $r_1 = (x_1, y_1)$ represents the light source plane (at the aperture) and $r_2 = (x_2, y_2)$ represents the observation plane. Given the source-plane optical field $U(x_1, y_1)$, light field $U(x_2, y_2)$ is given by the Fresnel diffraction integral using paraxial approximations

$$U(x_2, y_2) = \frac{e^{ik\Delta z}}{i\lambda\Delta z}\int_{-\infty}^{\infty}\int_{-\infty}^{\infty} U(x_1, y_1)e^{i\frac{k}{2\Delta z}[(x_2-x_1)^2+(y_2-y_1)^2]}dx_1dy_1. \tag{2.12}$$

When light propagates very far $\left(\Delta z > \frac{2D^2}{\lambda},\right)$ from its source aperture, the optical field in the observation plane is very closely approximated by the Fraunhofer diffraction integral,

$$U(x_2, y_2) = \frac{e^{ik\Delta z}e^{i\frac{k}{2\Delta z}(x_2^2+y_2^2)}}{i\lambda\Delta z}\int_{-\infty}^{\infty}\int_{-\infty}^{\infty} U(x_1, y_1)e^{-i\frac{k}{\Delta z}(x_2x_1+y_2y_1)}dx_1dy_1. \tag{2.13}$$

In general, these integrals do not have closed-form analytical solutions except for special aperture cases. Fourier optics based on Fourier transformations and convolutions provides effective means to analyze optical system behavior such as diffraction of an aperture and lens.

2.4.2 Fourier Optics

Fourier optics is the subfield of physical optics using Fourier transforms (FTs) and convolutions, in which the waveform being considered is regarded as made up of a combination, or superposition, of plane waves. Fourier optics focuses on analyzing optical systems in the frequency domain. It interprets optical signals in terms of their spatial frequency content, using Fourier transforms to switch between spatial and frequency representations. Physical optics is crucial for understanding fundamental wave behaviors and phenomena, whereas Fourier optics is essential for designing and analyzing imaging systems, especially in terms of their spatial frequency response.

In the case of Fraunhofer diffraction, the Fraunhofer diffraction integral equation can be represented as a Fourier transform of the aperture function (field at the source plane),

$$U\left(x_2, y_2\right) = \frac{e^{ik\Delta z}e^{i\frac{k}{2\Delta z}(x_2^2+y_2^2)}}{i\lambda\Delta z}\mathcal{F}\left\{U\left(x_1, y_1\right)\right\}\Bigg|_{f_{x_1}=\frac{x_2}{\lambda\Delta z},\ f_{y_1}=\frac{y_2}{\lambda\Delta z}} \tag{2.14}$$

With Fourier optics, the evaluation of a Fraunhofer diffraction integral is a simple matter of performing a Fourier transformation with the appropriate multipliers and spatial scaling. In the

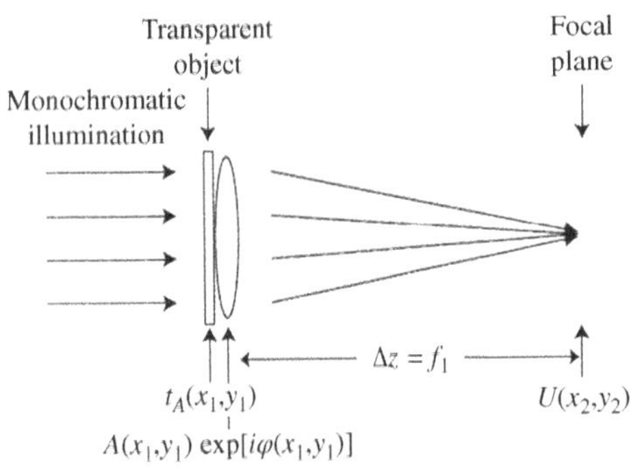

Figure 2.38 Diagram of a lens.

case of a light wave that encounters a lens, Fresnel diffraction integral in the paraxial approximation closely represents the optical lens.

In Figure 2.38, the light field at $U(x_1, y_1)$ is written as

$$U(x_1, y_1) = t_A(x_1, y_1)A(x_1, y_1)e^{-i\frac{k}{2f_l}(x_1^2+y_1^2)} \tag{2.15}$$

where $t_A(x_1, y_1)$ is the aperture transmittance of the object and $P(x_1, y_1) = A(x_1, y_1)e^{i\phi(x_1,y_1)}$ is the complex pupil function that represents the lens, , accounting for both amplitude apodization and phase modulation. The light at the focal plane becomes

$$\begin{aligned} U(x_2, y_2) = \frac{1}{i\lambda f_l} e^{i\frac{k}{2f_l}(x_2^2+y_2^2)} \int_{-\infty}^{\infty}\int_{-\infty}^{\infty} t_A(x_1, y_1) \\ \times A(x_1, y_1)e^{-i\frac{2\pi}{\lambda f_l}(x_2x_1+y_2y_1)}dx_1\,dy_1. \end{aligned} \tag{2.16}$$

This equation can be further expressed in terms of a Fourier transformation so that

$$U\left(x_2, y_2\right) = \frac{1}{i\lambda f_l} e^{i\frac{k}{2f_l}(x_2^2+y_2^2)} \mathscr{F}\left\{t_A\left(x_1, y_1\right)A\left(x_1, y_1\right)\right\}\Big|_{f_x=\frac{x_2}{\lambda f_l},\ f_y=\frac{y_2}{\lambda f_l}} \tag{2.17}$$

Eq. 2.17 shows that a lens performs a Fourier transform of the input field onto its back focal plane, accomplished by a phase factor, $e^{i\frac{k}{2f_l}(x_2^2+y_2^2)}$, and coordinate scaling factor, $\frac{1}{i\lambda f_l}$.

2.4.3 Amplitude Spread Function and Point Spread Function

The amplitude spread function (ASF) describes how the amplitude of an optical wave is distributed in the image plane when a point source is imaged by an optical system. It focuses on the amplitude of the wave, rather than its intensity. The point spread function (PSF) represents the intensity distribution in the image plane and is obtained by taking the squared magnitude of the ASF. ASF and PSF provide crucial insights into how an optical system performs. ASF observed at a distance Δz from the pupil can be written in terms of Fourier transformation such that

$$\begin{aligned} h(x_2, y_2) &= \frac{1}{\lambda \Delta z}\int_{-\infty}^{\infty}\int_{-\infty}^{\infty} t_A(x_1, y_1)A(x_1, y_1)e^{-i\frac{2\pi}{\lambda z_i}(x_2x_1+y_2y_1)}dx_1dy_1 \\ &= \frac{1}{\lambda \Delta z}\mathscr{F}\left\{t_A(x_1, y_1)\mathcal{P}(x, y)\right\}_{f_x=\frac{x_2}{\lambda z_i},\ f_y=\frac{y_2}{\lambda z_i}} \end{aligned} \tag{2.18}$$

Figure 2.39 illustrates how the intensity of light in the image plane from a point source can be computed using Fourier transformation.

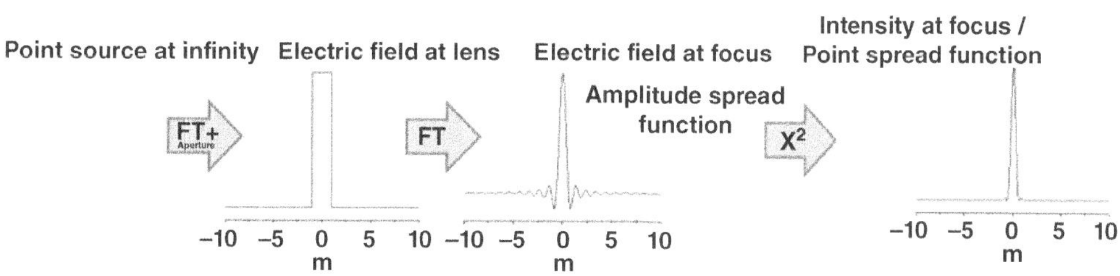

Figure 2.39 Fourier transformation relationship.

Figure 2.40 Incoherent imaging using convolutional.

PSF can be nonuniform across the focal plane and is field-angle-dependent since each field has its own optical path. It is often assumed that the PSF is invariant across the focal plane; however, PSF could be temporally variant such as imaging through the atmosphere.

2.4.4 Coherent and Incoherent Imaging System

In analogy to signal processing, where a system's response to a delta function characterizes its impulse response, the amplitude spread function (ASF) and point spread function (PSF) describe the optical system's response to a point source, effectively serving as its spatial impulse response. When the light is coherent, imaging systems are linear in the optical field. Accordingly, the image amplitude $U_i(x_2, y_2)$ is the convolution of the object's amplitude $U_o(x_2, y_2)$ with the amplitude spread function $h(x_2, y_2)$ according to

$$\begin{aligned} U_i(x_2, y_2) &= \int_{-\infty}^{\infty}\int_{-\infty}^{\infty} h\left(x_2 - \eta, y_2 - \xi\right) U_o\left(\eta, \xi\right) d\xi d\eta \\ &= h(x_2, y_2) \otimes U_o(x_2, y_2). \end{aligned} \tag{2.19}$$

Eq. 2.19 assumes a unit magnification imaging system; incorporating magnification simply requires scaling the object coordinates accordingly.

In incoherent imaging systems, interference does not affect image formation, and the system is linear in intensity rather than in the optical field. As a result, the image intensity is given by the convolution of the object's intensity with the system's PSF. Figure 2.40 illustrates a simulated imaging system in which a pristine image, free of optical aberrations, is convolved with the system's PSF to model the effects of optical aberrations.

2.4.5 Optical Transfer Function and Modulation Transfer Function

The optical transfer function (OTF) describes how well an optical system can transfer different spatial frequencies from the object to the image. It essentially measures the system's ability to reproduce the detail and contrast of the original scene in the resulting image. The OTF is defined as the Fourier transform of the PSF such that

$$\text{OTF}\left(f_x, f_y\right) = \mathcal{F}\left\{\text{PSF}\left(x, y\right)\right\} \tag{2.20}$$

The modulation transfer function (MTF) is given by the absolute value of the OTF such that

$$\text{MTF}\left(f_x, f_y\right) = \left|\text{OTF}\left(f_x, f_y\right)\right| \tag{2.21}$$

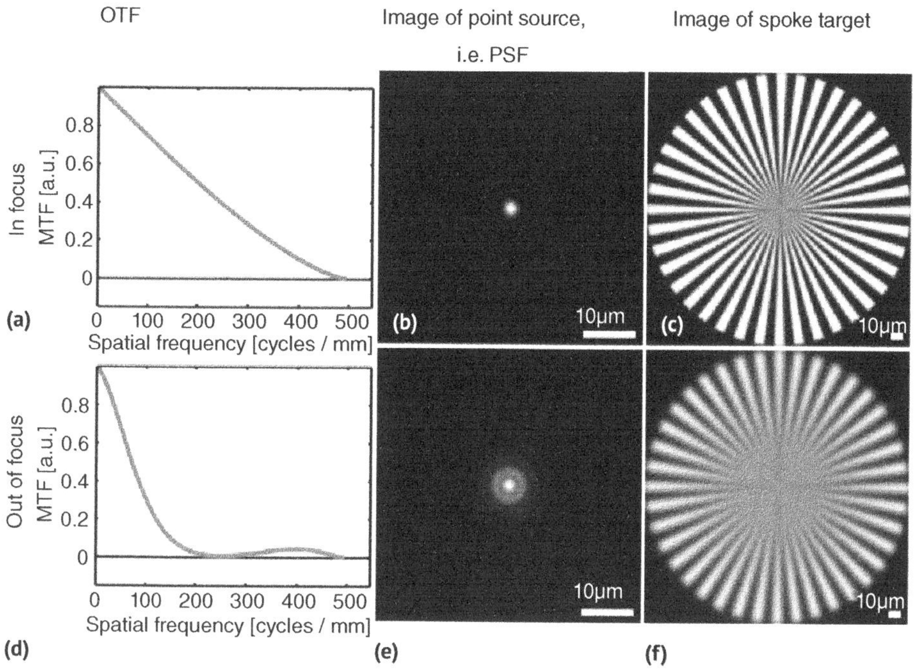

Figure 2.41 OTF and MTF for symmetric in-focus and out-of-focus beams, along with their corresponding PSF images and simulated spoke target images. *Source*: Tom.vettenburg/Wikimedia Commons/CC BY-SA 4.0.

Figure 2.41 shows the example OTF and MTF of in-focus and out-of-focus beams and corresponding PSF. While ASF and PSF measure the performance of an optical system in the spatial domain, OTF and MTF are commonly used to assess system performance in the frequency domain.

References

1 https://www.horticulture.red/en/the-spectrum-and-colors-of-light/.
2 https://en.wikipedia.org/wiki/File:Commercial_laser_lines.svg.
3 https://en.wikipedia.org/wiki/Refractive_index.
4 https://www.telescope-optics.net/diffraction.htm.
5 Adaptive Optics Class Notes, UC Santa Cruz.

3

Feedback Control

Feedback control system analysis begins with mathematical models, typically based on small perturbations around an operating point and described by ordinary differential equations. Direct manipulation of these equations can be cumbersome. To address this, systematic methods such as the Laplace transform, state-space, and optimal control are employed to simplify analysis and design. The following chapters will explore each of these methods, providing an introductory overview of their applications in control system design. The emphasis will be on concepts and practical usage, while detailed derivations will not be covered. Interested readers are encouraged to consult the references provided for a deeper understanding of mathematical foundations and advanced techniques.

3.1 Foundations of Classical Control

3.1.1 The Laplace Transform

The Laplace transform converts linear differential equations into the complex frequency or Laplace domain, simplifying integration into algebraic manipulations. The basic Laplace transform of a time signal $f(t)$ is

$$F(s) = \int_0^\infty f(t)e^{-st}dt$$

or more compactly

$$F(s) = \mathscr{L}[f(t)]. \tag{3.1}$$

It is assumed that $f(t)$ is zero for all times before $t = 0$. Here, $s = j\omega$ represents the complex frequency. The following properties of the Laplace transform are particularly useful in feedback control

$$\mathscr{L}[f_1(t) + f_2(t)] = \mathscr{L}[f_1(t)] + \mathscr{L}[f_2(t)] \tag{3.2}$$

$$\mathscr{L}[af(t)] = a\mathscr{L}[f(t)] \tag{3.3}$$

$$\mathscr{L}\left[\frac{d}{dt}f(t)\right] = sF(s) - f(0) \tag{3.4}$$

Optical Beam Control: Imaging Satellites and Laser Systems, First Edition.
Brij N. Agrawal, Jae Jun Kim and Sachin Agrawal.

$$\mathscr{L}\left[\frac{d^{\mathrm{n}}}{dt^{\mathrm{n}}}f(t)\right] = s^{\mathrm{n}}F(s) - \sum_{i=1}^{\mathrm{n}}\frac{d^{i-1}}{dt^{i-1}}f(0)s^{n-i}. \tag{3.5}$$

A table of Laplace transform operations can be found in most textbooks on feedback control [1].

3.1.2 The Inverse Laplace Transform

The inverse Laplace Transform converts $F(s)$ to the corresponding time function $f(t)$ and is given by

$$f(t) = \frac{1}{2\pi j}\int_{\sigma-j\infty}^{\sigma+j\infty} F(s)e^{st}ds$$

where $\sigma = Re(s)$ is the real number to the right of all poles of $F(s)$. In practice, these complex functions are broken into simpler components using tables. The Inverse Laplace Transform is more compactly expressed as

$$f(t) = \mathscr{L}^{-1}[F(s)]. \tag{3.6}$$

Example: Laplace transform of a lumped-parameter system.

Consider the system described by $\dfrac{d^2x}{dt^2} + 5\dfrac{dx}{dt} + 6x = 0$, with $x(0) = 0$ and $\dfrac{dx(0)}{dt} = 7$. Applying the Laplace transform to each term

$$\left(s^2X(s) - sx(0) - \frac{dx(0)}{dt}\right) + 5(sX(s) - x(0)) + 6X(s) = 0$$

which with some algebraic manipulation and substitution of the initial conditions

$$X(s) = \frac{7}{s^2 + 5s + 6} = \frac{7}{(s+3)(s+2)}. \tag{3.7}$$

The time-domain response, obtained from inverse Laplace transform tables is

$$x(t) = 7e^{-2t} - 7e^{-3t}.$$

3.1.3 Transfer Functions, Zeros, Poles

The transfer function, derived from linear differential equations of a dynamical system relates the Laplace-transformed output $Y(s)$ to the input $U(s)$

$$\frac{Y(s)}{U(s)} = H(s) \text{ or } Y(s) = H(s)U(s) \tag{3.8}$$

where $H(s)$ is called the transfer function because it shows how the input is transferred to the output. For physical systems, $H(s)$ is a rational polynomial such that

$$H(s) = \frac{N(s)}{D(s)}$$

where the order of $D(s) \geq$ the order of $N(s)$. The roots of $D(s)$ are called poles, values of s that make $H(s)$ infinite, while the roots of $N(s)$ are the zeros, values of s that make $H(s)$ zero. For example, Eq (3.7) has poles, $s = -2$ and $s = -3$ but no zeros, as there are no terms in s in the numerator. Complex poles and zeros appear as conjugate pairs, expressed as $s_{1,2} = \sigma \pm j\omega$.

For many design problems, $H(s)$ is often factored as

$$H(s) = \frac{K(s+z_1)(s+z_2)\cdots(s+z_m)}{(s+p_1)(s+p_2)\cdots(s+p_n)} = \frac{K\prod_{i=1}^{m}(s+z_i)}{\prod_{i=1}^{n}(s+p_i)} \tag{3.9}$$

where $m \leq n$ and $-z_i$ and $-p_i$ are the nonzero finite zeros and poles. The Bode form is another factored form [2] that is expressed as

$$H(s) = \frac{K_B \prod_{i=1}^{m}\left(1+\frac{s}{z_i}\right)}{\prod_{i=1}^{n}\left(1+\frac{s}{p_i}\right)} \tag{3.10}$$

where K_B is called the Bode gain.

3.1.3.1 DC Motor Dynamics Model

To obtain the transfer function, we take the Laplace transform of the motor's equations of motion, yielding a relationship between input and output. A DC motor combines the armature electric circuit and the rotor's rotational dynamics [8], as shown in Figure 3.1.

Using Newton's Law ($T = I\alpha$) and Kirchoff's Voltage Law, the motor's dynamics are

$$J_m \frac{d^2\theta_m(t)}{dt^2} + b\frac{d\theta_m(t)}{dt} = K_t i_a(t) + \omega(t),\ K_e\dot{\theta}_m + L_a i_a + R_a i_a = v_a. \tag{3.11}$$

Taking the Laplace transform (assuming zero initial conditions), we have

$$sJ_m Y(s) + bY(s) = K_t I_a(s) + W(s),\ K_e Y(s) + sL_a I_a(s) + R_a(s)\, I_a(s) = V_a(s).$$

Solving and rearranging these equations gives

$$Y(s) = \frac{A}{(\tau_1 s+1)(\tau_2 s+1)} V_a(s) + \frac{B}{(\tau_1 s+1)(\tau_2 s+1)} W(s)$$

where A, B, τ_1, and τ_2 are derived constants. Here, $Y(s)$ is the output, $V_a(s)$ is the input, and $W(s)$ is the disturbance. Block diagrams, such as Figures 3.2 and 3.3, graphically represent the open-loop and closed-loop dynamics of the DC motor [1]. Block diagrams simplify solving equations while retaining insights into system behavior.

To analyze the open-loop and closed-loop responses, we compute $y(t)$ for a step change in the disturbance, $w(t) = -0.1\ N \cdot m$.

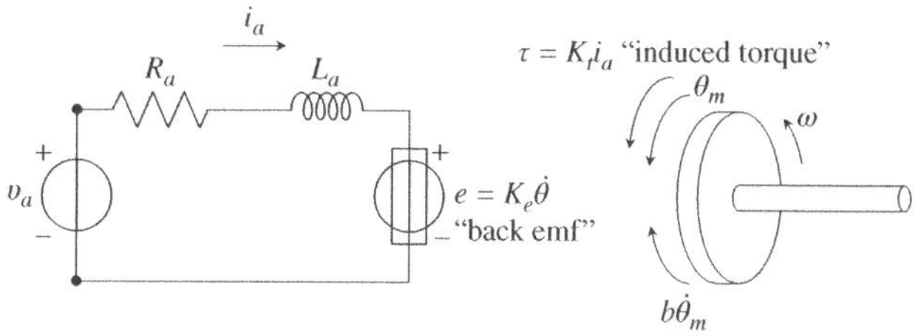

Figure 3.1 Combined armature electric circuit with the rotational dynamics of the rotor.

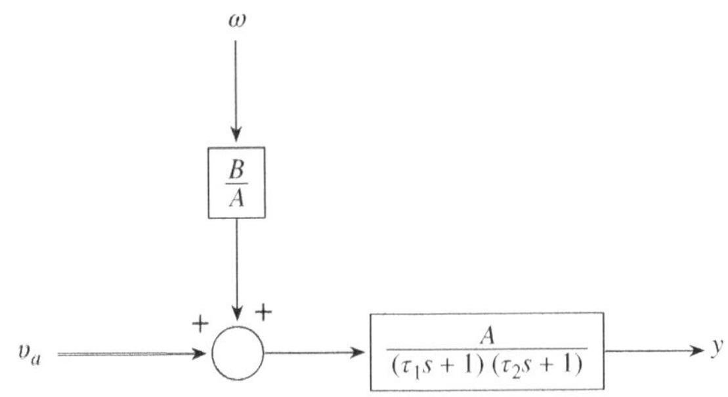

Figure 3.2 Open-loop block diagram of DC motor.

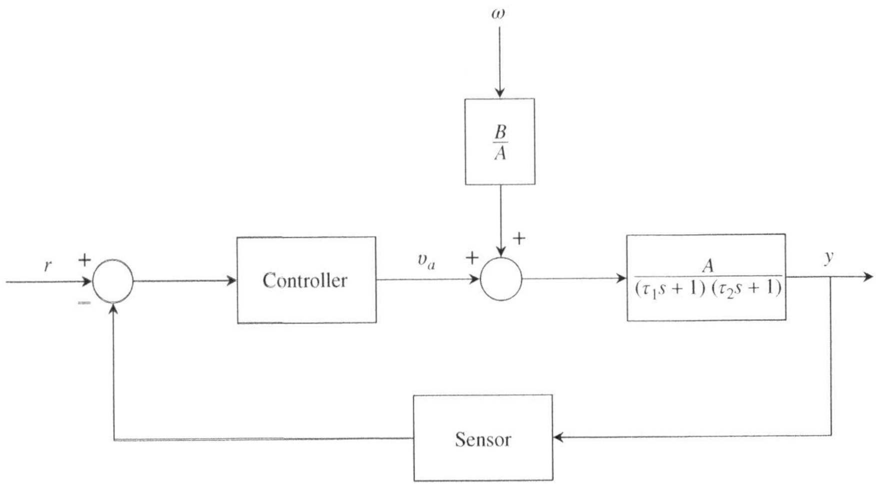

Figure 3.3 Closed-loop block diagram of DC motor.

Example: Open-loop dynamics of a DC motor:
The open-loop dynamics for the DC motor are

$$Y(s) = \frac{-0.1B}{s(\tau_1 s + 1)(\tau_2 s + 1)}$$

Using the partial fraction expansion

$$\frac{-0.1B}{s(\tau_1 s + 1)(\tau_2 s + 1)} = \frac{a}{(\tau_1 s + 1)} + \frac{b}{(\tau_2 s + 1)} + \frac{c}{s},$$

we can solve for a, b,c to compute $y(t)$.

Initial and Final Value Theorems

The initial value $f(0^+)$ of the function $f(t)$ whose Laplace Transform is $F(s)$ is

$$f(0^+) = \lim_{t \to 0} f(t) = \lim_{s \to \infty} sF(s),\ t > 0, \tag{3.12}$$

and final value $f(\infty)$, if it exists, is

$$f(\infty) = \lim_{t \to \infty} f(t) = \lim_{s \to 0} sF(s). \tag{3.13}$$

3.1.4 First-order Systems

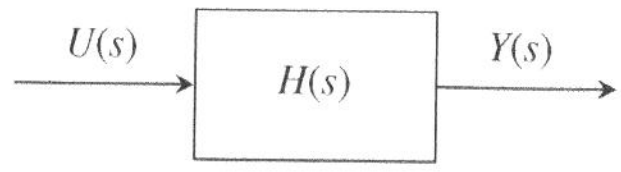

Figure 3.4 Input–output model of a system in the Laplace domain.

The system in Figure 3.4 is modeled in the Laplace domain, where $H(s)$ represents the transfer function. For a single-input single-output system, the response depends on the poles of $H(s)$.

For a first-order system (one real pole)

$$H(s) = \frac{1}{s+a} \rightarrow h(t) = \mathscr{L}^{-1}(H(s)) = e^{-at} \cdot 1(t) \text{ where } a \in \mathbb{R}. \tag{3.14}$$

Figures 3.5–3.7 illustrate the pole-zero maps and impulse responses for $a < 0$, $a > 0$, and $a = 0$. The time constant, $\tau = \frac{1}{a}$, is the time it takes the response to reach $\frac{1}{e}$ times the initial value [1].

3.1.5 Second-order Systems

For a general second-order system, the poles are complex conjugates

$$s_1, s_2 = -\zeta\omega_n \pm j\omega_n\sqrt{1-\zeta^2}. \tag{3.15}$$

Taking the inverse Laplace transform of a second-order system, we have

$$H(s) = \frac{\omega_n^2}{s^2 + 2\zeta\omega_n s + \omega_n^2} \xrightarrow{\mathscr{L}^{-1}} h(t) = \frac{\omega_n}{\sqrt{1-\zeta^2}} e^{-\zeta\omega_n t} \sin\omega_d t \cdot 1(t), \tag{3.16}$$

where ω_n= undamped natural frequency, ζ = damping ratio, and the damped natural frequency is $\omega_d = \omega_n\sqrt{1-\zeta^2}$. Figure 3.8 illustrates the pole-zero map of the impulse response for a second-order system.

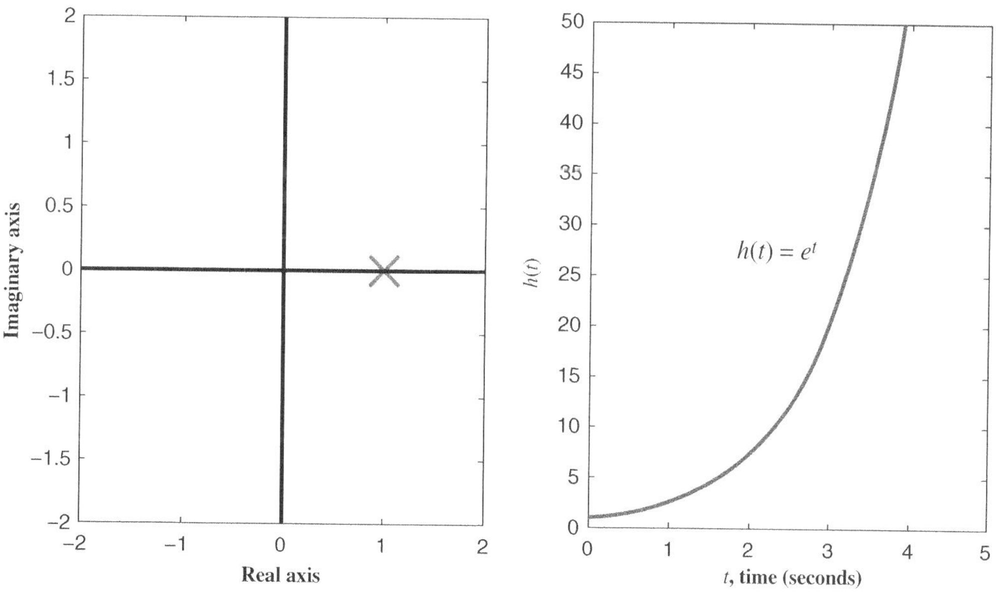

Figure 3.5 Pole-zero map and impulse response for $a = -1$.

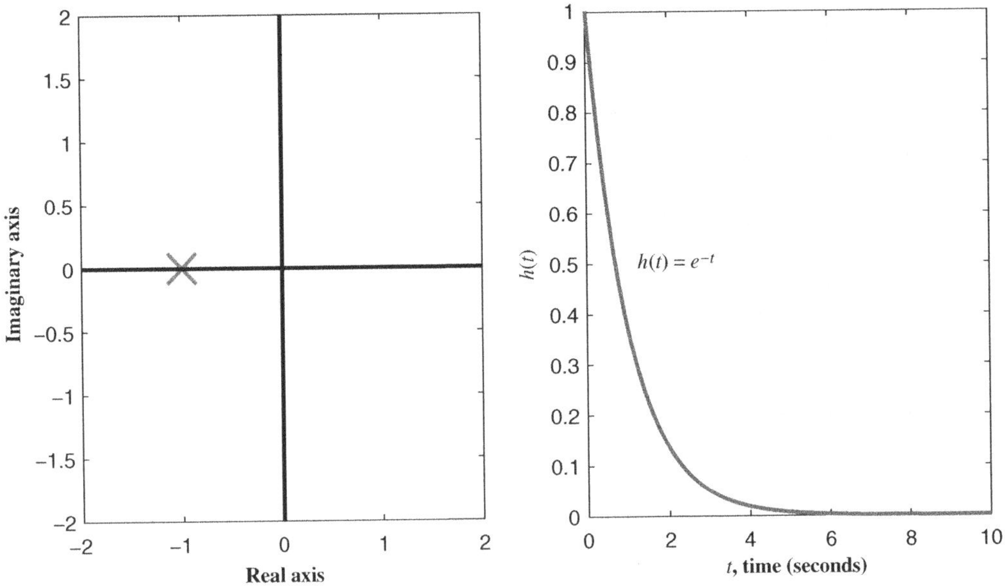

Figure 3.6 Pole-zero map and impulse response for $a = 1$.

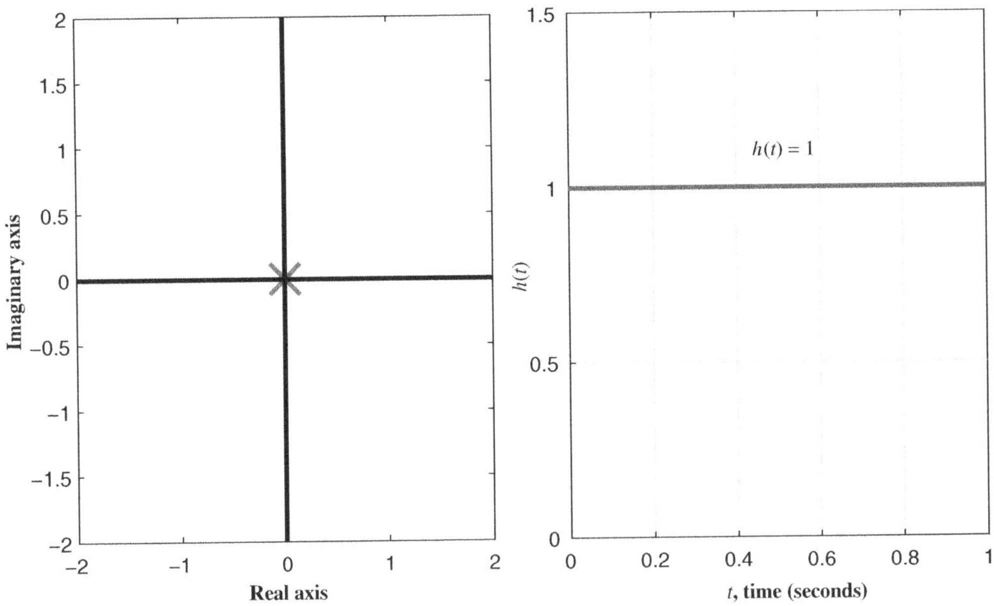

Figure 3.7 Pole-zero map and impulse response for $a = 0$.

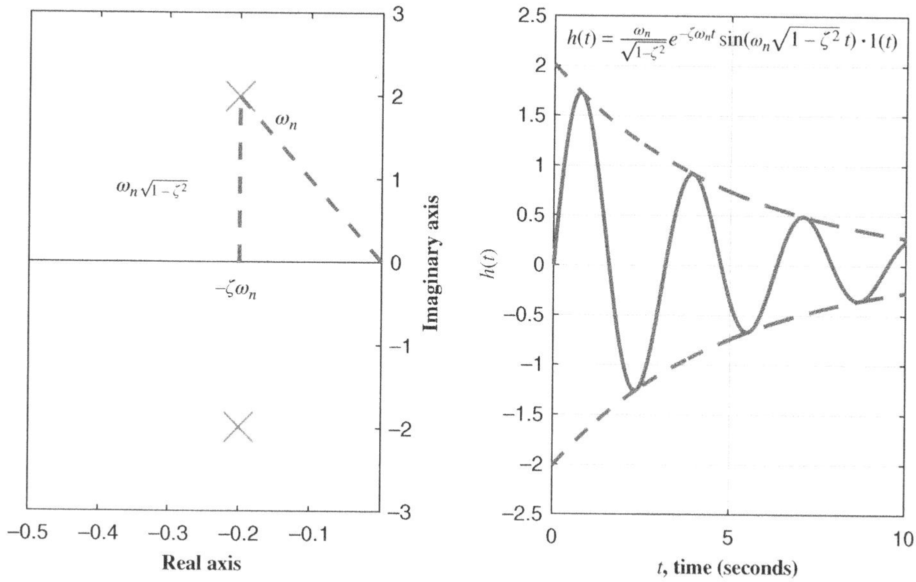

Figure 3.8 Pole-zero map and impulse response for $s_1, s_2 = -\zeta\omega_n \pm j\omega_n\sqrt{1-\zeta^2}$.

The step response of this second-order system is

$$Y(s) = H(s)\cdot\frac{1}{s} \overset{\mathcal{L}^{-1}}{\Longrightarrow} y(t) = 1 - e^{-\zeta\omega_n t}\left(\cos\omega_d t + \frac{\zeta\omega_n}{\omega_d}\sin\omega_d t\right). \tag{3.17}$$

From the step response of the second-order system, key performance metrics include [1]:

1. Rise time, t_r, is the time it takes the step response to rise from 10 to 90% of its final value. For $\zeta = 0.5$, $t_r \cong \frac{1.8}{\omega_n}$.
2. Settling time, t_s, is the time it takes transients to decay within 1% of the final value of the exponential part of the second-order response or,

$$e^{-\zeta\omega_n t_s} = 0.01 \Longrightarrow t_s = \frac{4.6}{\zeta\omega_n}. \tag{3.18}$$

3. Peak time, t_p, is the time to reach the maximum overshoot,

$$t_p = \frac{\pi}{\sqrt{1-\zeta^2}}. \tag{3.19}$$

4. The maximum overshoot relative to the final value, M_p, is illustrated in Figure 3.9, and can be calculated as

$$M_p = \frac{y(t_p) - 1}{1} = e^{-\zeta\omega_n\pi/\sqrt{1-\zeta^2}}. \tag{3.20}$$

Overshoot, M_p, is a decaying exponential function of ζ.

To meet minimum specifications on t_r, M_p, t_s, the poles of our system need to be on the s-plane such that

$$\omega_n \geq \frac{1.8}{t_p},\ \zeta \geq \zeta(M_p),\ \text{and}\ |\zeta\omega_n| \geq \frac{4.6}{t_s}.$$

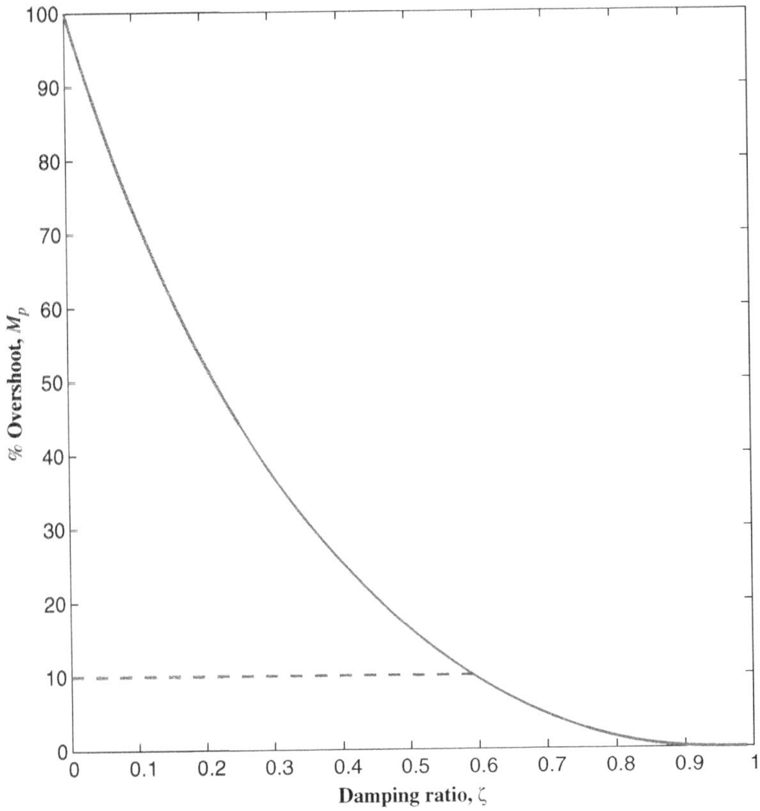

Figure 3.9 M_p as a function of ζ.

3.1.5.1 Effect of Zeros on the Response

Zeros influence the response by modifying the coefficients of the exponential functions, as seen in the following transfer functions $H_1(s)$ and $H_2(s)$

$$H_1(s) = \frac{2}{(s+1)(s+2)}, \quad H_2(s) = \frac{2(s+1.1)}{1.1(s+1)(s+2)}.$$

Partial fraction expansion of H_1 and H_2 yield

$$H_1(s) = \frac{2}{s+1} - \frac{2}{s+2}, \quad H_2(s) = \frac{0.18}{s+1} - \frac{1.64}{s+2}.$$

The zero at $s = -1.1$ in $H_2(s)$ reduces the coefficient of $(s+1)$. The reduction happens because the zero at $s = -1.1$ in $H_2(s)$ almost cancels the pole at $s = -1$. Let us add a zero at $s = -\alpha\zeta\omega_n$ and rearrange $H(s)$ with normalizing for time and zero locations such that

$$H(s) = \frac{\left(\frac{s}{\alpha\zeta\omega_n}\right) + 1}{\left(\frac{s}{\omega_n}\right)^2 + 2\zeta\left(\frac{s}{\omega_n}\right) + 1}.$$

As we increase α, the zero moves farther from the pole locations and has less effect on the response. Figure 3.10, plots the step response for $\zeta = 0.5$ and different values of α.

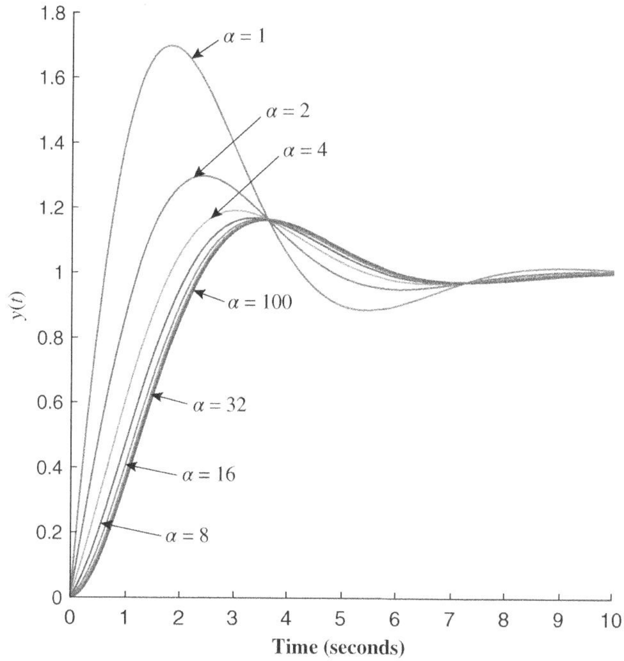

Figure 3.10 Step response of second-order system for different values of zero at α.

As illustrated in Figure 3.10, the main effect of the zero, or increasing α is on the overshoot M_p whereas it has little influence on the settling time.

3.1.5.2 Bounded-Input Bounded-Output (BIBO) Stability

A system is BIBO stable if every bounded input produces a bounded output. For a linear time-invariant (LTI) system, this condition is met when all poles of the system's transfer function, p_i, lie strictly in the left-half plane [1], where

$$\mathbb{R}(p_i) < 0. \tag{3.21}$$

3.1.6 Basic Equations for Feedback

For the open-loop system shown in Figure 3.11, there is no feedback to adjust the control effort based on the output. The output $Y(s)$ depends on the reference signal $R(s)$ and the plant disturbance $W(s)$ [1].

The open-loop output is

$$Y(s) = D(s)G(s)R(s) + G(s)W(s). \tag{3.22}$$

If we take the error as the difference between the reference input and system output, then

$$E(s) = R(s) - Y(s) = [1 - D(s)G(s)]R(s) - G(s)W(s). \tag{3.23}$$

The open-loop transfer function is

$$T_{OL}(s) = D(s)G(s). \tag{3.24}$$

Figure 3.12 is the basic unity feedback closed-loop structure with the sensor noise, $V(s)$.

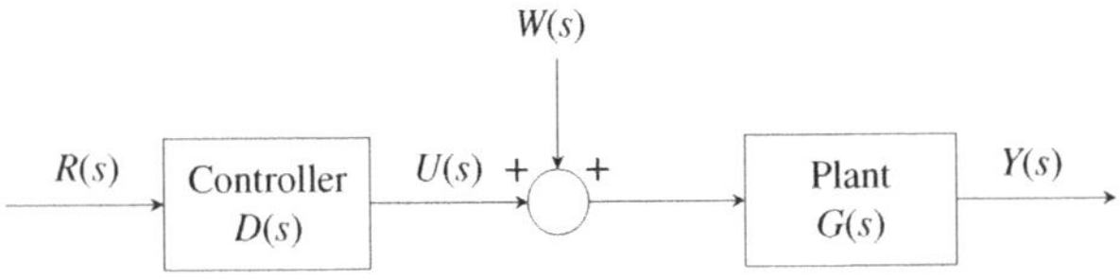

Figure 3.11 Open-loop system showing the reference $R(s)$, control $U(s)$, disturbance $W(s)$, and output $Y(s)$.

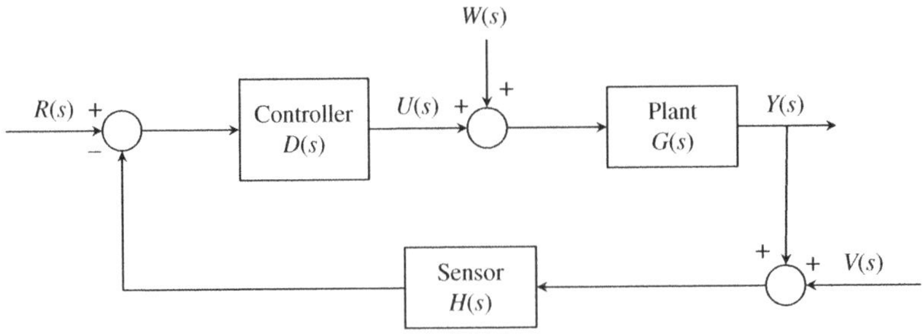

Figure 3.12 Basic unity feedback block diagram with sensor noise.

The equations for the output and control are given by the superposition of the responses of the three inputs individually (letting $H(s) = 1$)

$$Y(s) = \frac{D(s)G(s)}{1 + D(s)G(s)}R(s) + \frac{G(s)}{1 + D(s)G(s)}W(s) - \frac{D(s)G(s)}{1 + D(s)G(s)}V(s). \tag{3.25}$$

After some algebraic manipulation, we can find the error transfer function for the closed-loop system

$$E(s) = \frac{1}{1 + D(s)G(s)}R(s) - \frac{G(s)}{1 + D(s)G(s)}W(s) + \frac{D(s)G(s)}{1 + D(s)G(s)}V(s). \tag{3.26}$$

The closed-loop transfer function is ($W(s) = V(s) = 0$)

$$T_{CL}(s) = \frac{Y(s)}{R(s)} = \frac{D(s)G(s)}{1 + D(s)G(s)}. \tag{3.27}$$

Let the controller be a pure gain, $D(s) = K$, and drop the subscript so that the closed-loop transfer function is now

$$T(s) = \frac{KG(s)}{1 + KG(s)}.$$

To see how $T(s)$ changes with respect to K, we can define differential sensitivity

$$\frac{\partial T(s)/T(s)}{\partial K/K} = \frac{K}{T(s)}\frac{\partial T(s)}{\partial K}$$

which can be used to find the sensitivity function

$$S(s) = \frac{1}{1 + D(s)G(s)}. \tag{3.28}$$

Also, we can call $T(s)$ the complementary sensitivity function since

$$T(s) + S(s) = \frac{D(s)G(s)}{1 + D(s)G(s)} + \frac{1}{1 + D(s)G(s)} = 1. \tag{3.29}$$

If we set our output and error equations in terms of the sensitivity and complementary sensitivity functions, we get the following compact form

$$Y(s) = T(s)R(s) - T(s)V(s) + S(s)G(s)W(s) \tag{3.30}$$

$$E(s) = S(s)R(s) - S(s)G(s)W(s) + T(s)V(s). \tag{3.31}$$

3.1.6.1 Tracking, Regulation, and Noise

Control system tracking ensures that the system's output, $Y(s)$, closely follows a reference input, $R(s)$. Regulation aims to minimize errors due to disturbances, $W(s)$. For tracking, a stable plant without right-half-plane (RHP) poles or zeros allows the controller $D(s)$ to replace the plant's dynamics with the desired ones. Using Eq (3.30), the response to a disturbance is given by

$$S(s)G(s)W(s) = \frac{G(s)}{1 + D(s)G(s)} W(s). \tag{3.32}$$

Minimizing $S(s)$ requires that $D(s)G(s)$ be large, especially at low frequencies, which improves tracking and disturbance rejection. However, a large $D(s)$ at high frequencies amplifies noise, as seen in the complementary sensitivity function. To balance these effects, $D(s)$ should dominate at low frequencies for disturbance rejection while remaining small at high frequencies to suppress noise.

3.1.7 System Type

For the unity feedback system in Figure 3.12, the error is defined as (assuming $W(s) = V(s) = 0$)

$$E(s) = \frac{1}{1 + G(s)D(s)} R(s). \tag{3.33}$$

Considering a polynomial input $r(t) = \frac{t^k}{k!}1(t)$, with $R(s) = \frac{1}{s^{k+1}}$, and assuming the closed-loop system is stable, the Final Value theorem gives [1]

$$e_{ss} = \lim_{s\to 0} sE(s) = \lim_{s\to 0} \frac{s}{1 + G(s)D(s)} \frac{1}{s^{k+1}} = \lim_{s\to 0} \frac{1}{s^n + G(s)D_0(s)} \frac{s^n}{s^k}. \tag{3.34}$$

The integrators (poles at the origin) of the open-loop transfer function have been factored out so that $G(s)D(s) = \frac{G(s)D_0(s)}{s^n}$ has n poles at the origin. If a system's open-loop transfer function has k poles, then it is said to be a type k system.

- If type 0 ($n = k = 0$), with step input $R(s) = \frac{1}{s}$ the steady-state error is

$$e_{ss} = \lim_{s\to 0} sE(s) = \lim_{s\to 0} \frac{1}{s^n + G(s)D_0(s)} \frac{s^n}{s^k} = \frac{1}{1 + G(0)D_0(0)}.$$

- For $n > k$, $e_{ss} = 0$; for $n < k$, $e_{ss} = \infty$; and for $n = k$, then $e_{ss} = \frac{1}{G(0)D_0(0)}$.

The steady-state errors across system types (0, 1, and 2) are summarized in Table 3.1. Steady-state error for system types 0, 1, and 2.

Table 3.1 Steady-state error for system types 0, 1, and 2.

System type	Unit step (1/s)	Unit ramp (1/s^2)	Unit parabola (1/s^3)
Type 0	$\frac{1}{1+G(0)D_0(0)}$	∞	∞
Type 1	0	$\frac{1}{G(0)D_0(0)}$	∞
Type 2	0	0	$\frac{1}{G(0)D_0(0)}$

3.1.8 Basic Feedback Controllers

3.1.8.1 Proportional Control

Proportional control produces a response proportional to the system error [1]

$$u(t) = K_p e(t), \tag{3.35}$$

with the transfer function

$$\frac{U(s)}{E(s)} = D(s) = K_p. \tag{3.36}$$

The closed-loop response is

$$\frac{Y(s)}{R(s)} = \frac{K_p G(s)}{1 + K_p G(s)}.$$

Figure 3.13 shows responses for $K_p = 2$, $K_p = 6$, $r(t) = 1(t)$, and $G(s) = \frac{1}{s^2 + 1.4s + 1}$.

The output, $y(t)$, of the system exhibits decreasing steady-state error and improved tracking for increasing K_p. However, higher values of K_p, can worsen transient performance (longer settling times, more overshoot, and lead to instability). Any Type 0 system with proportional control will have a nonzero steady-state offset in the response.

3.1.8.2 Integral Control

Integral control addresses steady-state error by integrating the error over time, which results in a control signal proportional to the accumulated error [1]

$$u(t) = K_i \int_0^t e(\tau)d\tau, \tag{3.37}$$

with the transfer function

$$\frac{U(s)}{E(s)} = D(s) = \frac{K_I}{s}, \tag{3.38}$$

which has infinite gain at DC (when $s = 0$). Because $u(t)$ is a function of all past values of $e(t)$ rather than just the current value means that constant disturbances can be canceled with zero error. The closed-loop response is

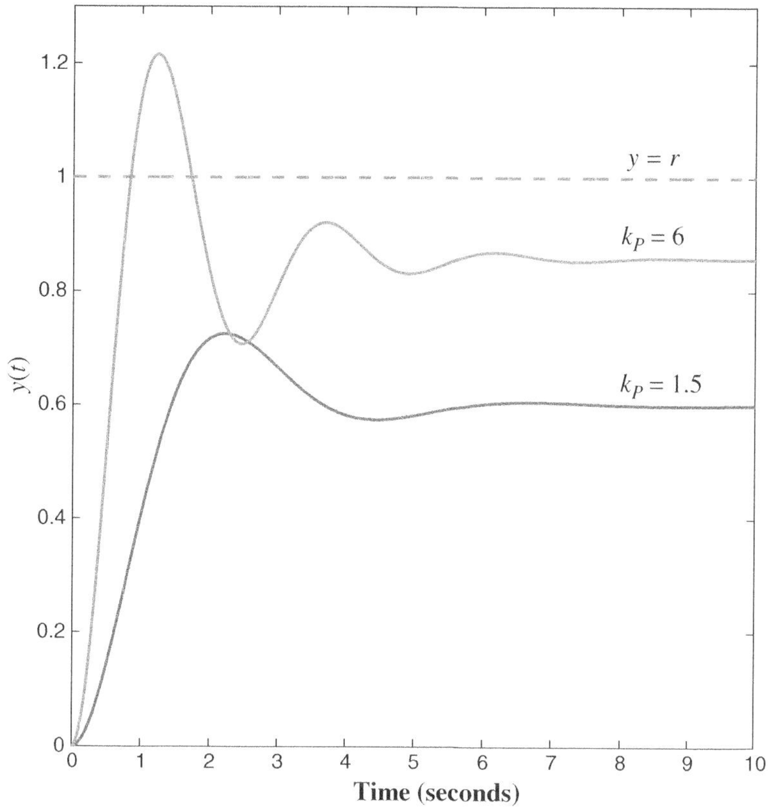

Figure 3.13 Steady-state tracking error and damping for increasing values of K_p.

$$\frac{Y(s)}{R(s)} = \frac{K_i G(s)}{s + K_i G(s)}.$$

Typically, integral control is combined with proportional control. Figure 3.14 illustrates the integral control response for $K_i = 0.5$.

Figure 3.14 shows that integral feedback results in zero steady-state error for both tracking and disturbance rejection. This property holds regardless of the value of K_i, as long as the closed-loop system remains stable. While integral control effectively eliminates steady-state error, it may compromise system stability. Integral control causes the closed-loop system to become Type 1. Type 1 systems, including those with integral control, exhibit a constant tracking error in response to a ramp input.

3.1.8.3 Derivative Control

Derivative control improves system stability and speeds up transient response by responding to the derivative or change in error [1]

$$u(t) = K_d \dot{e}(t), \tag{3.39}$$

with the transfer function

$$\frac{U(s)}{E(s)} = D(s) = K_d s. \tag{3.40}$$

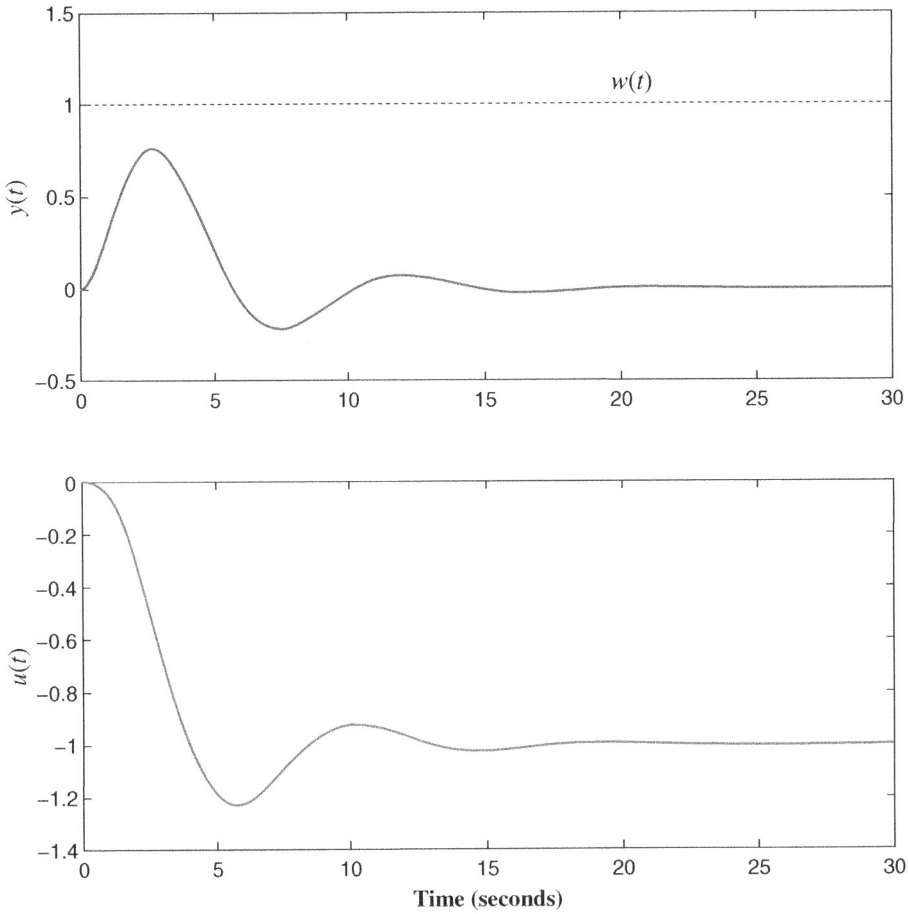

Figure 3.14 Constant error and disturbance rejection property using integral control.

Derivative control is not used by itself but often with proportional control. The derivative action must be used with care as it is sensitive to noise and can respond to rapidly changing signals. Derivate control often requires filtering.

3.1.8.4 Proportional-integral-derivative Control

When the three terms are combined, proportional–integral–derivative (PID) control achieves better transient and steady-state performance. The control law is [1]

$$u(t) = K_p e(t) + K_i \int_0^t e(\tau)d\tau + K_d \dot{e}(t), \tag{3.41}$$

with the transfer function

$$\frac{U(s)}{E(s)} = D(s) = K_p + \frac{K_i}{s} + K_d s. \tag{3.42}$$

A block diagram of a general PID controller is shown in Figure 3.15. Generally, K_p is selected for adequate rise time, K_i (units of gain per second) is selected for steady-state accuracy without

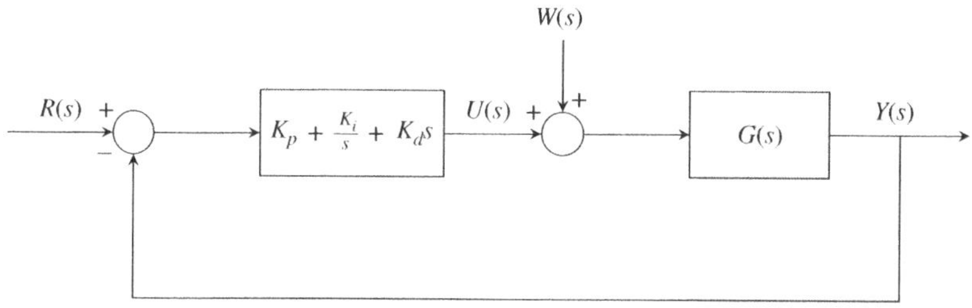

Figure 3.15 Block diagram with PID controller.

unacceptably degrading performance, K_d (units of gain · seconds) is selected to overcome excessive oscillation or too long a settling time.

If we have the second-order plant $G(s) = \dfrac{1}{s^2 + 1.4s + 1}$, then the characteristic equation is given by

$$1 + G(s)D(s) = s^3 + (1.4 + K_d)s^2 + K_p s + K_i,$$

where K_p, K_i, K_d can be tuned to achieve the desired performance. The control gains are tuned so that $K_p = 3$, $K_i = 15$, $K_d = 0.3$. The P, PI, and PID control responses to a step reference and disturbance input are shown below. Figure 3.16 shows that the integral term eliminates the steady-state error but introduces oscillations. Adding the derivative term for the PID reduces the oscillation while maintaining steady-state error.

Often in practice, the PID controller is rewritten as [2]

$$\frac{U(s)}{E(s)} = D(s) = K_p\left(1 + \frac{1}{sT_i} + sT_d\right), \tag{3.43}$$

where $T_i = \dfrac{K_p}{K_i}$ = integral action (or "reset time" in seconds), and $T_d = \dfrac{K_d}{K_i}$ = derivative action time (or "rate time" in seconds).

3.1.8.5 Heuristic Tuning of the PID Controller

In many practical systems, PID controllers are tuned without a closed-loop system model, particularly for open-loop stable plants that can safely undergo testing. The Ziegler–Nichols method offers a historical approach to transform step response characteristics into a practical PID design. It approximates the process response with a first-order lag (with gain K, time constant τ) and a pure delay T

$$G(s) \cong \frac{Ke^{-Ts}}{\tau s + 1}. \tag{3.44}$$

This method is suitable for plants without lightly damped complex poles or poles at the origin. The rules for Ziegler–Nichols PID tuning are summarized in Table 3.2.

If no pure time delay exists ($T = 0$), the proportional gain can get arbitrarily high. Any response other than a true first-order lag would potentially cause a measurable delay. Modern heuristic approaches have been developed based on the Internal Model Principle, namely the Simple Internal Model Control [5].

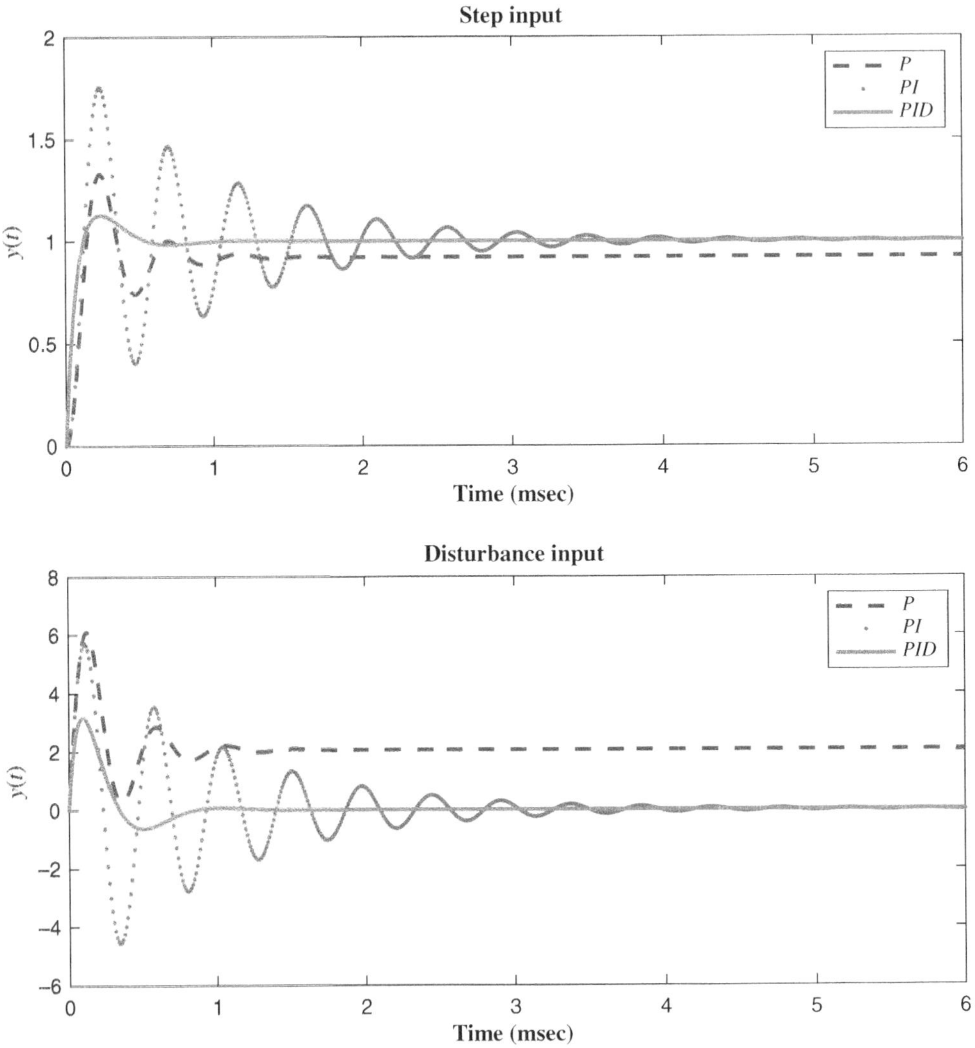

Figure 3.16 P, PI, and PID control to step reference input and step disturbance input.

Table 3.2 Rules for Ziegler–Nichols PID tuning.

P	$K_p = 1.0\tau/(KT)$		
PI	$K_p = 0.9\tau/(KT)$	$K_i = 0.27\tau/(KT^2)$	
PID	$K_p = 1.2\tau/(KT)$	$K_i = 0.60\tau/(KT^2)$	$K_d = 0.6\tau/K$

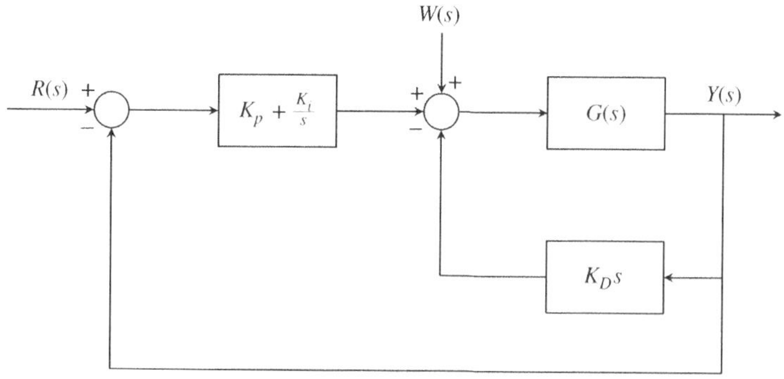

Figure 3.17 Block diagram of velocity feedback.

3.1.8.6 Velocity Feedback (Rate Feedback) Control

Derivative control can enhance rise time and minimize overshoot but is sensitive to noise and prone to "derivative kick," a sharp response caused by abrupt setpoint changes. This can lead to actuator wear and nonlinear behavior. To mitigate these effects, many systems use velocity feedback, where the derivative of the output, rather than the error, is fed back. A tachometer is a common example of a velocity feedback device [1].

Figure 3.17 illustrates a system that incorporates velocity feedback, where block $k_D s$ represents the tachometer, outputting a signal proportional to the derivative of $Y(s)$, which is the angular velocity of the motor shaft. The closed-loop transfer function for this configuration is

$$\frac{Y(s)}{R(s)} = \frac{D(s)G(s)}{1 + K_d sG(s) + D(s)G(s)}. \tag{3.45}$$

If $D(s)$ is a PI controller, the characteristic equation of the closed-loop system remains the same, regardless of whether the derivative term is in the forward or feedback path

$$s^3 + (1.4 + K_d)s^2 + (1 + K_p)s + K_i = 0.$$

When the derivative term is in the forward path, it introduces an extra zero in the transfer function. This can improve the transient response and speed, but the zero may also amplify noise. Placing the derivative term in the feedback path avoids this amplification, yielding a smoother response without the "kick," albeit at the cost of slightly slower system dynamics.

3.1.8.7 Feedforward Control

Feedforward control improves system response and tracking performance. As shown in Figure 3. 18, the feedforward path introduces a pre-compensating action, while the feedback path corrects deviations caused by disturbances or plant variations.

The system error response and output response transfer functions are

$$\frac{E(s)}{R(s)} = \frac{1 - G_{ff}(s)G(s)}{1 + D(s)G(s)}, \frac{Y(s)}{R(s)} = \frac{G_{ff}(s)G(s) + D(s)G(s)}{1 + D(s)G(s)}. \tag{3.46}$$

If $G_{ff}(s)G(s) = 1$, then the output tracks the input with zero error since $E(s)/R(s) = 0$ and $Y(s)/R(s) = 1$. However, plant parameters are often imprecisely known, so $G_{ff}(s)$ is typically an approximation. The feedback loop compensates for parameter uncertainties and disturbances, ensuring robustness.

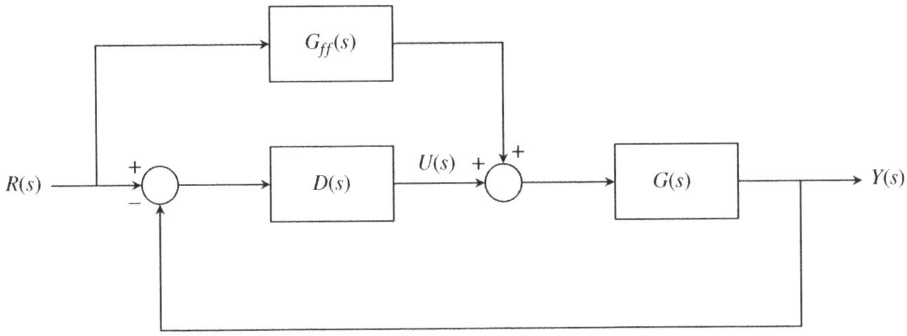

Figure 3.18 Feedforward control with error-correcting feedback path.

Example: A simple feedforward.

For the system in Figure 3.18, with $D(s) = K_d s + K_p$, $G(s) = \dfrac{A}{s^2 + as}$, and $G_{ff}(s) = \alpha(s)$, then the output response after some algebraic manipulation is

$$Y(s) = \frac{AK_d s + AK_p + A\alpha(s)}{s^2 + A(a + k_D)s + AK_p} R(s).$$

- If $\alpha(s) = 0$, the system tracks step inputs with zero error.
- If $\alpha(s) = (a * s)/A$, where $a* = a$, the system tracks ramp inputs with zero error.

The accuracy of feedforward control depends on precise knowledge of plant parameters, such as $a*$.

3.2 The Root Locus

3.2.1 The Root Locus Technique

The root-locus technique provides a design method based on the system's open-loop transfer function, which will give the closed-loop pole positions based on variations of the open-loop gain, K. The open-loop transfer function also called "loop gain" is $L(s) = D(s)G(s)$.

The closed-loop characteristic equation in root-locus form can be expressed as

$$1 + KL(s) = 0. \tag{3.47}$$

The root-locus of the closed-loop system is the set of all complex numbers s such that

$$|KL(s)| = 1 \text{ and } \angle L(s) = \pm 180^\circ \tag{3.48}$$

as K varies from 0 up to but not including infinity [8].

Example: DC motor root locus.

For the DC motor from voltage to position for $D(s) = 1$ the transfer function is

$$\frac{\Theta(s)}{V_a(s)} = \frac{Y(s)}{U(s)} = \frac{A}{s(s + c)}.$$

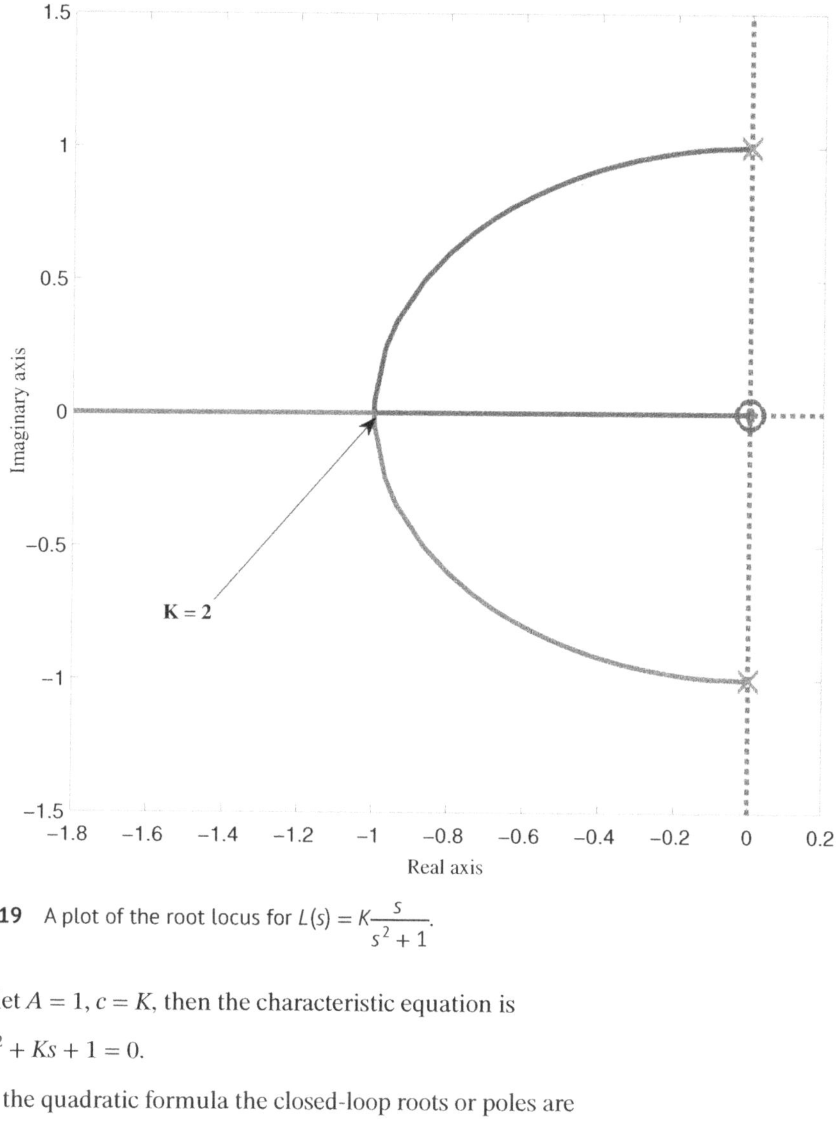

Figure 3.19 A plot of the root locus for $L(s) = K\frac{s}{s^2+1}$.

If we let $A = 1, c = K$, then the characteristic equation is

$$s^2 + Ks + 1 = 0.$$

Using the quadratic formula the closed-loop roots or poles are

$$s_{1,2} = -\frac{K}{2} \pm \sqrt{\frac{1}{4}K^2 - 1}.$$

The locus of solutions is shown below in Figure 3.19 with zeros indicated by x's and poles indicated by o's of the loop transfer function, $L(s)$.

When $K = 0$, the roots lie on the imaginary axis, resulting in an oscillatory response. The damping ratio ζ grows as K increases from 0. At $K = 2$, there are two roots at $s = -1$ and the two locus segments abruptly change directions and move in opposite directions along the real axis. The point where multiple roots come into the real axis is called a break-in point. The point when two or more roots move toward each other and break away is called the break-away point.

The procedure for drawing the root locus is to search for points s^* such that $\angle L(s^*) = \pm 180°$ and find a suitable K such that $|KL(s^*)| = 1$. Normally, root locus plots like Figure 3.20 are made using

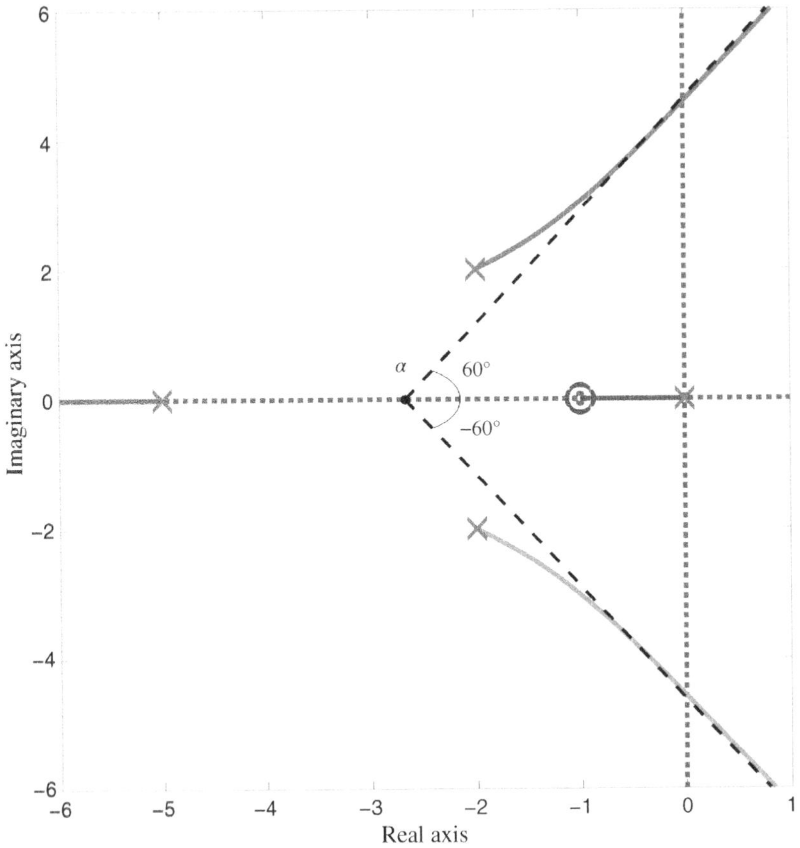

Figure 3.20 Root locus for $L(s) = \frac{s+1}{s(s+5)((s+2)^2+4)}$, with asymptotes at 60°, −60°.

software like Matlab, but it is also possible to get insight into the salient features of the system by quickly sketching the general shape of the root locus using the following rules [2].

1. **Asymptotes**: For $n-m$ branches where $K \to \infty$, the asymptotes' angles ϕ_i radiating out from the center point $s = \alpha$ on the real axis where

$$\phi_i = \frac{180°+360°(i-1)}{(n-m)}, \; i = 1, 2, \cdots, n-m,$$

$$\alpha = \frac{\sum p_i - \sum z_i}{n-m}, \text{ where } p_i \text{ are the poles, and } z_i \text{ are the zeros of } L(s).$$

For example, if $L(s) = \frac{s+1}{s(s+5)((s+2)^2+4)}$, we have $n-m = 4-1 = 3$,

$$\alpha = \frac{\sum_i p_i - \sum_i z_i}{n-m} = \frac{0-2+2j-2-2j-5-(-1)}{3} = -\frac{8}{3} = -2.67$$

$$\phi_i = \frac{180°+360°(i-1)}{(n-m)}, \; i = 1, 2, 3 \Longrightarrow \phi_1 = 60°, \; \phi_2 = 180°, \; \phi_2 = 300° = -60°.$$

2. **Real-axis Locus**: The locus lies to the left of an odd number of poles and zeros on the real axis. Complex conjugate roots are symmetric about the real-axis and so they will cancel.
3. **Angles of Departure/Arrival**: The angles of departure of a branch of a locus from a pole of multiplicity q are given by

$$q\phi_{l,dep} = \sum\psi_i - \sum\phi_i - 180° - 360°(l-1),$$

where $l = 1, 2, \cdots, q$ and the angles(s) of the arrival of a branch at a zero of multiplicity q is given by

$$q\phi_{l,arr} = \sum\phi_i - \sum\psi_i + 180° + 360°(l-1).$$

Most of the time we only calculate the angles of departure. For our example, if we choose a test point, s^*, close to the pole $-2+2j$ that is labeled $*$ in Figure 3.21.

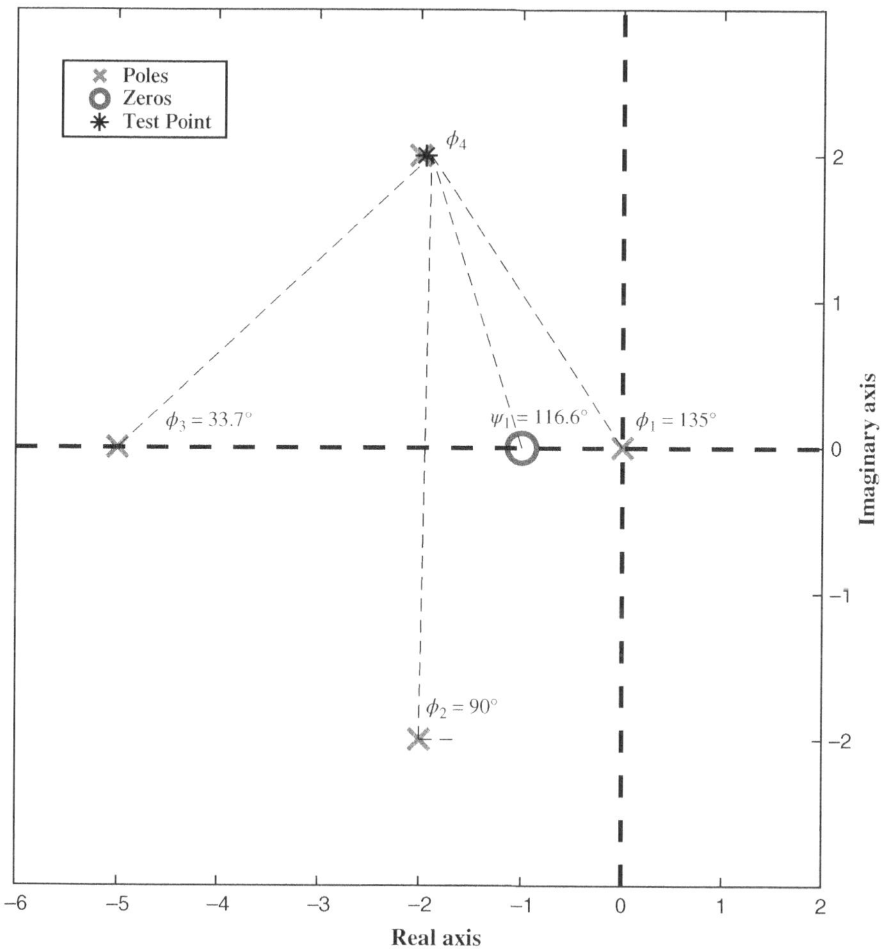

Figure 3.21 Angles of departures for root locus.

$$\begin{aligned}\phi_4 &= \psi_1 - \phi_1 - \phi_2 - \phi_3 - 180^\circ \\ &= a\tan2(2,\ -1) - a\tan2(2,\ -2) - a\tan2(4, 0) - (180^\circ - a\tan2(2,\ -3)) - 180^\circ \\ &= 116.6^\circ - 135^\circ - 90^\circ - 33.7^\circ - 180^\circ \\ &= 37.9^\circ.\end{aligned}$$

Therefore, the root locus leaves pole #4 at an angle of 37.9°.

4. **Multiplicity**: Multiple roots at points on the locus of multiplicity q. The branches will approach a point of q roots at angles separated by

$$\frac{180^\circ + 360^\circ(i-1)}{q}$$

and will depart at angles with the same separation.

To design specific closed-loop pole locations that meet desired specifications (e.g. transient response), we select an appropriate value of K. The process involves

1. Placing the dominant poles at the desired location, $s = s_{des}$.
2. Using the magnitude condition to compute K,

$$|KL(s_{des})| = 1 \Longrightarrow K = \frac{1}{|L(s_{des})|}.$$

For example, in Figure 3.22, K is chosen to achieve a damping ratio of $\zeta = 0.5$, with the root locus illustrating radial lines corresponding to this specification.

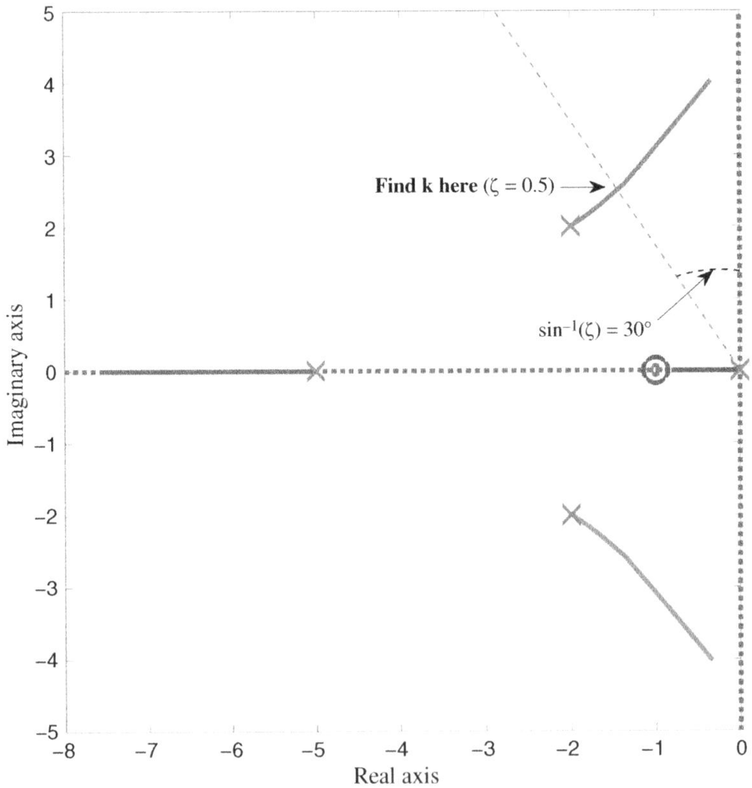

Figure 3.22 Root locus of $L(s) = \dfrac{s+1}{s(s+5)((s+2)^2+4)}$ with radial lines corresponding to $\zeta = 0.5$.

3.3 Frequency Response Methods

3.3.1 Bode Plots

This section explores the relationship between the Laplace and the frequency domain, establishing conditions for closed-loop stability to evaluate frequency performance metrics. Consider the unity feedback closed loop system in Figure 3.23.

Recall, for $s = j\omega$ to be a closed loop pole, the conditions are

$$|KG_{OL}(s)| = 1 \text{ and } \angle G_{OL}(s) = \pm 180^\circ. \tag{3.49}$$

The system is critically stable when a ω^* satisfies these conditions, and the closed-loop system has a pole on the $j\omega$ – axis. A Bode plot is a graphical method to assess stability by showing $|KG_{OL}(j\omega)|$ and $\angle KG_{OL}(j\omega)$ across ω [2].

3.3.1.1 Bode Plot Construction

Bode plots express magnitude in decibels (dB) and phase in degrees, both as functions of the logarithm of frequency. The steps for creating Bode plots are as follows [8]

1. Put the transfer function into Bode form

$$KG_{OL}(j\omega) = K\frac{(1 + j\omega\tau_1)(1 + j\omega\tau_2)\cdots}{(1 + j\omega\tau_a)(1 + j\omega\tau_b)\cdots}.$$

2. Plot gain $|KG_{OL}(j\omega)|$ on a log scale and phase $\angle G_{OL}(j\omega)$ on a linear scale. The magnitude is plotted on the log scale since we can add contributions such that

$$|KG_{OL}(j\omega)| = \frac{|K|\,|1 + j\omega\tau_1|\,|1 + j\omega\tau_2|\cdots}{|1 + j\omega\tau_a|\,|1 + j\omega\tau_b|\cdots}$$

and taking the base 10 log of both sides

$$\begin{aligned}\log_{10}|KG_{OL}(j\omega)| &= \log_{10}|K| + \log_{10}|1 + j\omega\tau_1| + \log_{10}|1 + j\omega\tau_2| + \\ &\quad -\log_{10}|1 + j\omega\tau_a| - \log_{10}|1 + j\omega\tau_b| - \cdots.\end{aligned}$$

More commonly, the magnitude is expressed in decibels (dB), which is

$$|KG_{OL}(j\omega)|_{dB} = 20\log_{10}|KG_{OL}(j\omega)|. \tag{3.50}$$

The phase is already in additive form

$$\begin{aligned}\angle KG_{OL}(j\omega) &= \angle K + \angle(1 + j\omega\tau_1) + \angle(1 + j\omega\tau_2) + \cdots \\ &\quad -\angle(1 + j\omega\tau_a) - \angle(1 + j\omega\tau_b) - \cdots.\end{aligned} \tag{3.51}$$

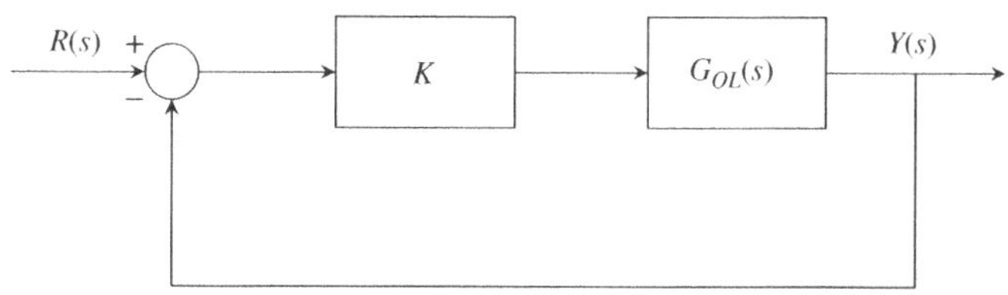

Figure 3.23 Unity feedback closed loop system.

Knowing that we can add magnitudes and phases in this way, we can study the individual contributions of each elementary term of the transfer function

$$KG_{OL}(j\omega) = \frac{K(1 + j\omega\tau_1)}{(j\omega)^2(1 + j\omega\tau_a)\left(\left(\frac{j\omega}{\omega_n}\right)^2 + 2\zeta\frac{j\omega}{\omega_n} + 1\right)} \tag{3.52}$$

Let us look at the Bode plots for each of these elementary terms in Eq (3.52)

1. $K(j\omega)^{\pm n}$:
   - Magnitude: $20\log_{10}(K|(j\omega)^n|) = 20\log_{10}K + 20n\log_{10}|j\omega|$
   - Phase: $\angle K(j\omega)_n = \underbrace{(j\omega) \times (j\omega) \times \cdots \times (j\omega)}_{\text{n times}}$

     $= n \cdot 90°$
   - A Bode plot of $K(j\omega)^{\pm n}$ for $n = -1, 1, 2$ is shown in Figure 3.24.

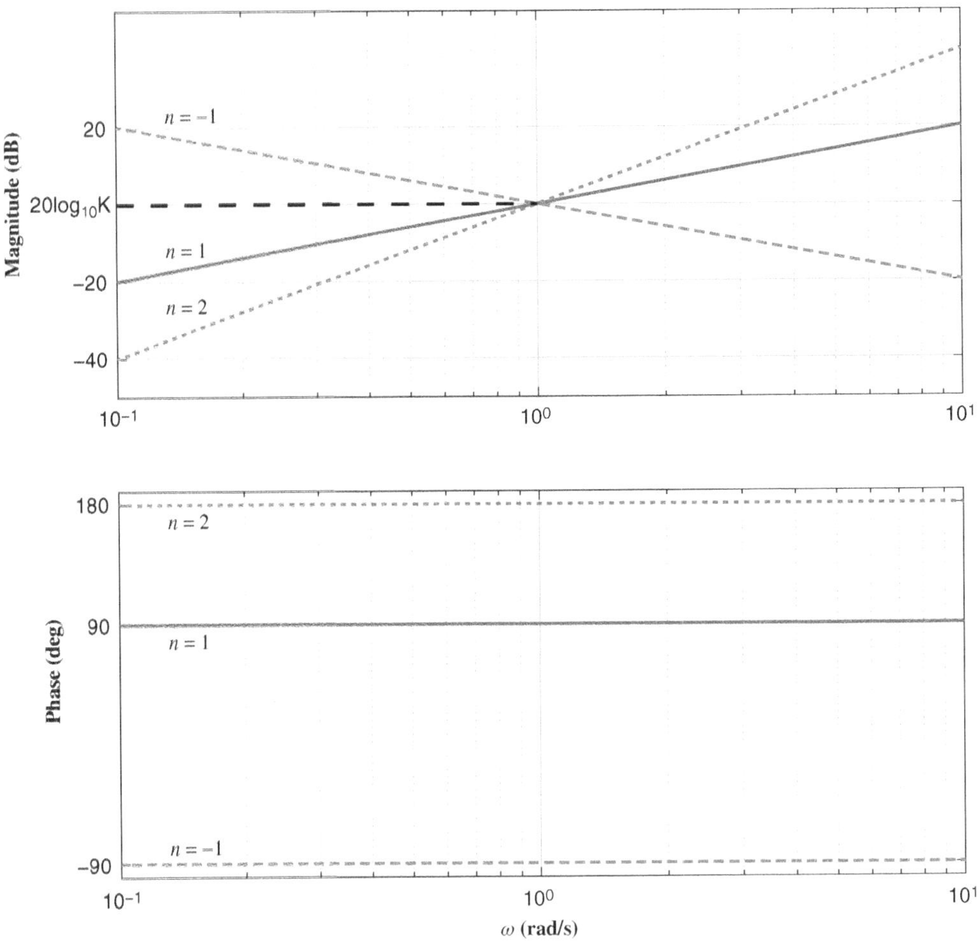

Figure 3.24 Bode plot for $n = -1, 1, 2$.

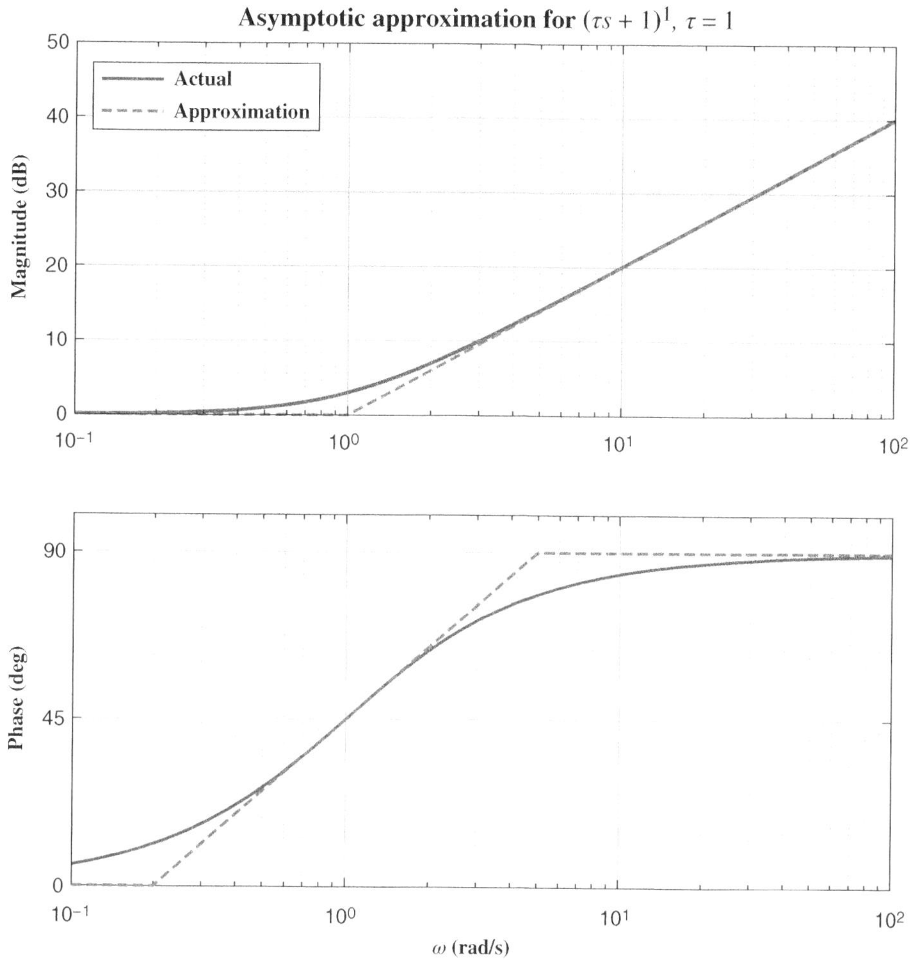

Figure 3.25 Bode plot for $(1+j\omega\tau)^1$, with approximation and actual curves.

2. $(1+j\omega\tau)^1$:

- Magnitude: $20\log_{10}(|1+j\omega\tau|) = \begin{cases} 0 \text{ dB}, \omega \to 0 \\ 3 \text{ dB}, \omega = 1/\tau \\ 20 \text{ dB}, \omega\tau = 10 \\ 40 \text{ dB}, \omega\tau = 100 \end{cases}$
- Phase: $\angle(1+j\omega\tau) = \begin{cases} 0^\circ, \omega \to 0 \\ 45^\circ, \omega = 1/\tau \\ 90^\circ, \omega\tau > 10 \end{cases}$
- A Bode plot of $(1+j\omega\tau)^1$ is shown in Figure 3.25.

For $(1+j\omega\tau)^{-1}$ we can just flip the above Bode plot about the ω axis.

3. $\left(\left(\frac{j\omega}{\omega_n}\right)^2 + 2\zeta\left(\frac{j\omega}{\omega_n}\right) + 1\right)^{\pm 1}$

- Breakpoint is $\omega = \omega_n$ with 40 dB/decade magnitude and $\pm 180^\circ$ phase.

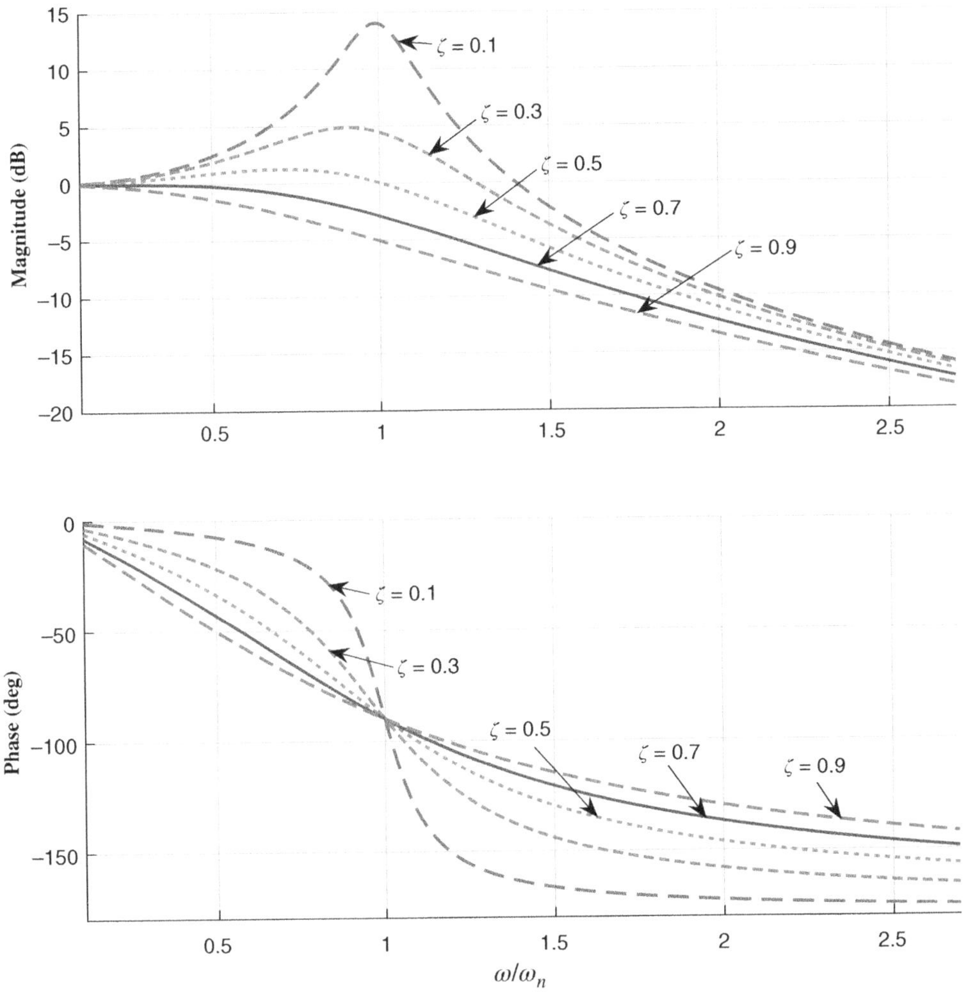

Figure 3.26 Normalized Bode plot for $\left(\left(\frac{j\omega}{\omega_n}\right)^2 + 2\zeta\left(\frac{j\omega}{\omega_n}\right) + 1\right)^1$.

- The size of the magnitude approximation near the corner frequency depends on ζ. The peak value occurs at $\omega = \omega_n\sqrt{1-2\zeta^2}$ with the magnitude of $20\log_{10}\left(\frac{1}{2\zeta\sqrt{1-\zeta^2}}\right)$.
- A Bode plot of $\left(\left(\frac{j\omega}{\omega_n}\right)^2 + 2\zeta\left(\frac{j\omega}{\omega_n}\right) + 1\right)^1$ is shown in Figure 3.26.

3.3.1.2 Non-minimum Phase Transfer Functions

Transfer functions with poles or zeros in the right-hand s-plane are non-minimum phase [1]. Consider

$$G_1(s) = \frac{(s+1)}{(s+2)(s+3)} \text{ and } G_2(s) = \frac{(s-1)}{(s+2)(s+3)}.$$

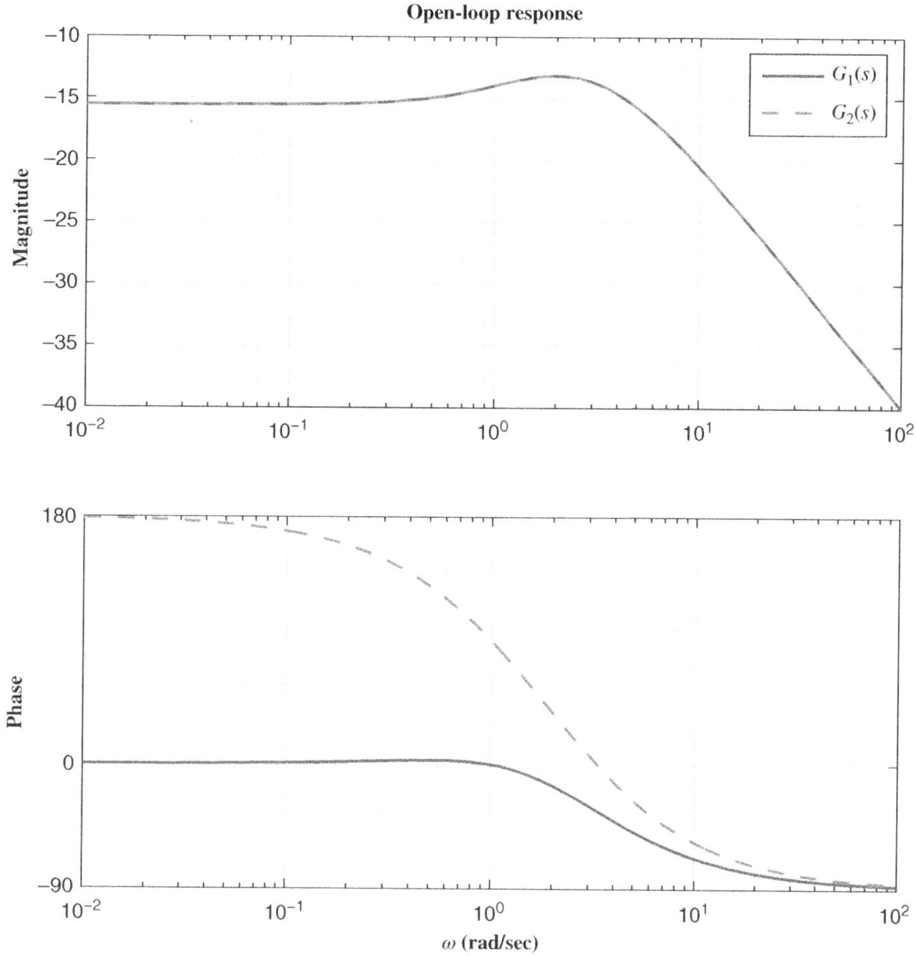

Figure 3.27 Open-loop Bode plot for $G_1(s) = \dfrac{K(s+1)}{(s+2)(s+3)}$ and $G_2(s) = \dfrac{K(s-1)}{(s+2)(s+3)}$.

Both systems are type 0 systems and will have a steady-state error to a unit step input. The open-loop Bode plot for $G_1(s)$ and $G_2(s)$ in Figure 3.27 shows identical magnitudes. However, the non-minimum phase zero in $G_2(s)$ produces a phase plot, which at low frequencies approaches 180° and for high frequencies approaches −90°.

Figure 3.28 shows the closed loop Bode plot for $G_1(s)$ and $G_2(s)$. The closed-loop Bode plot for $G_1(s)$ resembles a heavily damped second-order system. At low frequencies the closed-loop magnitude for $G_1(s)$ is 16.9 dB below the 0 dB line. In contrast, $G_2(s)$ has a magnitude of approximately −14 dB below the 0 dB. The closed-loop phase plot for $G_2(s)$ shows the system is non-minimum phase but gives no indication of the step response characteristic shown in Figure 3.29. The step response plots show that $G_2(s)$ when subjected to a positive unit step of 0.1, initially responds in a positive direction but ends at a negative steady-state value of −0.2. This behavior by a

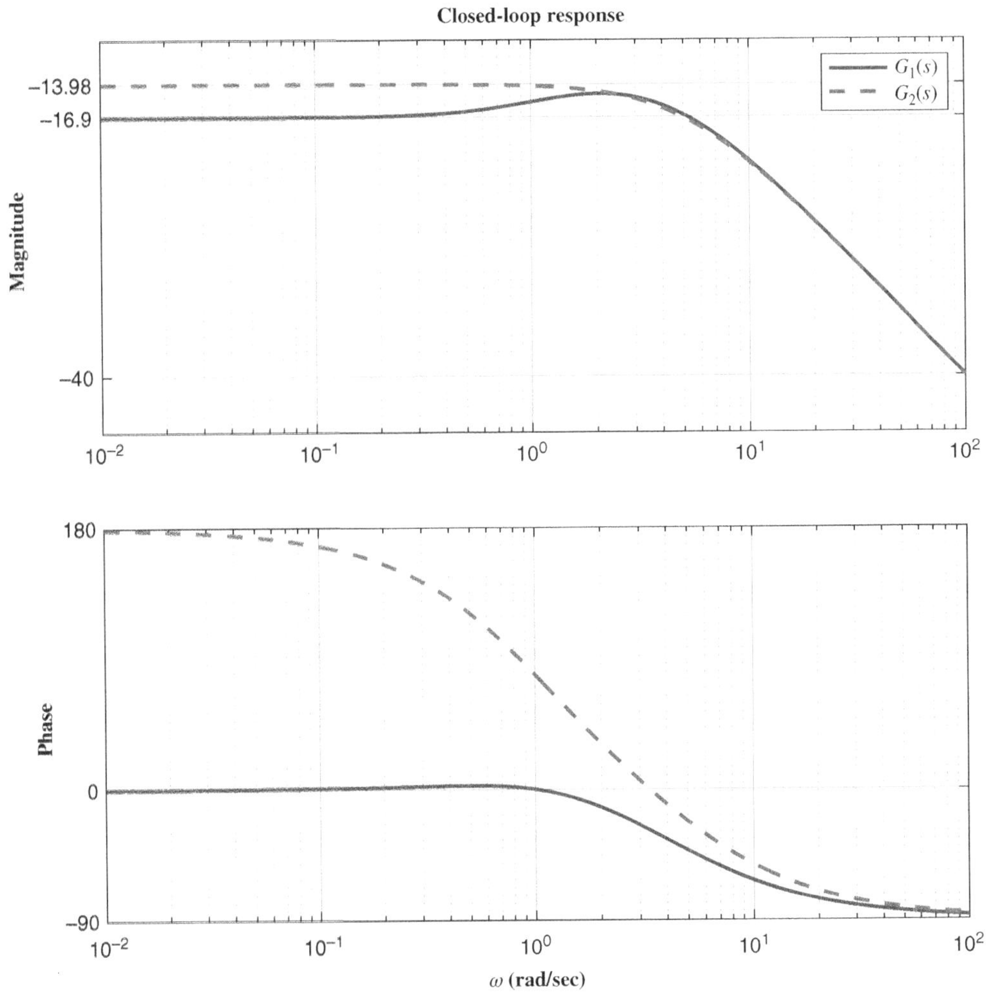

Figure 3.28 Closed-loop Bode plot for $G_1(s)$ and $G_2(s)$.

non-minimum phase step response makes the control more difficult since the controller must know that the system may initially move in the wrong direction.

3.3.1.3 Relating Root Locus to Bode Plots

For the system, $G(s) = \frac{1}{s(s+1)^2}$, we analyze the effect of increasing the gain K on the system's stability using Bode plots for $K = 0.1, 2, 10$. Figure 3.30 shows at $K = 0.1$, the system is stable since the magnitude crosses 0 dB line when the phase is 80°. At $K = 2$, the magnitude crosses the 0 dB when the phase is 0° and so the system is critically stable. If we increase the gain further to $K = 10$, the phase becomes −35° and so the system is unstable.

The root locus of the poles for these values of K is shown in Figure 3.31.

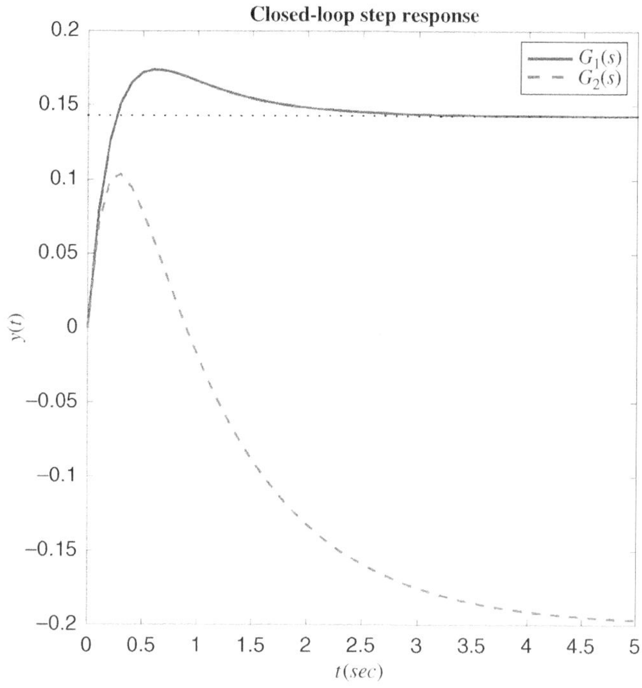

Figure 3.29 Closed-loop step response for $G_1(s)$ and $G_2(s)$.

The root locus for $K = 2$, shows the poles are neutrally stable since they are on the imaginary axis of the root locus. If we increase the gain to $K = 10$, then the poles move to the right-half plane and the system is now unstable. Therefore, as we increase the gain, our Bode plot shows a higher magnitude but at the expense of stability as verified by the root locus plot.

3.3.2 Gain and Phase Margin

For systems in which increasing gain leads to instability, and $|KG_{OL}(j\omega)|$ crosses the 0 dB line once, the stability condition based on the open-loop frequency response is

$$|KG(j\omega)| < 1 \text{ at } \angle G(j\omega) = -180^\circ. \tag{3.53}$$

The gain margin, GM, is the factor by which K can be increased before instability, and the phase margin, PM, is the amount by which the phase of $G(j\omega)$ exceeds -180° when $|KG(j\omega)| = 1$. Also, the crossover frequency is the frequency ω_c at which $|KG(j\omega)| = 1 = 0$ dB [1]. A Bode plot illustrating the gain margin, phase margin, and crossover frequency is shown in Figure 3.32.

The bandwidth, ω_{BW}, is the frequency of the input at which the output is attenuated to -3 dB [1]. It is the frequency up to which the output will track the input in a satisfactory manner. It is usually defined with respect to the closed loop Bode plot as shown in Figure 3.33.

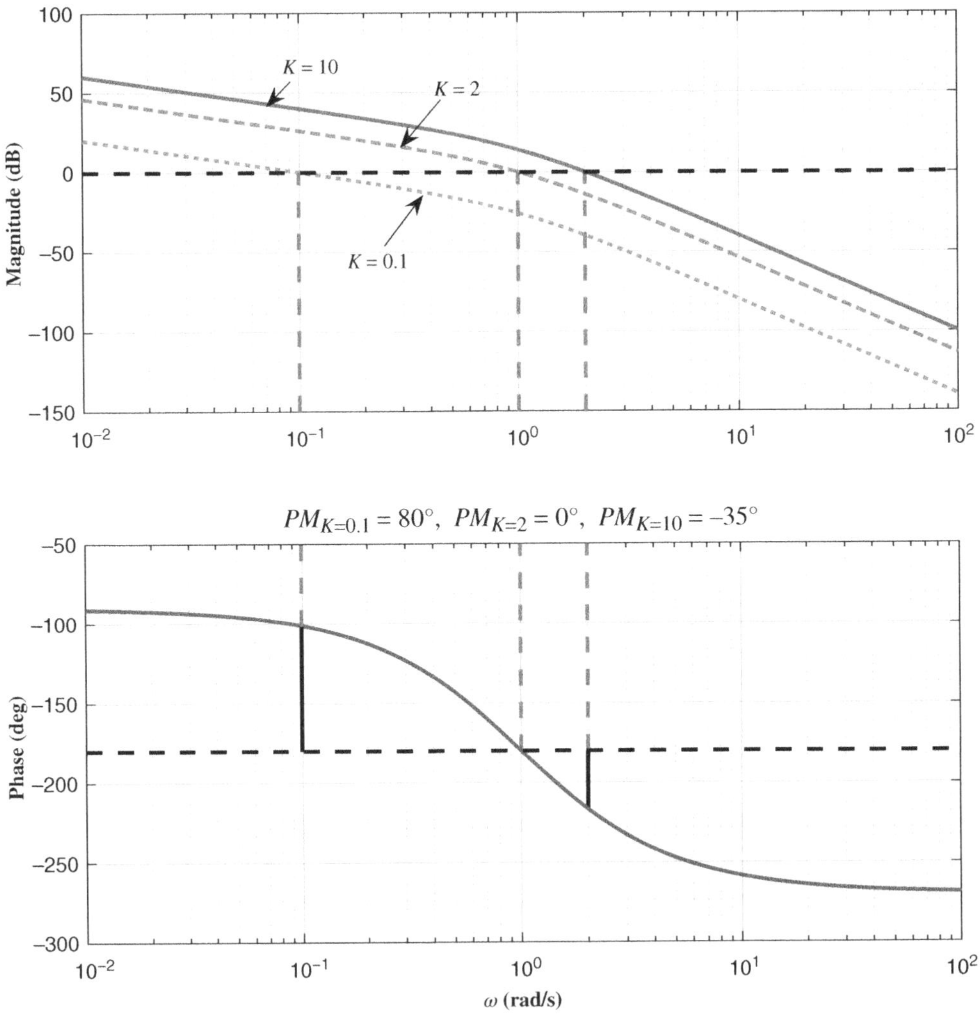

Figure 3.30 Bode plots for $K = 0.1, 2, 10$.

3.3.3 Lead and Lag Compensator Design

When adjusting the system's closed-loop gain is insufficient, dynamic compensation is necessary. Lead, lag, and lead-lag compensators can be used to adjust the system response to meet performance requirements [8].

3.3.3.1 Pure Gain Compensation

Consider $K \cdot G_{OL}(j\omega) = \dfrac{K}{j\omega\left(1+\dfrac{j\omega}{2}\right)}$. We would like a phase margin of 45°. What is the maximum bandwidth I can achieve with just a pure gain compensation?

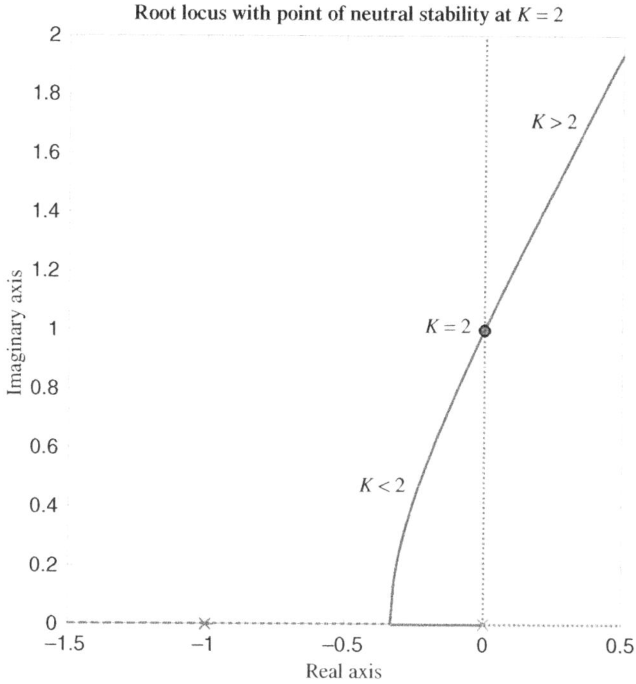

Figure 3.31 Root locus plot for $K = 0.1, 2, 10$.

From the Bode plot shown in Figure 3.34, pure gain compensation does not alter the open-loop phase characteristics. The phase, $\angle G_{OL}(j\omega)$, attains a value of $-180^\circ + 45^\circ = -135^\circ$ at $\omega = 2\,\text{rad/s}$. To meet the gain crossover condition, $20\log_{10}|G_{OL}(j\omega)| = 0$ dB, an increase of 9 dB is required for the magnitude $20\log_{10}|G_{OL}(\omega)|$ to intersect the 0 dB line at $\omega = 2\,\text{rad/s}$. Calculating from $20\log_{10}|K| = 9$, we find the necessary gain compensation, K, to be approximately 2.82. This gain, when applied, will adjust the open-loop system's response to intersecting the 0 dB line at the new crossover frequency, ensuring the desired phase margin.

3.3.3.2 Lead Compensation

Lead compensation is normally used to improve the speed of response of stable systems of type 1 and higher [8]. The transfer function is

$$KD(j\omega) = K_{\text{lead}} \cdot \frac{\left(\frac{j\omega}{a} + 1\right)}{\left(\frac{j\omega}{b} + 1\right)}, \text{ where } b > a. \tag{3.54}$$

When designing lead compensation, the pole (red x) is picked at a larger frequency than the zero (blue o). A pole-zero plot for lead compensation is shown in Figure 3.35.

The phase is maximized at $\omega_{\max} = \sqrt{ab}$, and the maximum phase lead is

$$\phi_{\max} = \sin^{-1}\left(\frac{1 - \frac{a}{b}}{1 + \frac{a}{b}}\right) \Rightarrow \frac{a}{b} = \frac{1 - sin\phi_{\max}}{1 + sin\phi_{\max}}. \tag{3.55}$$

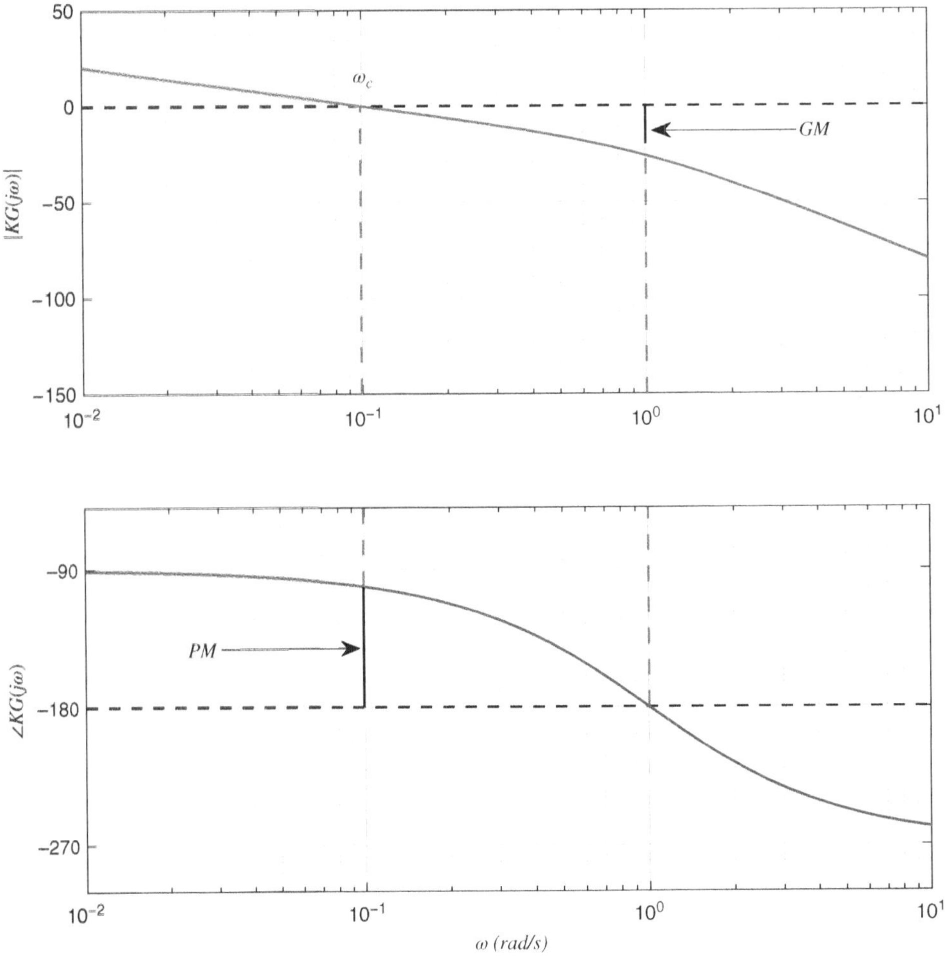

Figure 3.32 Bode Plot with *GM* (dB), *PM* (deg), and crossover frequency, ω_C (rad/s).

Figure 3.36 shows the lead compensator Bode plot with $\frac{a}{b} = 0.1$, resulting in $\phi_{\max} \approx 60°$ phase lead and total gain increase of 20 dB. Figure 3.37 shows Bode plots for a lead compensator $KD(j\omega) = \frac{a}{b} \cdot \frac{(j\omega/a + 1)}{(j\omega/b + 1)}$ for various lead ratios $\frac{a}{b}$.

As evident in Figure 3.37, smaller lead ratios result in changes in 20 dB/decade over a larger frequency range and an overall lower $\phi_{\max}$. Lead compensation can be used to increase gain margin/phase margin and bandwidth. The general procedure for designing a lead compensator is

1. Use the series gain, K, to achieve the desired ω_c. To do this, we find the additional gain that is needed to move the response up so that the magnitude crosses 0 db at the desired ω_c.
2. Compute $\phi_{\max}$ for desired phase margin at ω_c and select $\frac{a}{b}$.
3. Iterate K_{lead} to ensure $20\log_{10}|K_{\text{lead}}| = 0$ dB at $\omega = \sqrt{ab}$.

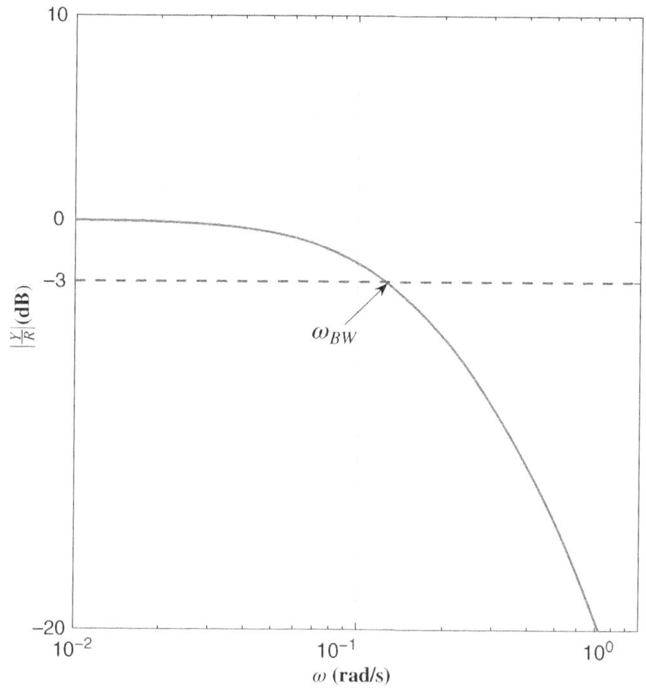

Figure 3.33 Bode of closed-loop system, $|Y/R|$.

Example: Lead compensation
Design a lead compensator for

$$G(s) = \frac{34381}{s(s+15.5)(s+35.5)},$$

to achieve a phase margin of 52° and a gain margin of 12.9 dB. A Bode plot of the uncompensated system, $G(s)$, is shown in Figure 3.38.

The GM is −1.76 dB and the phase margin is −5.235° for the uncompensated system. Therefore, the closed-loop system is unstable. We can design the compensator and leave the crossover frequency at $\omega_c = 25.91$ rad/s. To achieve a phase margin of 52° then we need a phase increase of $\phi = 52° - (-5.235) = 57.2°$. We want $\omega_c = \sqrt{ab} = 25.91$ rad/s, so we can use $\frac{a}{b} = \frac{1 - sin\phi_{\max}}{1 + sin\phi_{\max}}$ to get $a = 7.56$ and $b = 88.8$. Next, we evaluate $|KD(s)| = 1$ at $\omega_c = \sqrt{ab}$, to get the following result

$$KD(s) = 3.4 \cdot \frac{s + 7.62}{s + 88.1}.$$

The Bode plot summary of the uncompensated and compensated system with the lead control is shown in Figure 3.39. The compensated system has our target phase margin of 52° and a gain margin of 12.9 dB.

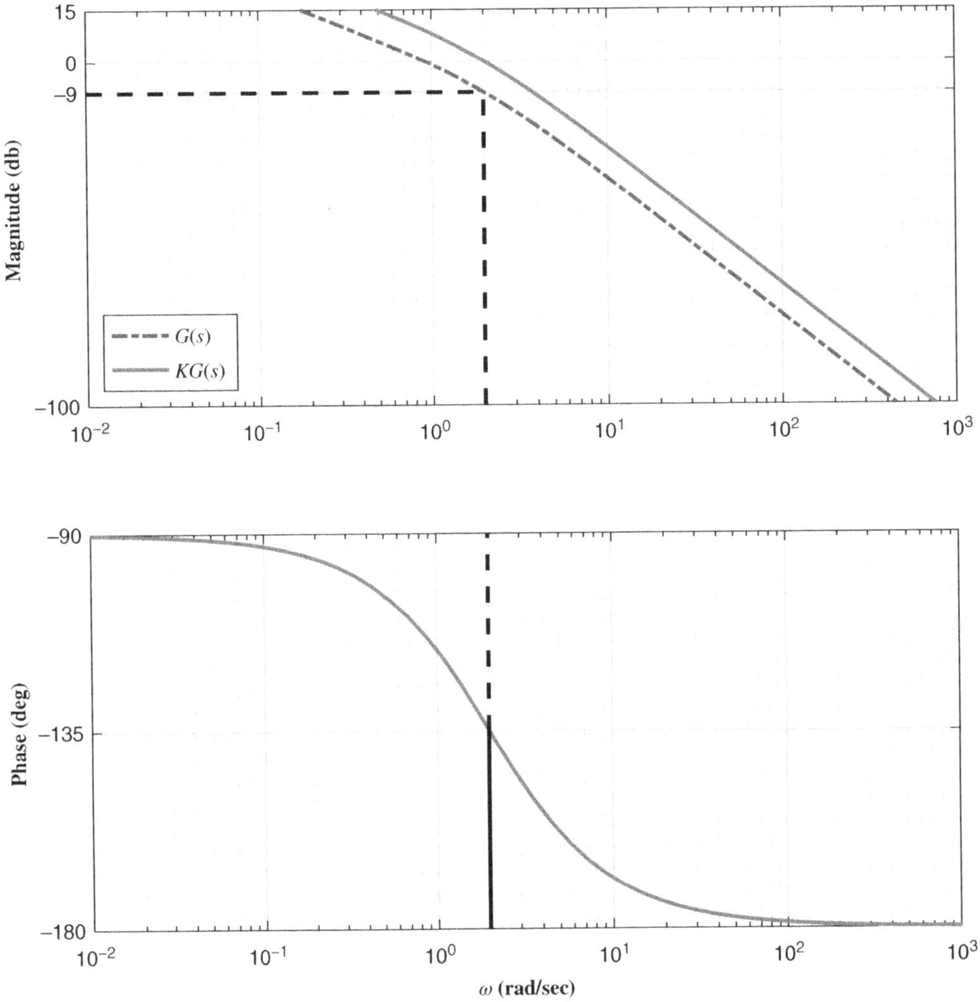

Figure 3.34 Bode plot for open-loop system $G_{OL}(s) = \dfrac{1}{j\omega\left(1 + \dfrac{j\omega}{2}\right)}$ and $KG_{OL}(s) = \dfrac{2.82}{j\omega\left(1 + \dfrac{j\omega}{2}\right)}$.

3.3.3.3 Lag Compensation

Lag Compensation can improve a system that has a satisfactory transient response but a large steady-state error [8]. The transfer function is

$$KD(j\omega) = \frac{b}{a} \cdot \frac{\left(\dfrac{j\omega}{b} + 1\right)}{\left(\dfrac{j\omega}{a} + 1\right)}, \text{ where } b > a. \tag{3.56}$$

When designing lag compensation, the zero is picked at a greater frequency than the pole. A pole-zero plot is shown in Figure 3.40.

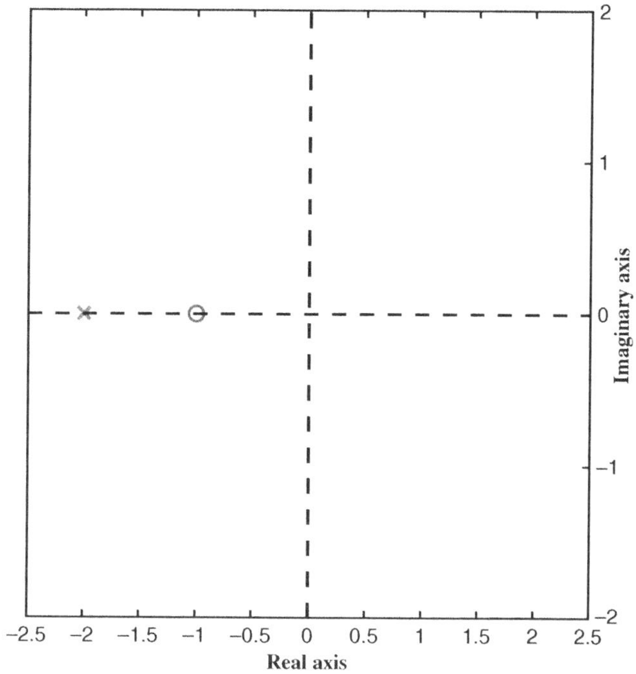

Figure 3.35 Pole-zero plot for lead compensation.

The Bode plot of the lag compensation is shown in Figure 3.41 and uses $\frac{b}{a} = 10$. This results in a total gain decrease of 20 dB and a phase lag of $\phi_{max} \approx 60°$. Bode plots for a lag compensator $KD(j\omega) = \frac{1 + j\omega/b}{1 + j\omega/a}$ for various lag ratios $\frac{b}{a}$ is shown in Figure 3.42.

The procedure for lag compensator design is as follows:

1. Choose b to be ~ $\frac{\omega_c}{50}$ so the phase lag does not significantly affect the phase margin.
2. Choose $\frac{b}{a}$ ratio to achieve the low-frequency gain to satisfy steady-state error specs:

$$e_{ss} = \lim_{s\to 0} s \frac{1}{\left[1 + \frac{b}{a}\frac{\left(\frac{s}{b}+1\right)}{\left(\frac{s}{a}+1\right)}G(s)\right]} R(s) \leq \text{steady-state error specification.}$$

3.3.3.4 Lead-lag Compensation

Another common compensation structure is the lead-lag compensation, which is shown in Figure 3.43.

The lead and lag compensators are in series and combine the best of both lead and lag compensation [1].

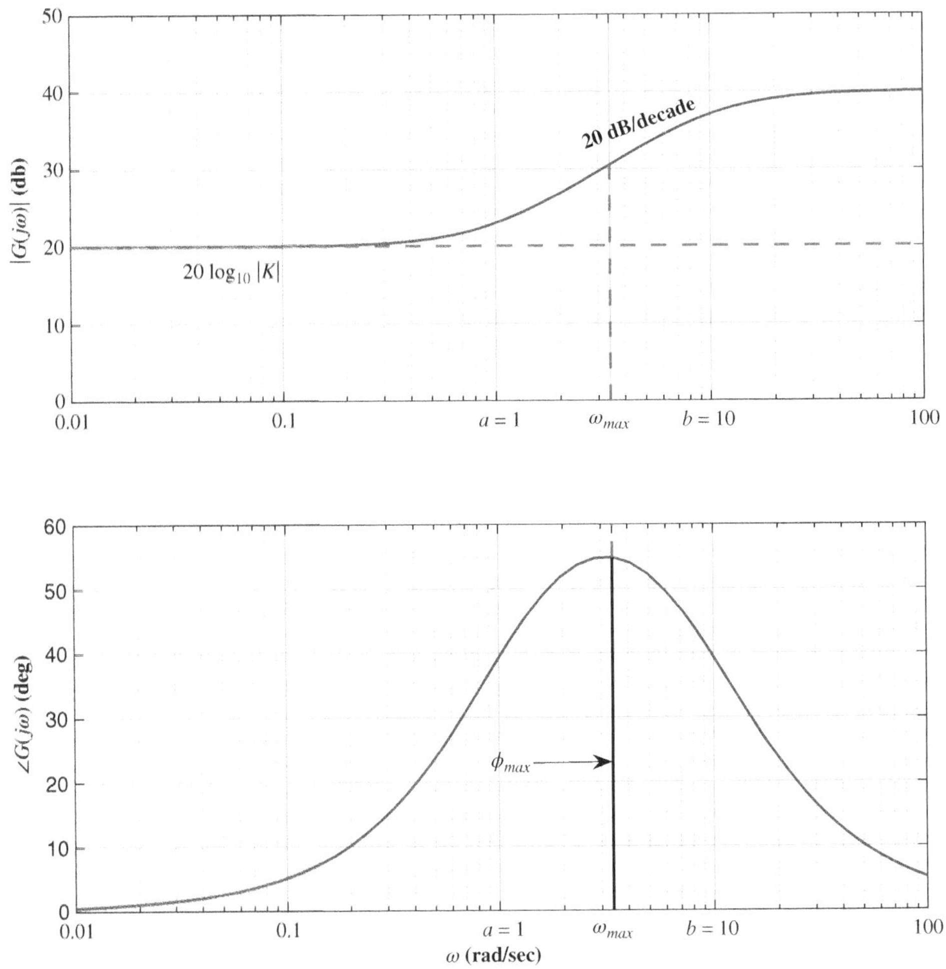

Figure 3.36 Bode plot for lead compensation.

$$KD(s) = K\left(K_{\text{lead}} \cdot \frac{\frac{s}{a}+1}{\frac{s}{b}+1}\right)\left(\frac{b}{a} \cdot \frac{\frac{s}{b}+1}{\frac{s}{a}+1}\right). \tag{3.57}$$

The procedure for lead-lag compensator design is as follows:

1. Use lead compensation and series gain, K, to achieve the desired ω_c and phase margin.
2. Use lag compensation to achieve the desired low-frequency gain to satisfy steady-state error specifications.

3.3.3.5 Summary of Lead and Lag Control Techniques

Table 3.3 compares lead and lag compensators across key categories. This comparison can help designers understand when to use each compensator based on specific design goals, such as improving stability, enhancing transient response, or optimizing steady-state behavior.

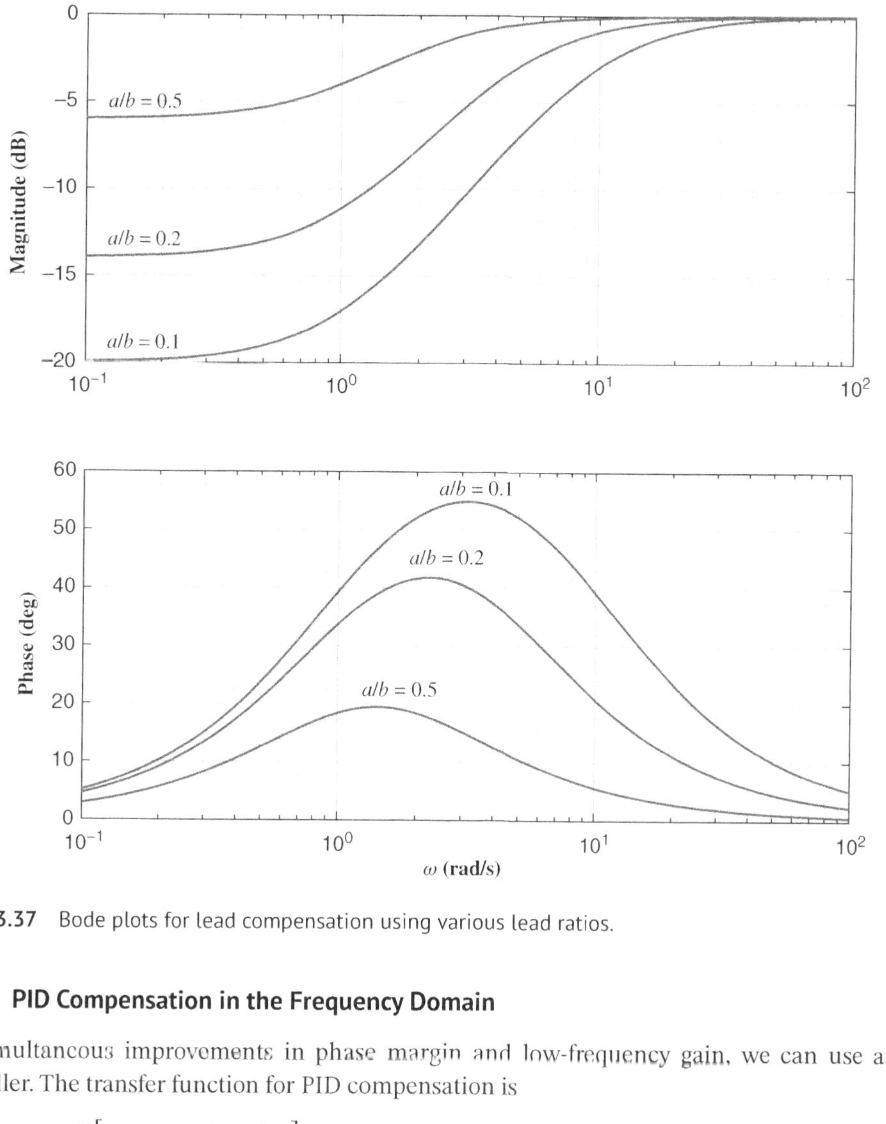

Figure 3.37 Bode plots for lead compensation using various lead ratios.

3.3.4 PID Compensation in the Frequency Domain

For simultaneous improvements in phase margin and low-frequency gain, we can use a PID controller. The transfer function for PID compensation is

$$D(s) = \frac{K}{s}\left[(T_D s + 1)\left(s + \frac{1}{T_I}\right)\right]. \tag{3.58}$$

The frequency response for $T_D = 0.1$ and $T_I = 1$ is shown in Figure 3.44.

PID compensation is practically equivalent to lead-lag compensation and it provides improvement of the transient and steady-state response [1].

3.3.5 Time Delay on a Bode Plot

The Laplace transform of a pure time delay, T_{delay}, is

$$G_{\text{delay}}(s) = e^{-sT_{\text{delay}}}. \tag{3.59}$$

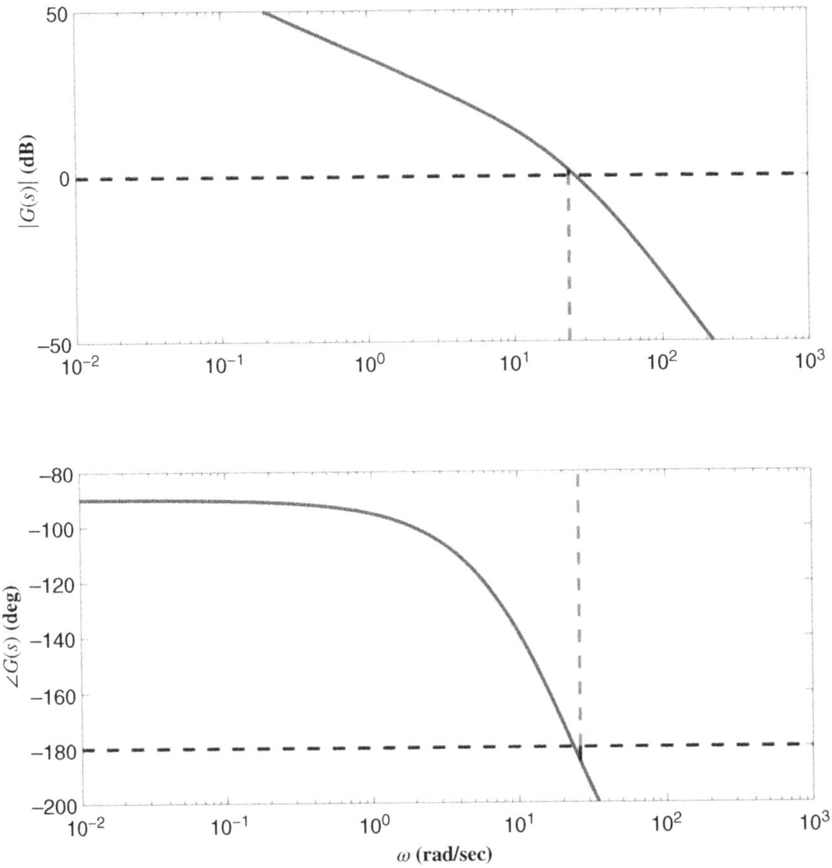

Figure 3.38 Bode plot of $G(s)$ for the lead compensation example.

This can be approximated using a Padé approximation

$$e^{-sT_{\text{delay}}} \cong \frac{1 - (T_{\text{delay}}s/2)}{1 + (T_{\text{delay}}s/2)}. \tag{3.60}$$

Time delays can also be analyzed with Bode plots. The magnitude of the time delay is

$$|G_{\text{delay}}(j\omega)| = \left|e^{-j\omega T_{\text{delay}}}\right| = 1 \text{ for all } \omega, \tag{3.61}$$

which has no effect on the magnitude. However, the phase shift is

$$\angle G_{\text{delay}}(j\omega) = -\omega T_{\text{delay}}, \tag{3.62}$$

which becomes increasingly negative as the frequency increases. Figure 3.45 shows the phase plot for $T_{\text{delay}} = 1$.

For $\omega T_{\text{delay}} > 5$ rad/s, the phase lag exceeds 270°, making it nearly impossible to stabilize this system with a crossover greater than 5 rad/s since it will be difficult to maintain a positive phase

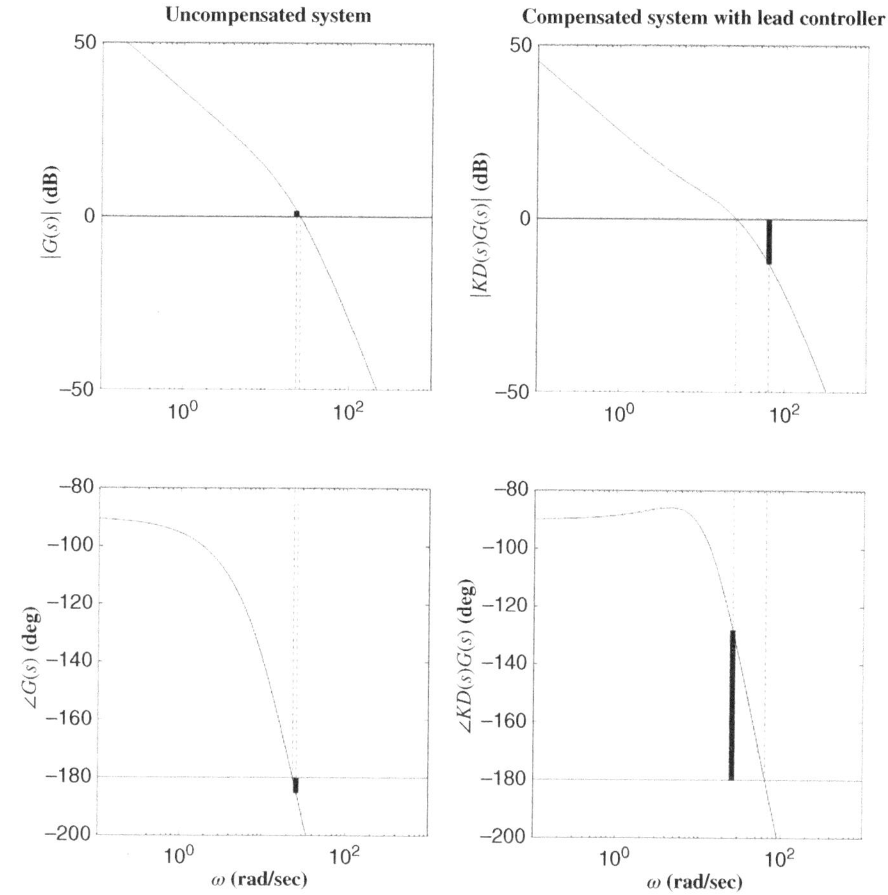

Figure 3.39 Bode plot of uncompensated and lead compensated system.

margin at frequencies greater than $\omega = 5/T_{\text{delay}}$. Maintaining a positive phase margin becomes challenging, effectively constraining the bandwidth in systems with time delays. In sampled systems, the effective delay varies between 0 and one sampling period T_s. On average the sampling period introduces a delay of

$$\angle G_{\text{delay}}(j\omega) = -\omega \frac{T_s}{2}.$$

To limit phase lag at the crossover frequency, the sampling time T_S must be carefully chosen. Increasing T_S reduces phase margin, potentially increasing overshoot, M_p, in the closed-loop system [1].

3.3.6 Summary of Classical Design Techniques

Table 3.4 compares the Root Locus and Bode methods based on key performance metrics, such as stability, transient performance, and tracking accuracy. It highlights the strengths and limitations of

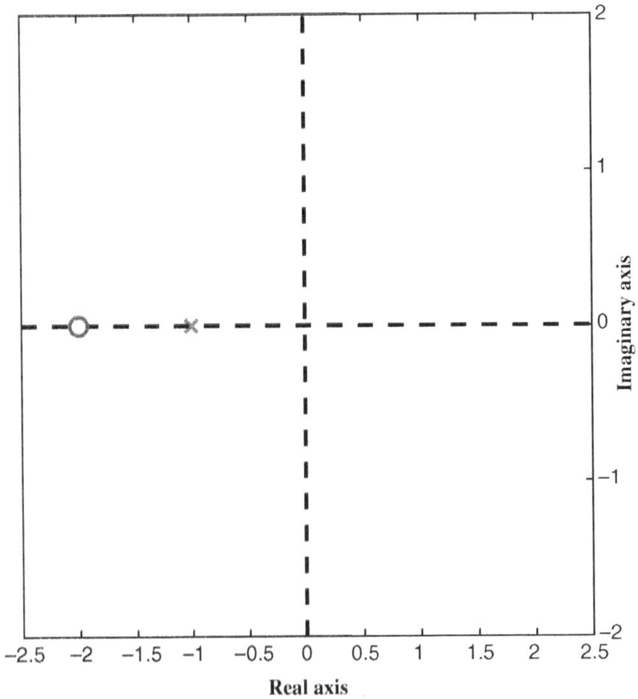

Figure 3.40 Pole-zero plot for lag compensation.

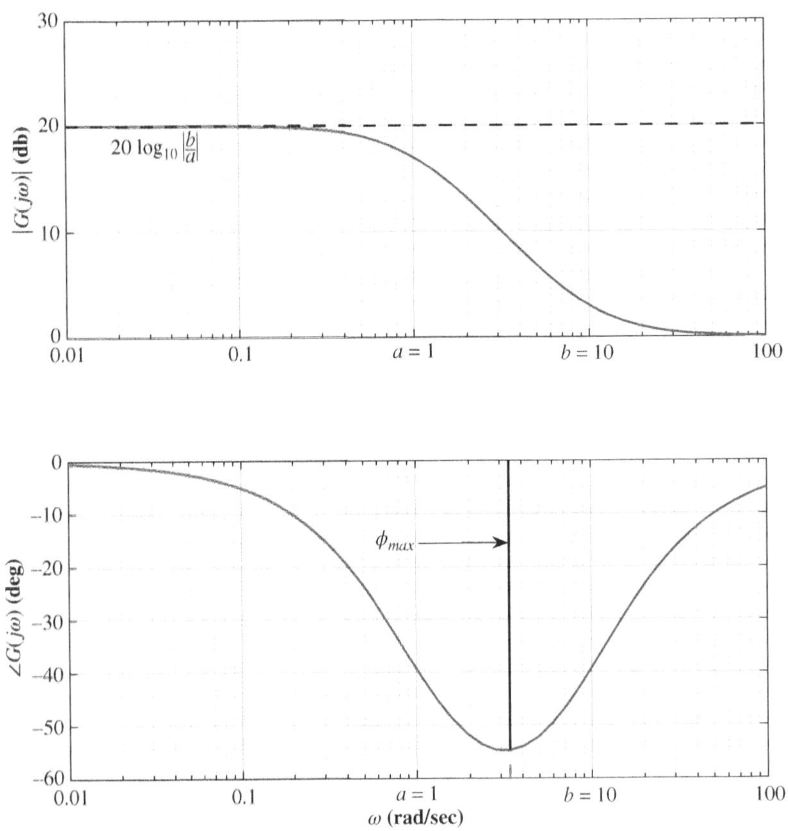

Figure 3.41 Bode plot for lag compensation.

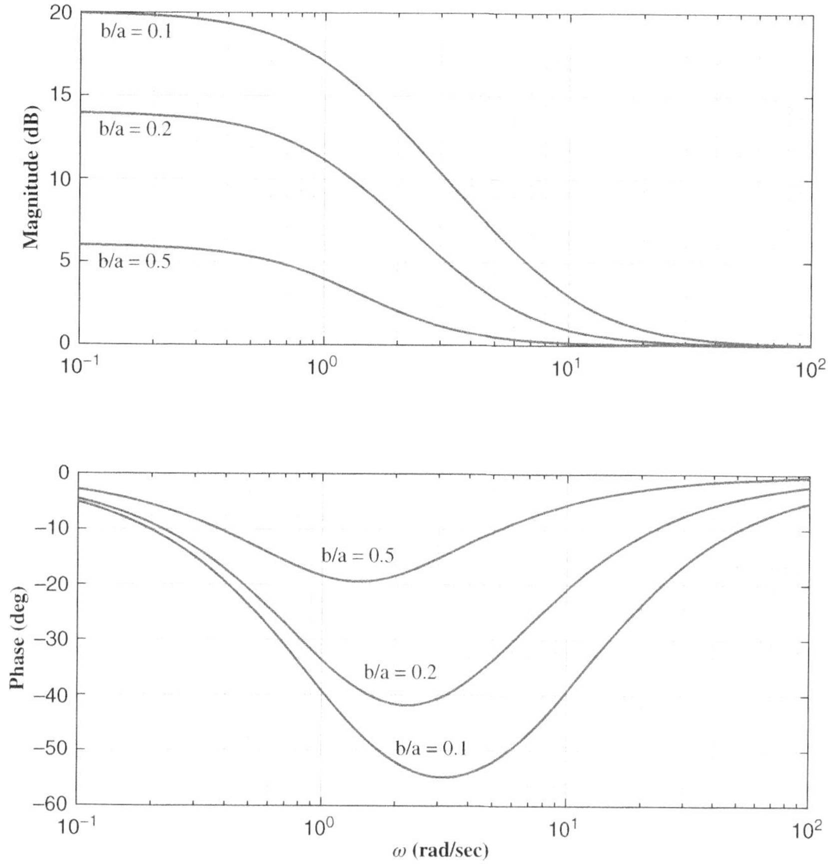

Figure 3.42 Bode plots for lag compensation using various lag ratios.

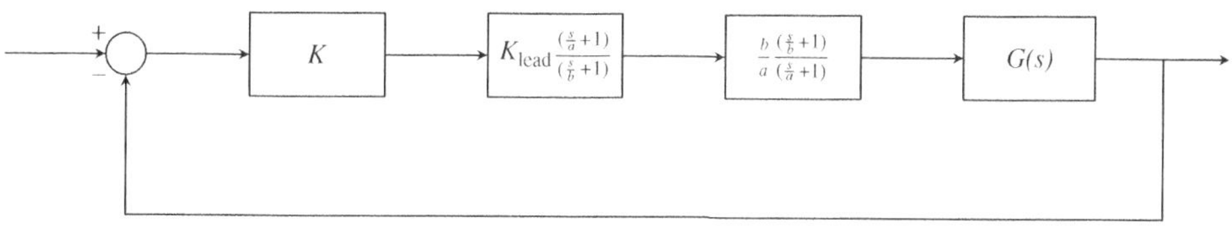

Figure 3.43 Lead-lag compensation block diagram.

each technique, aiding control engineers in selecting the most suitable method for their application. Understanding these distinctions helps tailor control strategies to specific system requirements.

3.4 Flexible Control

Gain and phase stabilization are essential techniques to ensure stability in control systems, particularly for those with high-frequency dynamics caused by flexible structures. These high-frequency dynamics, even when modeled and compensated for, may introduce instability due to modeling errors or uncertainties as depicted in Figure 3.46.

Table 3.3 Comparison of lead and lag control.

	Lead control	Lag control
Approach	• Add phase angle near crossover frequency to increase PM • Add lead to yield desired dominant poles in the s-plane	• Add phase lag at a low frequency while maintaining desired dominant roots
Results	• Increase bandwidth • Increase high-frequency gain	• Decrease bandwidth • Reduce steady-state error
Advantages	• Yields desired transient response • Used when fast transient response is required	• Suppress high-frequency gain • Used when error constant is specified
Disadvantages	• Need to re-adjust gain • Increase bandwidth, susceptible to noise	• Slow down transient response
Application	• Used when fast transient response is required	• Used when error constant is specified

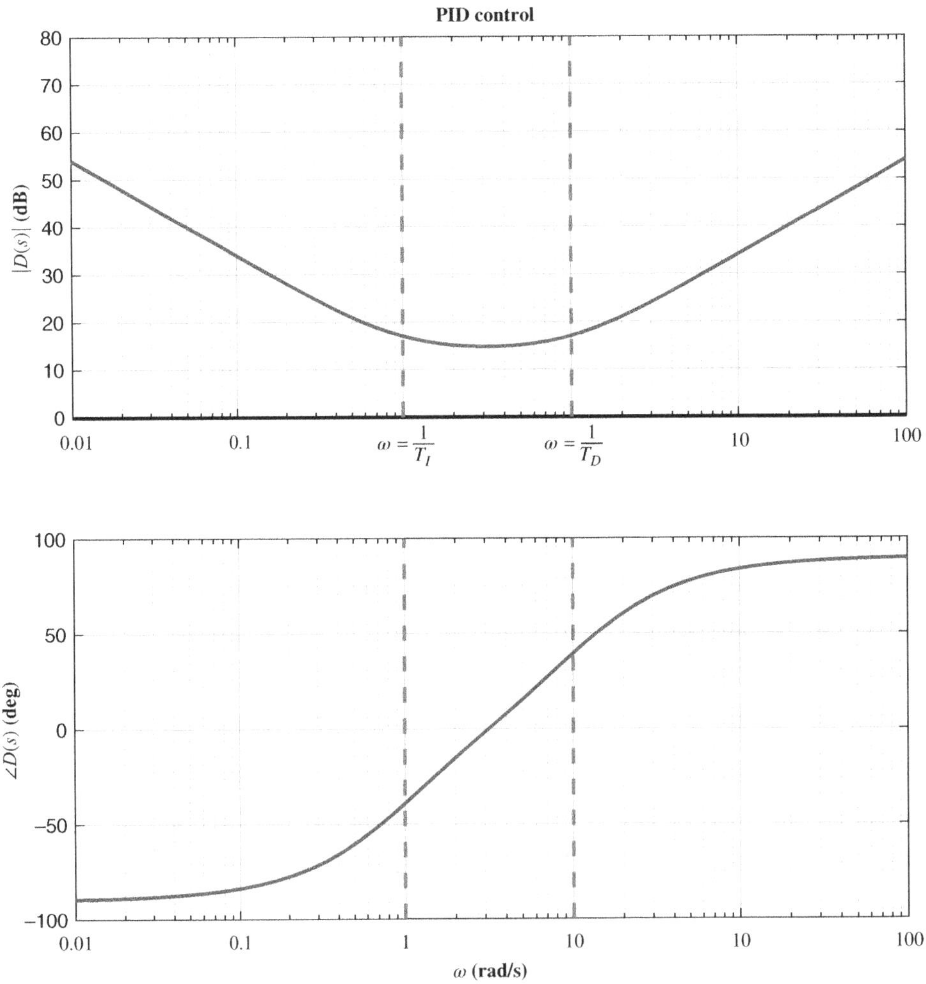

Figure 3.44 Bode plot of PID Compensation, $D(s) = \frac{K}{s}\left[(T_D s + 1)\left(s + \frac{1}{T_I}\right)\right]$.

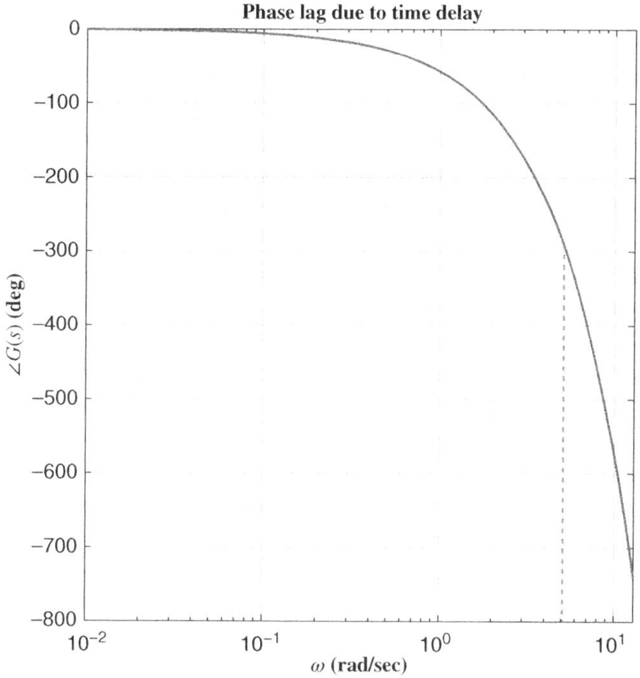

Figure 3.45 Phase plot of time-delay $\angle G_{\text{delay}}(j\omega) = -\omega T_{\text{delay}}$.

Table 3.4 Comparison of classical design techniques.

Properties	Root locus	Bode plot
Stability	✓	✓
Transient performance	✓	✓
Tracking		✓
Disturbance rejection		✓
Sensor noise rejection		✓
Gain margin		✓
Phase margin		✓

Gain stabilization is achieved by attenuating the control loop gain at specific frequencies, ensuring closed-loop stability even when phase uncertainty exists [4]. A flexible mode is a gain stabilized if it remains stable at the selected loop gain but can become unstable if the gain increases or damping decreases. The primary goal of gain stabilization is to maintain the gain margin while avoiding instability from high-frequency dynamics.

Phase stabilization focuses on adjusting the phase at the desired frequency to obtain a closed-loop damping that is greater than the passing damping of that mode [4]. A flexible mode is phase stabilized if it is closed-loop stable for arbitrarily small damping. A phase stabilized mode has a finite phase margin but is closed-loop stable regardless of loop gain uncertainty.

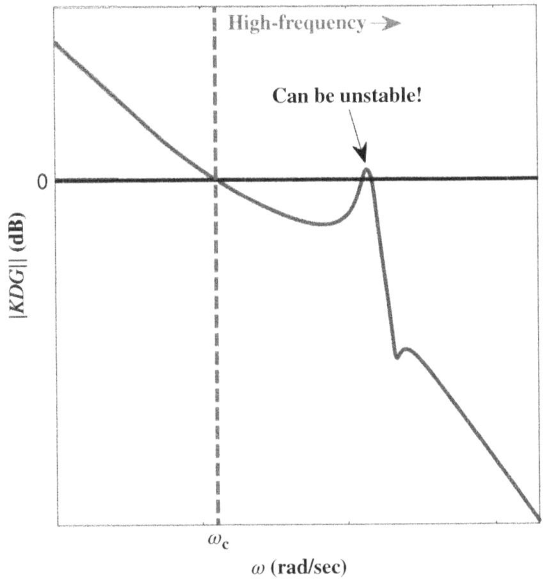

Figure 3.46 Effect of high-frequency plant uncertainty.

A rigid body or flexible mode is said to be **gain-phase stabilized** if it is closed-loop stable with finite gain and phase margin. Notch filtering is a way to gain stabilize unwanted flexible modes in the control loop [4]. To further mitigate the effects of unmodeled high-frequency dynamics and sensor noise, the control loop gain is typically "rolled off" at frequencies above the system's control bandwidth. This approach minimizes the risk of destabilizing unmodeled modes and attenuates high-frequency noise.

3.4.1 Second-order Filters

Second-order filters in the frequency domain are indispensable for solving flexible control problems. These filters are extensions of the classical design techniques and are represented by

$$D_f(s) = \frac{\left(\frac{s^2}{\omega_z^2}\right) + \frac{2\zeta_z s}{\omega_z} + 1}{\left(\frac{s^2}{\omega_p^2}\right) + \frac{2\zeta_p s}{\omega_p} + 1} \tag{3.63}$$

where ω_z, ζ_z, ω_p, ζ_p are tunable parameters [4]. By varying these parameters, several well-known filters can be realized such as minimum-phase and non-minimum-phase filters. For a minimum-phase notch (also known as the bandpass filter), where $\omega_c = \omega_z = \omega_p$, the filter simplifies to

$$D_{\text{notch}}(s) = \frac{\left(\frac{s^2}{\omega_c^2}\right) + \frac{2\zeta_z s}{\omega_c} + 1}{\left(\frac{s^2}{\omega_c^2}\right) + \frac{2\zeta_p s}{\omega_c} + 1}. \tag{3.64}$$

When $\zeta_p > \zeta_z$, the minimum gain at ω_c is $-20\,\log_{10}(\zeta_p/\zeta_z)$. The frequency response of the notch filter in Figure 3.47 shows alternating minimum phase and maximum phases, which can be tuned by adjusting ζ_p, ζ_z, and ω_c.

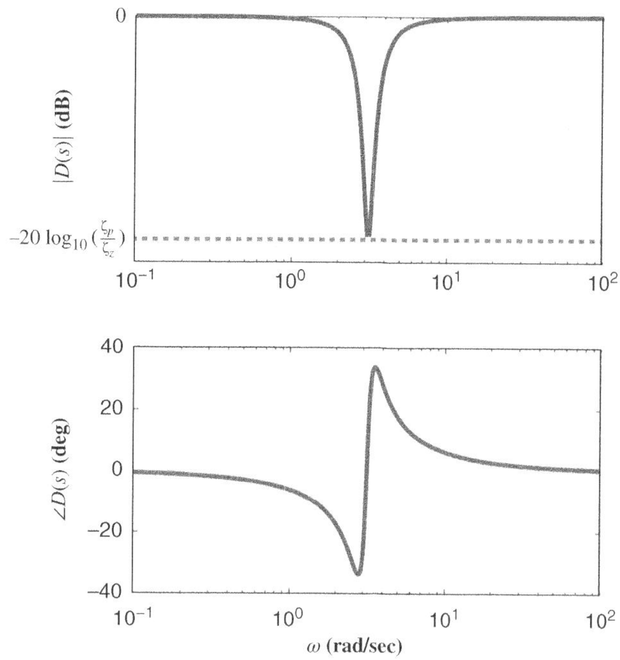

Figure 3.47 Notch filter with $\zeta_p = 0.7, \zeta_z = 0.2$, and $\omega_{\text{notch}} = 10$ rad/s.

A non-minimum-phase all-pass filter has $\omega_c = \omega_z = \omega_p$, $\zeta_c = \zeta_p = |\zeta_z|$, with $\zeta_z < 0$ due to a non-minimum phase zero. This filter takes the form

$$D_{ap}(s) = \frac{\left(\frac{s^2}{\omega_c^2}\right) - \frac{2\zeta_c s}{\omega_c} + 1}{\left(\frac{s^2}{\omega_c^2}\right) + \frac{2\zeta_c s}{\omega_c} + 1}. \tag{3.65}$$

Figure 3.48 demonstrates these filters can provide large phase lag while maintaining constant gain, with the phase varying from 0° to −360°.

Typical values for ζ_c range from 0.3 and 0.7 depending on the application. The non-minimum phase all-pass filter can be used for flexible mode stabilization by maintaining the loop gain at all frequencies while providing phase change (lead or lag) to increase the closed-loop damping ratio of the flexible modes [4]. There are more second-order filter structures that can be enumerated but we have focused on a few that are pertinent to flexible control.

Example: Colocated and noncolocated control design.
Figure 3.49 shows a flexible three-disk plant. The single input for this system is a torque motor that acts on the lowest disk. There are encoders that measure the orientation of the top and bottom disks, θ_3 and θ_1, respectively.

3.4.2 Colocated Control Design

Design a simple lead controller that uses a measurement of the rotation of the lowest disk and a torque input on the same disk. This type of control law is called "collocated" since the measurement

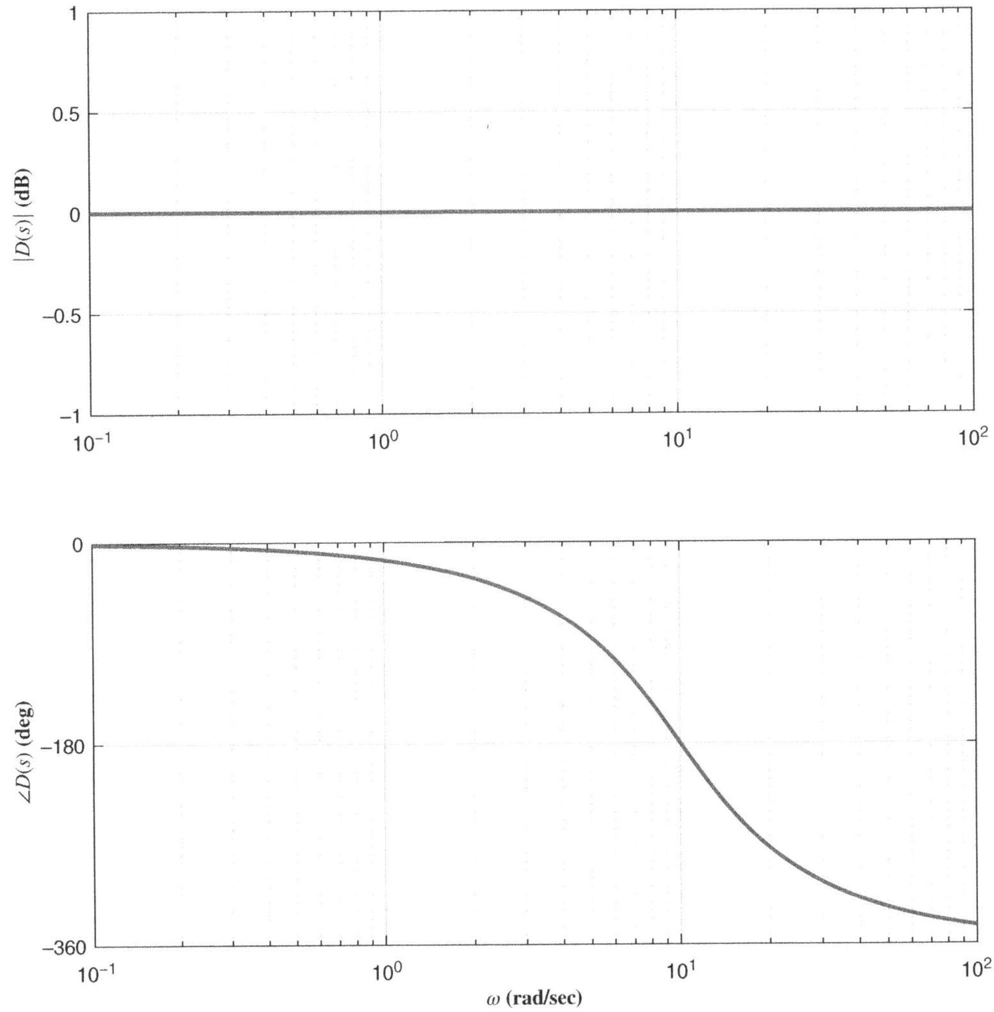

Figure 3.48 Non-minimum-phase all-pass filter with $\zeta_c = 0.7, \omega_c = 10$ rad/s.

and control input happen at the same location. The transfer function relating the torque input u to the output θ_1 is

$$G_1(s) = \frac{\Theta_1(s)}{U(s)} = \frac{K_1\left(s^2 + 2\zeta_{z1}\omega_{z1}s + \omega_{z1}^2\right)\left(s^2 + 2\zeta_{z2}\omega_{z2}s + \omega_{z2}^2\right)}{s(s+p)\left(s^2 + 2\zeta_{p1}\omega_{p1}s + \omega_{p1}^2\right)\left(s^2 + 2\zeta_{p2}\omega_{p2}s + \omega_{p2}^2\right)}$$

where $K_1 = 30.20$, $p = 0.5$ rad/s, $\zeta_{z1} = 0.026$, $\omega_{z1} = 10.68$ rad/s, $\zeta_{z2} = 0.018$, $\omega_{z2} = 40.83$ rad/s, $\zeta_{p1} = 0.020$, $\omega_{p1} = 14.95$ rad/s, $\zeta_{p2} = 0.018$, and $\omega_{p2} = 41.97$ rad/s.

a. Design a simple lead controller, $KD(s) = K_{\text{lead}}\dfrac{s+a}{s+b}$ that yields a closed loop bandwidth of 2π rad/s with a phase margin of about 50°.

b. Plot the Bode diagram of your system. What is your crossover frequency? What is your phase margin?

c. Plot the step response ($\frac{\pi}{2}$ rad command) of your closed-loop system.

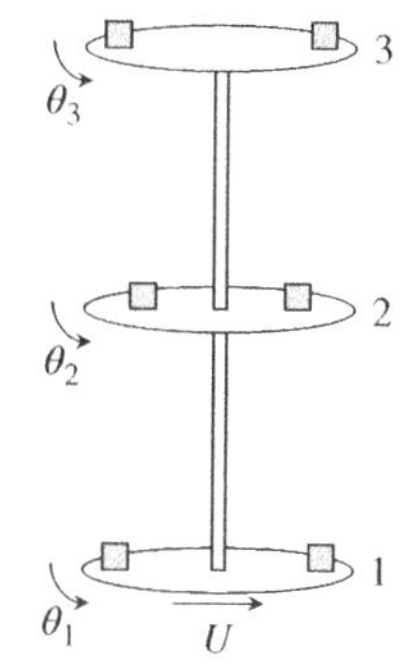

Figure 3.49 Flexible three-disk plant with torque motor.

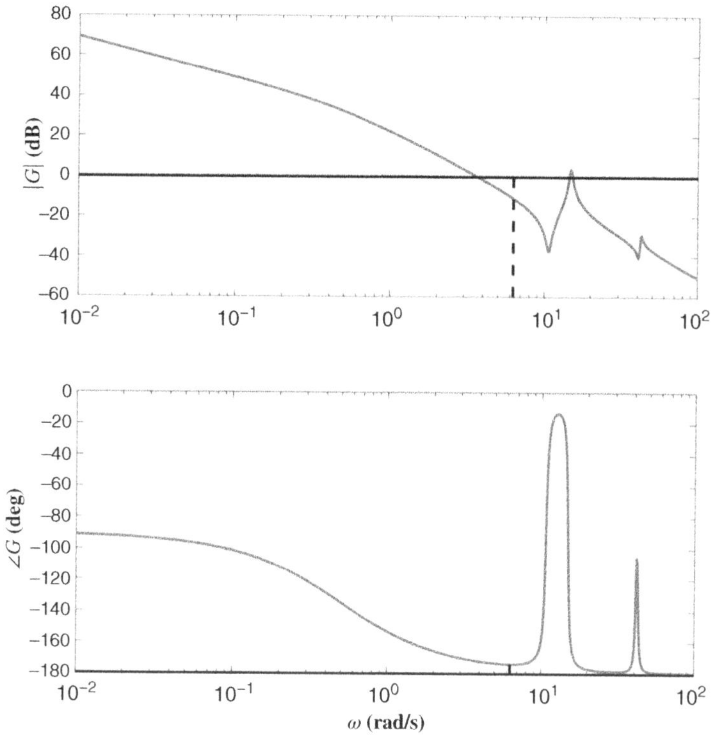

Figure 3.50 Uncompensated colocated system.

Solution:

a. The Bode diagram of the uncompensated system is plotted in Figure 3.50. The initial phase margin, *PM*, at a frequency of 2π was found to be about 2°.

The lead compensator is designed using equations $\omega = \sqrt{ab}$ and $\frac{a}{b} = \frac{1 - sin\phi_{\max}}{1 + sin\phi_{\max}}$. The lead ratio is determined to be $\frac{a}{b} = 0.13$ resulting in the lead compensator

$$KD(s) = 4.25 \cdot \frac{(s + 1.64)}{(s + 12.36)}.$$

This compensator produces 49° of phase at frequency of 2π rad/s.

b. The gain was adjusted to move the crossover frequency to 2π. However, the closed-loop system initially showed a bandwidth exceeding the desired value; therefore, the compensator was iteratively tuned to reduce the bandwidth to the target 2π frequency (-3 dB crossing of the closed-loop Bode diagram). A Bode plot of the compensated open-loop transfer function is shown in Figure 3.51.

The lead compensator $KD(s) = 4.25 \cdot \frac{(s + 1.64)}{(s + 12.36)}$, achieves an open-loop phase margin of 57° and crossover frequency of 4.5 rad/s.

Figure 3.52 illustrates that the bandwidth of the compensated closed-loop system is 2π rad/s.

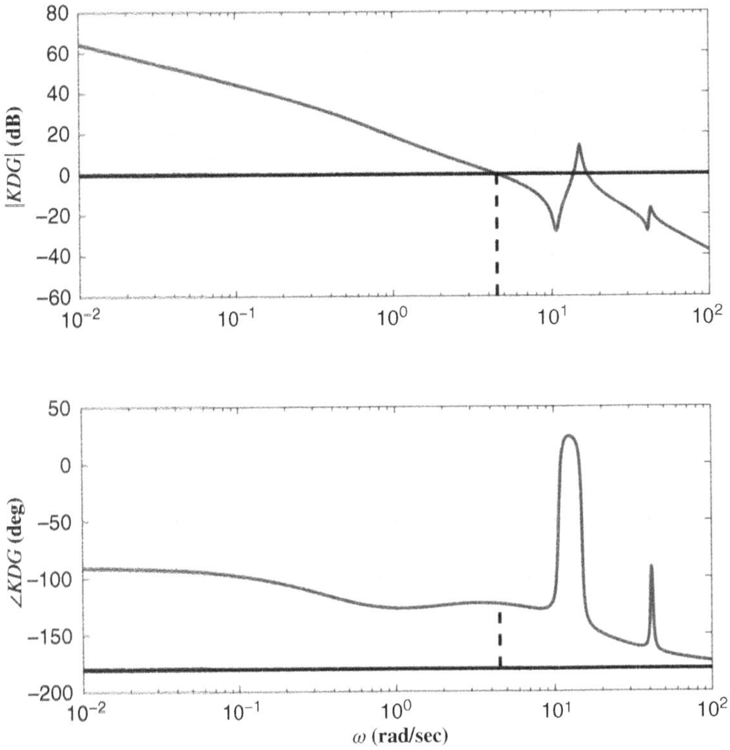

Figure 3.51 Bode diagram of the compensated open-loop transfer function.

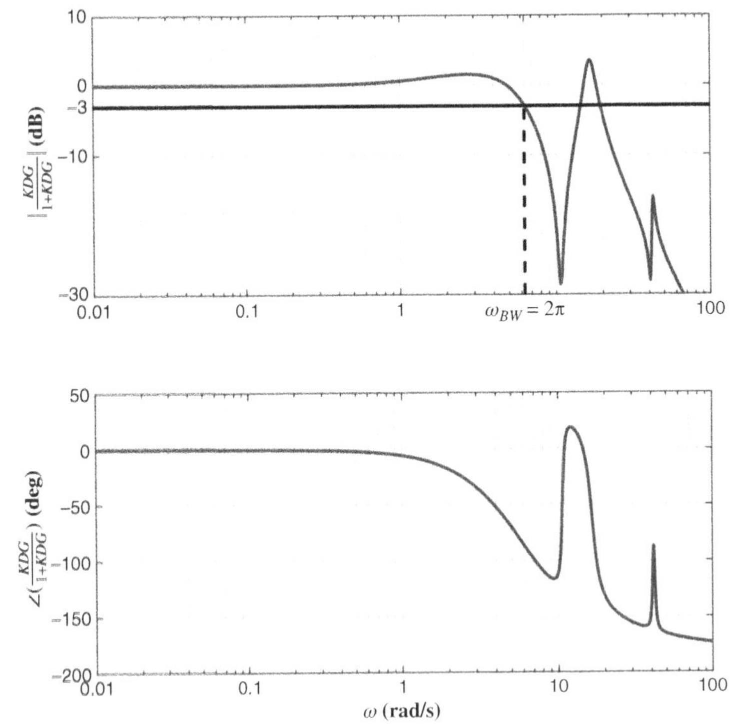

Figure 3.52 Bode diagram of the compensated closed-loop transfer function.

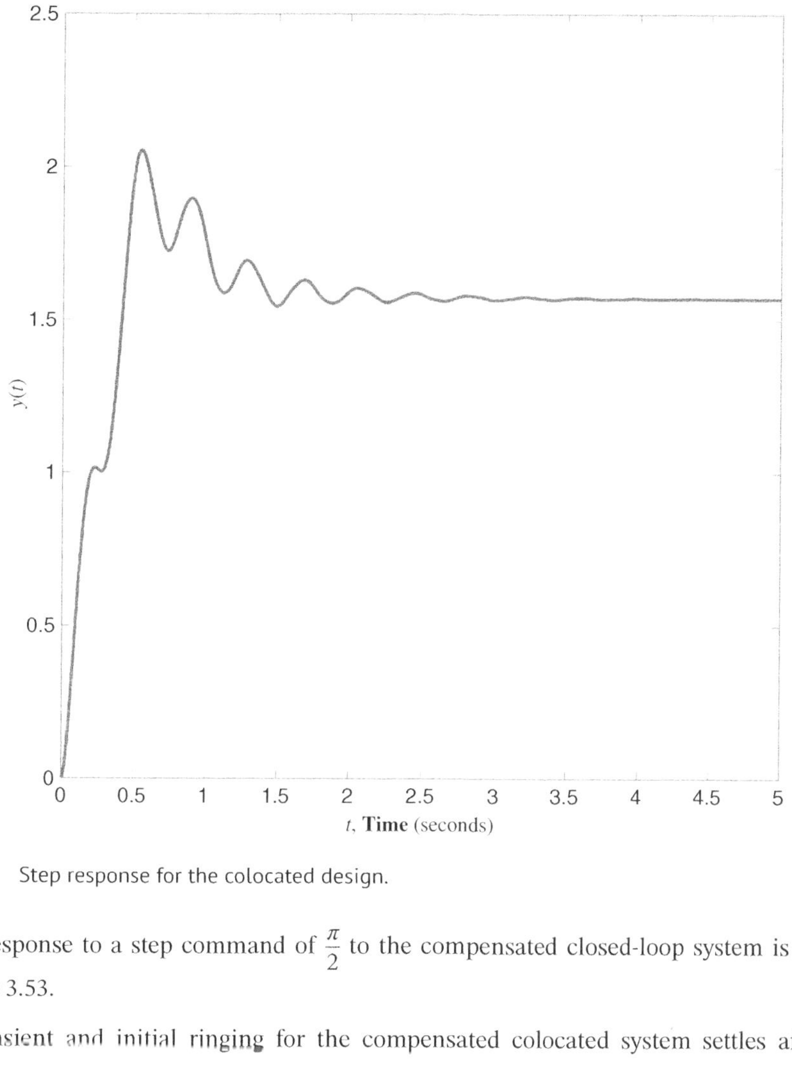

Figure 3.53 Step response for the colocated design.

c. The response to a step command of $\frac{\pi}{2}$ to the compensated closed-loop system is shown in Figure 3.53.

The transient and initial ringing for the compensated colocated system settles after about 4 seconds.

3.4.3 Non-colocated Control Design

What happens if you try to control the position of the top disk using an actuator that drives the bottom disk? This is called a "non-colocated" design, and it requires an additional lag or notch filter to stabilize the high-frequency flexible modes. The transfer function relating the torque input u to the output θ_3 for this system is

$$G_3(s) = \frac{\Theta_3(s)}{U(s)} = \frac{K_2}{s(s+p)\left(s^2 + 2\zeta_{p1}\omega_{p1}s + \omega_{p1}^2\right)\left(s^2 + 2\zeta_{p2}\omega_{p2}s + \omega_{p2}^2\right)}$$

where $K_2 = 5.7465 \times 10^6$, $p = 0.5$ rad/s, $\zeta_{p1} = 0.020$, $\omega_{p1} = 14.95$ rad/s, $\zeta_{p2} = 0.018$, and $\omega_{p2} = 41.97$ rad/s.

d. Design a simple lead controller, $KD(s) = K_{\text{lead}}\frac{s+a}{s+b}$ for the $G_3(s)$ plant. In the simulation, show the maximum bandwidth you can achieve with this type of design? Be sure to remember that there is some uncertainty in both the frequency and the damping of the high-frequency flex modes.

e. Now design a lead compensator plus notch filter, $D_{notch}(s)$, for the $G_3(s)$ system. The notch zeros should be "near" the poles of the first resonant mode (i.e. ω_{p1}). You should experiment in simulation with where to put them. Try to achieve the same performance as you did for the colocated design (closed-loop bandwidth of 2π rad/s with a phase margin of about 55°). Note that this might not be possible.

3.4.4 Non-colocated Control Design Solution

d. A simple lead compensator cannot modify the mode's resonant peak shape but can attenuate it by adjusting the bandwidth. An initial crossover frequency of 2 rad/s is chosen because the magnitude of the system is the same as the resonant peak. Without a phase margin specification, the maximum phase of 60° was targeted at this frequency. A gain was chosen to achieve a crossover frequency that slightly attenuates at 1 rad/s. The resulting lead compensator is

$$KD(s) = 0.52 \cdot \frac{(s + 0.54)}{(s + 7.6)}.$$

Figure 3.54 shows the Bode diagram for the system with and without lead compensation.

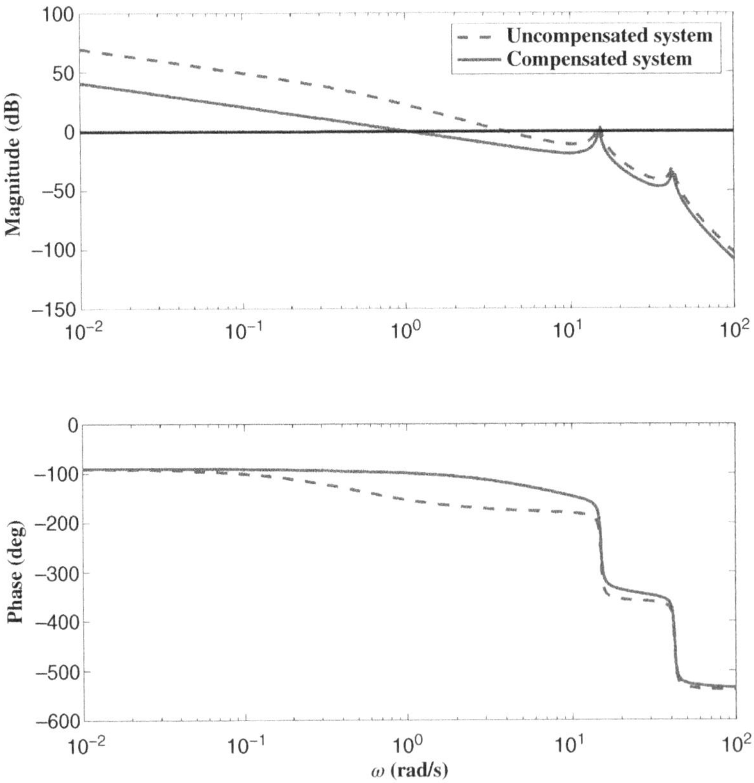

Figure 3.54 Bode of the open-loop system with and without lead control.

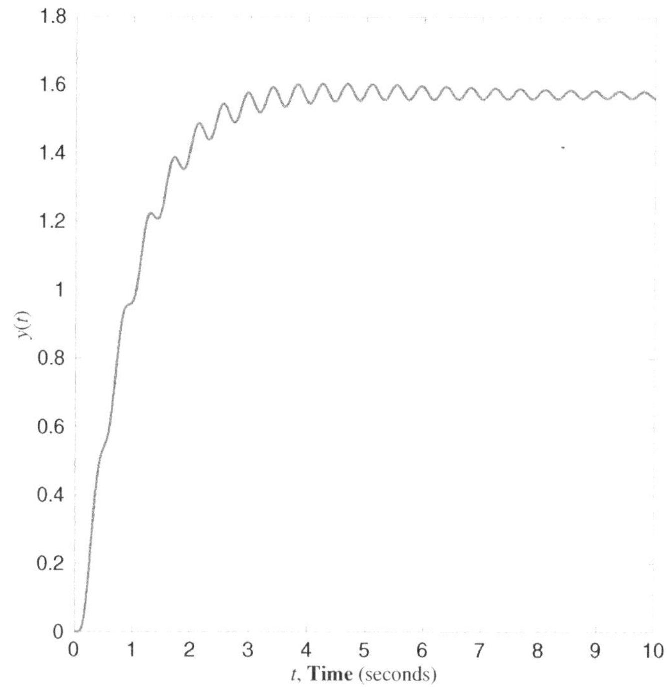

Figure 3.55 Step response of non-colocated system with lead control.

Bandwidth is limited by the high-frequency resonant peaks due to the flexible modes that cannot be fully attenuated using a simple lead compensator. To mitigate this, the bandwidth is reduced to ensure the resonant peak magnitude remains below 0 dB. Here, the bandwidth is closely approximated by the crossover frequency, which must be ≤ 1 rad/s. To account for uncertainties in the plant dynamics, a conservative crossover frequency well below 1 rad/s is chosen. The step response is shown in Figure 3.55.

e. The zeros of the notch compensator are placed to cancel the plant's low-frequency poles. The poles of the notch filter were chosen to be near the compensator zeros. The notch filter gain is adjusted to achieve a bandwidth of 2π, resulting in a crossover frequency of approximate 3.7 rad/s. The final notch filter is

$$D_{\text{notch}}(s) = 0.1822 \cdot \frac{s^2 + 0.598s + 223.5}{s^2 + 29.9s + 223.5}$$

and the frequency response is shown in Figure 3.56.

A lead compensator was added to provide 60° of phase at the crossover frequency of 3.7 rad/s, yielding a phase margin of 40°. Reducing the gain increased the phase margin of 50° but lowered the bandwidth to 2.9 rad/s as shown in Figure 3.57.

Figure 3.58 shows the Bode plot of the compensated system with a 2.9 rad/s bandwidth. This design highlights the tradeoff between achieving the specified bandwidth and phase margin, which was not fully possible for the non-colocated plant model.

The step response, shown in Figure 3.59, demonstrates that the notch and lead control reduces ringing compared to lead control alone, with the transient settling in about 5 seconds.

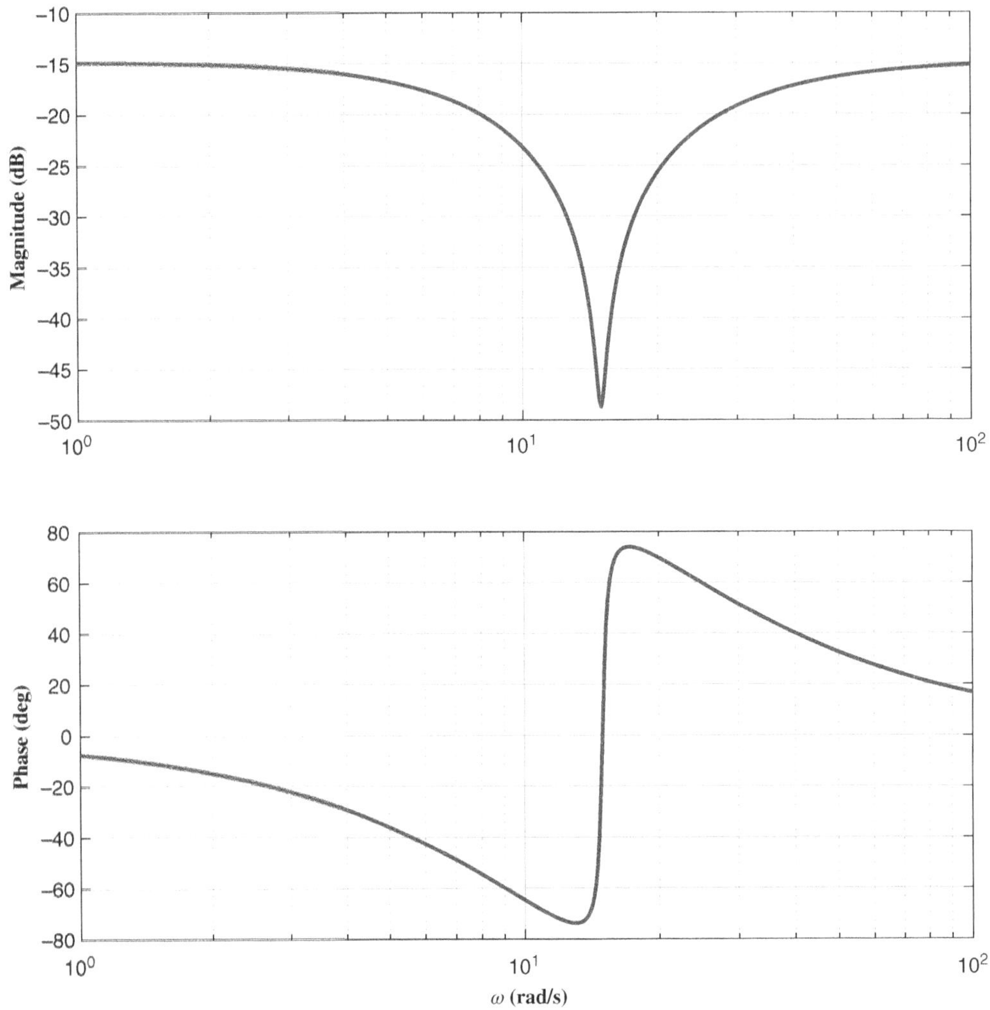

Figure 3.56 Frequency response of the notch filter, $D_{\text{notch}}(s)$.

3.5 State-space Models

State-space models are a fundamental framework for representing systems, particularly linear, time-invariant (LTI) systems, which are central in control theory. A system $H : U \rightarrow Y$ is said to be linear if and only if superposition applies. For example, if $u = \alpha_1 u_1 + \alpha_2 u_2$ with $y_1 = H_1(u)$ and $y_2 = H_2(u)$ then $H(u) = \alpha_1 y_1 + \alpha_2 y_2$. Also, H is time-invariant if and only if the time history of the output is independent of the "starting time" [8]. These properties enable tools like eigenvalue-based stability analysis, superposition, and frequency-domain methods. The DC motor is an example of an LTI system and is modeled as

$$J_m \ddot{\theta}_m(t) + b\dot{\theta}_m(t) = K_t i_a(t) + \omega(t), \tag{3.66}$$

$$K_e \dot{\theta}_m(t) + L_a \dot{i}_a(t) + R_a i_a(t) = v_a(t). \tag{3.67}$$

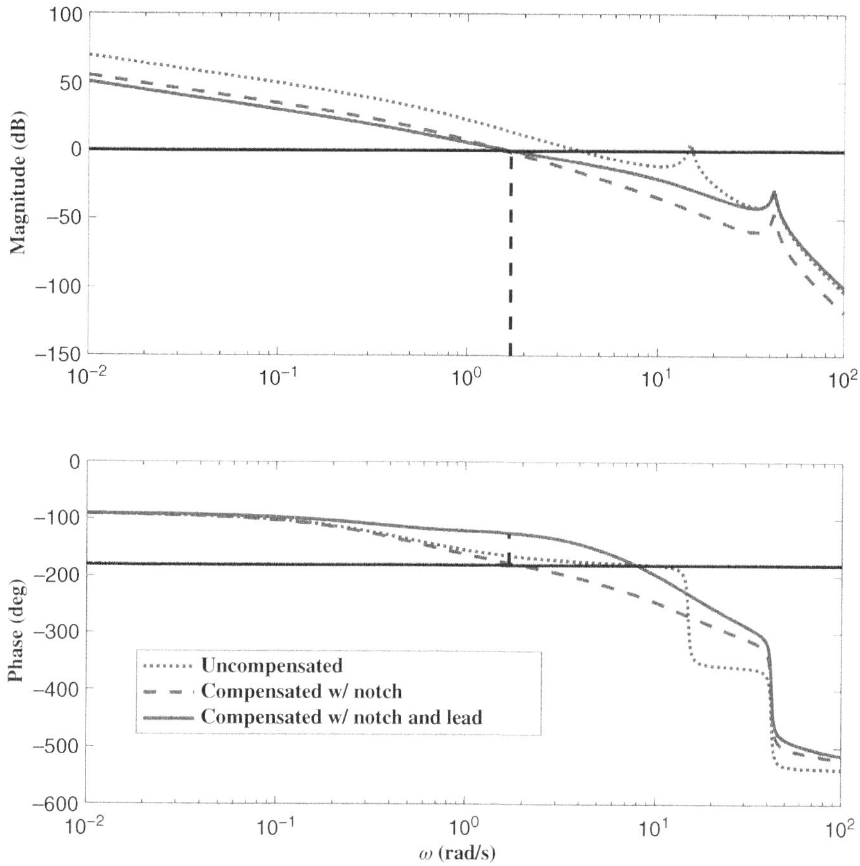

Figure 3.57 Bode plot of the uncompensated system, the compensated system with a notch filter, and the compensated system with a notch filter and lead controller.

To understand the properties of linear systems, let us write the model in "canonical form". Define a vector $\boldsymbol{x}(t) \in \mathbb{R}^3$ (a 3×1 vector) as $\boldsymbol{x}(t) = [\theta_m(t) \;\; \dot{\theta}_m(t) \;\; i_a(t)]^T$ where $\theta_m(t)$ is the shaft angle, $\dot{\theta}_m(t)$ is the shaft angular velocity, $i_a(t)$ is the armature current. Define another vector $\boldsymbol{u}(t) \in \mathbb{R}^2$ as $\boldsymbol{u}(t) = [v_a(t) \;\; \omega(t)]^T$ where $v_a(t)$ is the armature voltage, and $\omega(t)$ is the load torque disturbance. The vector $\boldsymbol{x}(t)$ is known as the state vector and its elements $\theta_m(t)$, $\dot{\theta}_m(t)$, $i_a(t)$ are the states; and the vector $\boldsymbol{u}(t)$ is known as the input vector with the elements $v_a(t)$ (control input), and $\omega(t)$ (disturbance input). For simplicity, the explicit time dependence in the notation will be omitted in the following expressions. We can rewrite Eqs (3.66) and (3.67) in what is known as state-space form as

$$\frac{d\theta_m}{dt} = \dot{\theta}_m$$

$$\frac{d\dot{\theta}_m}{dt} = -\frac{b}{J_m}\dot{\theta}_m + \frac{K_t}{J_m}i_a + \frac{\omega}{J_m}$$

$$\frac{di_a}{dt} = -\frac{K_e}{L_a}\dot{\theta}_m - \frac{R_a}{L_a}i_a + \frac{v_a}{L_a}$$

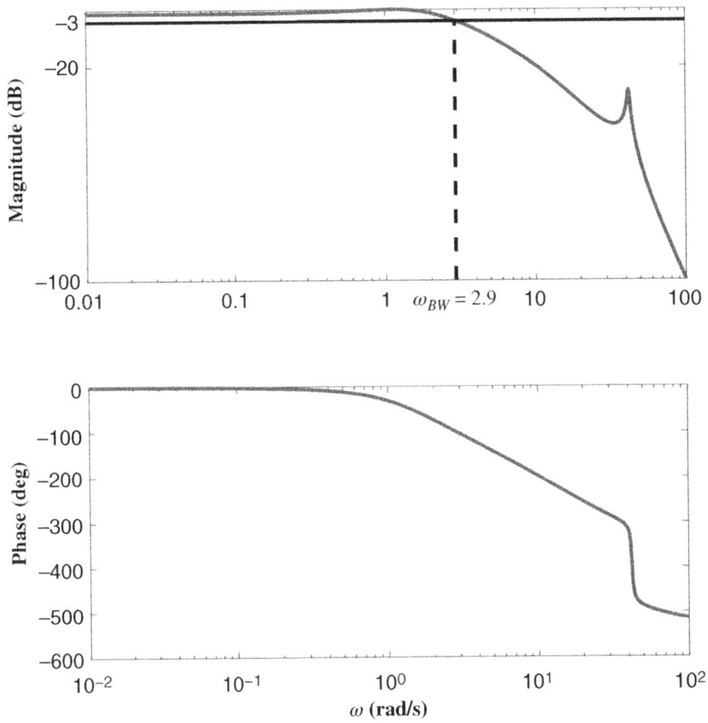

Figure 3.58 Bode plot of the compensated closed-loop system with a bandwidth of 2.9 rad/s.

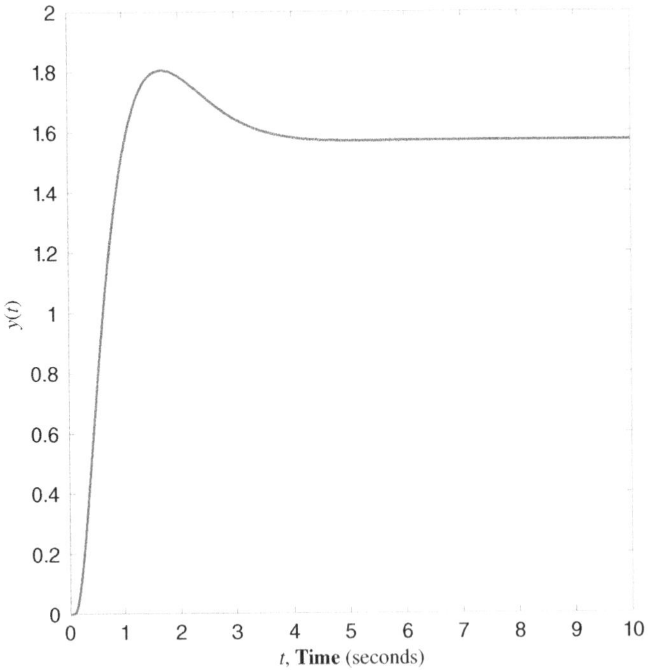

Figure 3.59 Step response of non-colocated system with notch and lead control.

which can be written more compactly as

$$\dot{x} = \begin{bmatrix} 0 & 1 & 0 \\ 0 & -\frac{b}{J_m} & \frac{K_t}{J_m} \\ 0 & -\frac{K_e}{L_a} & -\frac{R_a}{L_a} \end{bmatrix} x + \begin{bmatrix} 0 & 0 \\ 0 & \frac{1}{J_m} \\ \frac{1}{L_a} & 0 \end{bmatrix} u$$

with $x = [\theta_m(t) \;\; \dot{\theta}_m(t) \;\; i_a(t)]^T$ and $u = \begin{bmatrix} v_a \\ \omega \end{bmatrix}$.

The state update equation is

$$\dot{x}(t) = Ax(t) + Bu(t) \tag{3.68}$$

where A is a $n \times n$ matrix and B is a $n \times m$ matrix. For the DC motor, the output is defined as $y(t) = \dot{\theta}_m(t)$, which can be written as

$$y(t) = [0 \;\; 1 \;\; 0]x(t).$$

More generally, the output equation is

$$y(t) = Cx(t) + Du(t) \tag{3.69}$$

where $y(t) \in \mathbb{R}^p$. Finally, we must specify the initial condition or initial state, $x(t_0) = x_0$ where $x_0 \in \mathbb{R}^n$. The number of state variables equals the initial conditions needed to describe the system's energy storage, with a state variable assigned to each variable whose derivative appears in the equations of motion. The state-space model provides detailed information, including all state values over time and initial state data. However, the input-output model is often our primary focus [8]. Starting with the general state-space equations

$$\begin{aligned} \dot{x}(t) &= Ax(t) + Bu(t), \\ y(t) &= Cx(t) + Du(t), \\ x(t_0) &= x_0, \end{aligned} \tag{3.70}$$

we want to derive the transfer function, $G(s) = \frac{Y(s)}{U(s)}$. Taking the Laplace transform of the system, we get

$$sX(s) - x(0) = AX(s) + BU(s) \text{ and } Y(s) = CX(s) + DU(s).$$

Assuming $x(0) = 0$, we can solve for $X(s)$ and substitute the result in $Y(s)$ to get

$$G(s) = C(sI - A)^{-1}B + D. \tag{3.71}$$

If the system has m inputs and p outputs, then we would expect $m \times p$ transfer functions, one for each individual input/output relationship.

3.5.1 Linearization of a System

Suppose $\dot{x}(t) = f(x(t), u(t))$ and we have $\dot{x}_e(t) = f(x_e(t), u_e(t)) = 0$ then $(x_e(t), u_e(t))$ are called the equilibrium states and inputs.

We can linearize a system around an equilibrium $(x_e(t), y_e(t), u_e(t))$ or more generally, around an operating point or "nominal" trajectory [8]. Let $\delta x(t)$, $\delta u(t)$, $\delta y(t)$ be deviations from the nominal trajectory then

$$\boldsymbol{x} = \boldsymbol{x}_e + \delta\boldsymbol{x},$$
$$\boldsymbol{u} = \boldsymbol{u}_e + \delta\boldsymbol{u},$$
$$\boldsymbol{y} = \boldsymbol{y}_e + \delta\boldsymbol{y}.$$

Using a Taylor Series expansion on $f(\boldsymbol{x}, \boldsymbol{u})$ and $g(\boldsymbol{x}, \boldsymbol{u})$ while neglecting higher-order terms (h.o.t) we have

$$\dot{\boldsymbol{x}}(t) = f(\boldsymbol{x}, \boldsymbol{u})$$
$$= f(\boldsymbol{x}_e, \boldsymbol{u}_e) + \left.\frac{\partial f(\boldsymbol{x}, \boldsymbol{u})}{\partial \boldsymbol{x}}\right|_{(\boldsymbol{x}_e, \boldsymbol{u}_e)} \cdot (\boldsymbol{x} - \boldsymbol{x}_e) + \left.\frac{\partial f(\boldsymbol{x}, \boldsymbol{u})}{\partial \boldsymbol{u}}\right|_{(\boldsymbol{x}_e, \boldsymbol{u}_e)} \cdot (\boldsymbol{u} - \boldsymbol{u}_e) + (h.o.t) \tag{3.72}$$

where $\frac{\partial f(\boldsymbol{x}, \boldsymbol{u})}{\partial \boldsymbol{x}} = \begin{bmatrix} \frac{\partial f_1(\boldsymbol{x}, \boldsymbol{u})}{\partial x_1} & \cdots & \frac{\partial f_1(\boldsymbol{x}, \boldsymbol{u})}{\partial x_n} \\ \vdots & \ddots & \vdots \\ \frac{\partial f_n(\boldsymbol{x}, \boldsymbol{u})}{\partial x_1} & \cdots & \frac{\partial f_n(\boldsymbol{x}, \boldsymbol{u})}{\partial x_n} \end{bmatrix}$ is a $n \times n$ matrix.

Similarly, for $g(\boldsymbol{x}, \boldsymbol{u})$

$$\boldsymbol{y} = g(\boldsymbol{x}, \boldsymbol{u}) = g(\boldsymbol{x}_e, \boldsymbol{u}_e) + \left.\frac{\partial g(\boldsymbol{x}, \boldsymbol{u})}{\partial \boldsymbol{x}}\right|_{(\boldsymbol{x}_e, \boldsymbol{u}_e)} \cdot (\boldsymbol{x} - \boldsymbol{x}_e) + \left.\frac{\partial g(\boldsymbol{x}, \boldsymbol{u})}{\partial \boldsymbol{u}}\right|_{(\boldsymbol{x}_e, \boldsymbol{u}_e)} \cdot (\boldsymbol{u} - \boldsymbol{u}_e) + (h.o.t). \tag{3.73}$$

Now

$$\dot{\boldsymbol{x}} = \dot{\boldsymbol{x}}_e + \delta\dot{\boldsymbol{x}} = f(\boldsymbol{x}_e, \boldsymbol{u}_e) + \left.\frac{\partial f(\boldsymbol{x}, \boldsymbol{u})}{\partial \boldsymbol{x}}\right|_{(\boldsymbol{x}_e, \boldsymbol{u}_e)} \cdot (\boldsymbol{x} - \boldsymbol{x}_e) + \left.\frac{\partial f(\boldsymbol{x}, \boldsymbol{u})}{\partial \boldsymbol{u}}\right|_{(\boldsymbol{x}_e, \boldsymbol{u}_e)} \cdot (\boldsymbol{u} - \boldsymbol{u}_e)$$
$$= f(\boldsymbol{x}_e, \boldsymbol{u}_e) + \left.\frac{\partial f}{\partial \boldsymbol{x}}\right|_{(\boldsymbol{x}_e, \boldsymbol{u}_e)} \cdot \delta\boldsymbol{x} + \left.\frac{\partial f}{\partial \boldsymbol{u}}\right|_{(\boldsymbol{x}_e, \boldsymbol{u}_e)} \cdot \delta\boldsymbol{u} \tag{3.74}$$

$$\boldsymbol{y} = \boldsymbol{y}_e + \delta\boldsymbol{y} = g(\boldsymbol{x}_e, \boldsymbol{u}_e) + \left.\frac{\partial g(\boldsymbol{x}, \boldsymbol{u})}{\partial \boldsymbol{x}}\right|_{(\boldsymbol{x}_e, u_e)} \cdot (\boldsymbol{x} - \boldsymbol{x}_e) + \left.\frac{\partial g(\boldsymbol{x}, \boldsymbol{u})}{\partial \boldsymbol{u}}\right|_{(\boldsymbol{x}_e, \boldsymbol{u}_e)} \cdot (\boldsymbol{u} - \boldsymbol{u}_e)$$
$$= g(\boldsymbol{x}_e, \boldsymbol{u}_e) + \left.\frac{\partial g}{\partial \boldsymbol{x}}\right|_{(\boldsymbol{x}_e, \boldsymbol{u}_e)} \cdot \delta\boldsymbol{x} + \left.\frac{\partial g}{\partial \boldsymbol{u}}\right|_{(\boldsymbol{x}_e, \boldsymbol{u}_e)} \cdot \delta\boldsymbol{u}.$$

Letting $A = \left.\frac{\partial f(\boldsymbol{x}, \boldsymbol{u})}{\partial \boldsymbol{x}}\right|_{(\boldsymbol{x}_e, \boldsymbol{u}_e)}$, $B = \left.\frac{\partial f(\boldsymbol{x}, \boldsymbol{u})}{\partial \boldsymbol{u}}\right|_{(\boldsymbol{x}_e, \boldsymbol{u}_e)}$, $C = \left.\frac{\partial g(\boldsymbol{x}, \boldsymbol{u})}{\partial \boldsymbol{x}}\right|_{(\boldsymbol{x}_e, \boldsymbol{u}_e)}$, $D = \left.\frac{\partial g(\boldsymbol{x}, \boldsymbol{u})}{\partial \boldsymbol{u}}\right|_{(\boldsymbol{x}_e, \boldsymbol{u}_e)}$ this linearization around an operating point, $(\boldsymbol{x}_e, \boldsymbol{u}_e)$, results in the following linearized state-space model

$$\begin{aligned} \delta\dot{\boldsymbol{x}} &= A \cdot \delta\boldsymbol{x} + B \cdot \delta\boldsymbol{u} \\ \delta\boldsymbol{y} &= C \cdot \delta\boldsymbol{x} + D \cdot \delta\boldsymbol{u}. \end{aligned} \tag{3.75}$$

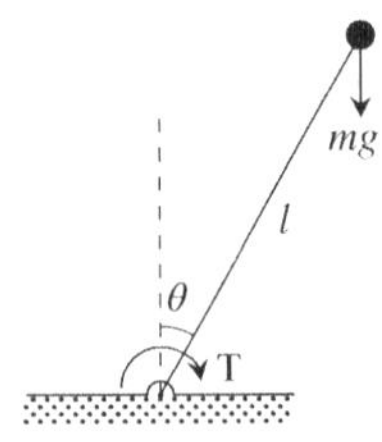

Figure 3.60 Simple inverted pendulum.

Example: Linearization of inverted pendulum dynamics.
For the simple inverted pendulum shown in Figure 3.60, the equations of motion are

$$ml^2\ddot{\theta} - mgl\sin\theta = T.$$

Let $\theta_1 = \theta$ and $\theta_2 = \dot{\theta}$, then the nonlinear model is

$$\dot{\theta}_1 = \theta_2$$
$$\dot{\theta}_2 = \frac{g}{l}\sin\theta_1 + \frac{T}{ml^2}.$$

Linearizing around the equilibrium point $(\theta_e, T_e) = (\mathbf{0}, 0)$ gives

$$\delta\dot{\theta}_1 = \delta\theta_2,$$
$$\delta\dot{\theta}_2 = \frac{g}{l}\delta\theta + \frac{1}{ml^2}\cdot\delta T.$$

The linearized state-space model is

$$\delta\dot{\boldsymbol{\theta}} = A\cdot\delta\boldsymbol{\theta} + B\cdot\delta T = \begin{bmatrix} 0 & 1 \\ \frac{g}{l} & 0 \end{bmatrix}\begin{bmatrix} \delta\theta_1 \\ \delta\theta_2 \end{bmatrix} + \begin{bmatrix} 0 \\ \frac{1}{ml^2} \end{bmatrix}\delta T.$$

The linearized state-space model provides a simplified approximation of the inverted pendulum dynamics, valid near the equilibrium point $(\theta_e, T_e) = (\mathbf{0}, 0)$. This model can now be used for control design.

Example: Inverted pendulum pole cancellation.
The inverted pendulum dynamic equation is repeated

$$ml^2\ddot{\theta} - mgl\sin\theta = T.$$

Using the small angle approximation, $\sin\theta \approx \theta$, the equation can be reduced to

$$\ddot{\theta} = \Omega^2\theta + u,$$

where $\Omega^2 = g/l$ and $u = T/ml^2$. If we define $y = \theta$, the open-loop transfer function is

$$\frac{Y(s)}{U(s)} = G(s) = \frac{1}{s^2 - \Omega^2} = \frac{1}{(s-\Omega)(s+\Omega)}.$$

Let us design a compensator such that

$$D(s) = \frac{s - \overline{\Omega}}{s}, \text{ where } \overline{\Omega} = \Omega.$$

This compensator cancels the unstable pole at Ω and replaces it with an open-loop pole at the origin. The new open-loop transfer function is

$$D(s)G(s) = \frac{1}{s(s+\Omega)}.$$

The closed-loop step response indicates that our compensation performs satisfactorily. But let us take a closer look at what is really going on. The open-loop state-space system with the compensator block diagram is shown in Figure 3.61.

We can define $x_1 = \theta$, $x_2 = \dot{\theta}$, and add a new state $x_3 = \overline{u} - u$ so the compensated system becomes

$$\dot{x}_1 = x_2$$
$$\dot{x}_2 = \Omega^2 x_1 - x_3 + \overline{u}$$
$$\dot{x}_3 = \Omega\overline{u}$$

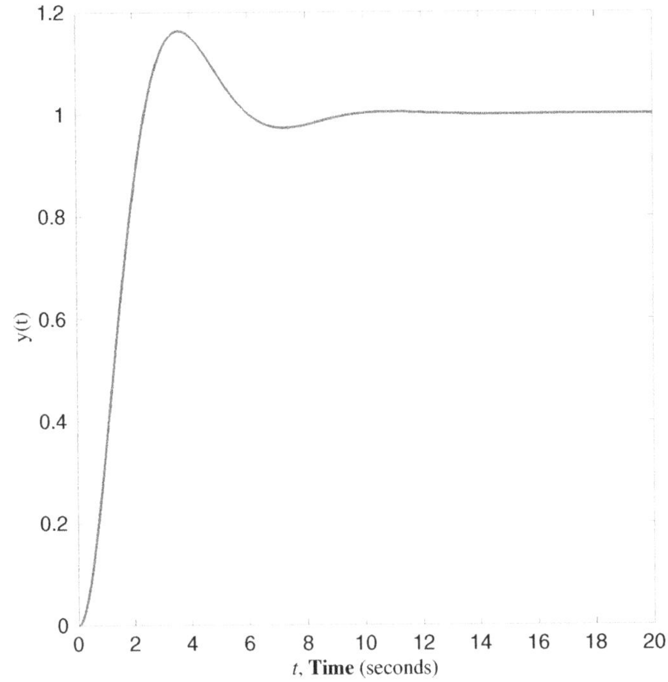

Figure 3.61 Compensated closed-loop step response of inverted pendulum.

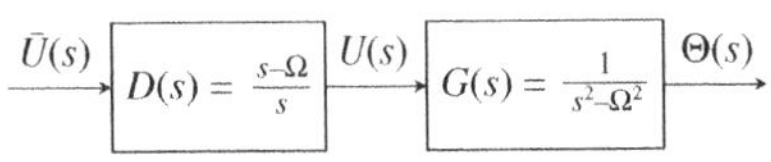

Figure 3.62 Block diagram of state-space system.

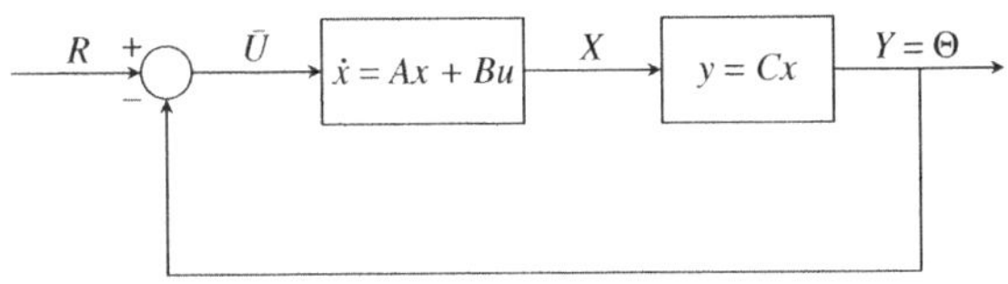

Figure 3.63 State-space block diagram of the closed-loop system.

where the state update and output equations are

$$\dot{\boldsymbol{x}} = A\boldsymbol{x} + Bu = \begin{bmatrix} 0 & 1 & 0 \\ \Omega^2 & 0 & -1 \\ 0 & 0 & 0 \end{bmatrix} \boldsymbol{x} + \begin{bmatrix} 0 \\ 1 \\ \Omega \end{bmatrix} u$$
$$\boldsymbol{y} = C\boldsymbol{x} = [1 \quad 0 \quad 0]\boldsymbol{x}.$$

A block diagram of the closed-loop system is shown in Figure 3.63.
With feedback, $\bar{u} = \boldsymbol{r} - \boldsymbol{y} = \boldsymbol{r} - \boldsymbol{Cx}$ the closed loop system is

$$\begin{aligned} \dot{\boldsymbol{x}} &= A\boldsymbol{x} + B(\boldsymbol{r} - \boldsymbol{Cx}) = (A - BC)\boldsymbol{x} + B\boldsymbol{r} \\ \boldsymbol{y} &= C\boldsymbol{x}. \end{aligned} \tag{3.76}$$

Figure 3.64 shows the closed-loop step response of the state-space system for $\Omega = 1$, which is the same as before.

But the state response when $\boldsymbol{r}(\boldsymbol{t}) = 0$ and the initial condition is $[x_1(0) \quad x_2(0) \quad x_3(0)]^T = [1 \quad 0 \quad 0]^T$ in Figure 3.65 is unstable.

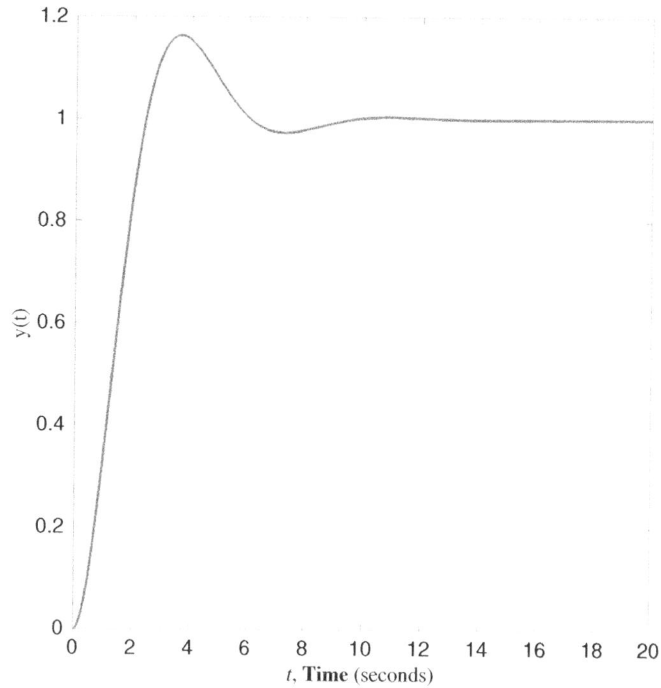

Figure 3.64 State-space step response for inverted pendulum.

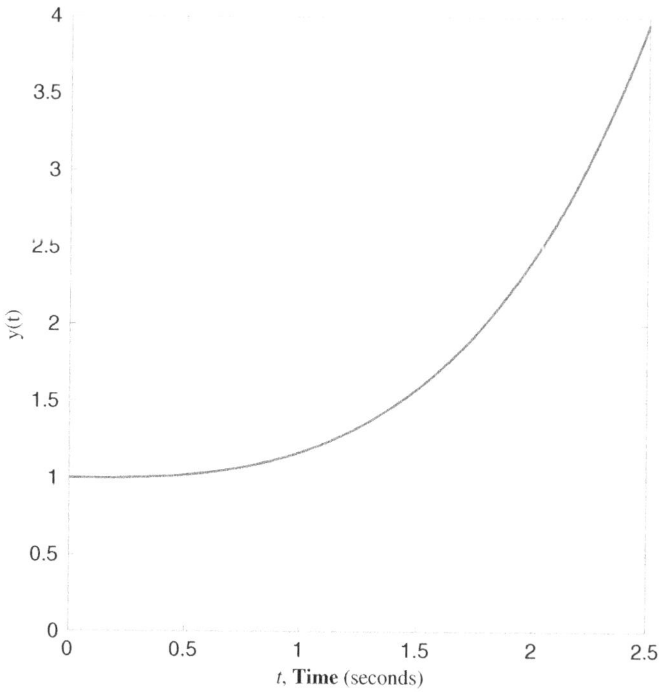

Figure 3.65 Step response with initial condition $[x_1(0) \quad x_2(0) \quad x_3(0)]^T = [1 \quad 0 \quad 0]^T$.

This step response is unstable because although the inclusion of the compensator $D(s) = \frac{s-\Omega}{s}$, breaks the connection between the input $\bar{u}$ and the output θ, it did not remove the internal instability in θ from the system. The unstable pole is still there and may be excited by:

- nonzero initial conditions
- disturbances or plant uncertainty.

State-space models reveal internal system dynamics, including those hidden in input/output transfer functions. In this example, unstable modes could have been identified by analyzing eigenvalues of A and $(A - BC)$. Compared to transfer functions, state-space models offer several advantages [8]:

1. State-space models expose unstable dynamics that are uncontrollable (cannot be influenced by inputs) or unobservable (not visible at outputs).
2. These are more convenient for multi-input/multi-output (MIMO) systems.
3. They account for initial state responses, unlike transfer functions.
4. These are also applicable to nonlinear systems.

3.5.2 The Matrix Exponential

To solve the system of equations

$$\begin{aligned} \dot{\boldsymbol{x}} &= A\boldsymbol{x} + B\boldsymbol{u} \\ \boldsymbol{y} &= C\boldsymbol{x} + D\boldsymbol{u} \\ \boldsymbol{x}(t_0) &= \boldsymbol{x}_0 \end{aligned}$$

where $A \in \mathbb{R}^{n\times n}$, $B \in \mathbb{R}^{n\times n}$, $C \in \mathbb{R}^{m\times n}$, $D \in \mathbb{R}^{m\times n}$, $\boldsymbol{x} \in \mathbb{R}^{\boldsymbol{n}}$, $\boldsymbol{u} \in \mathbb{R}^{n}$, and $\boldsymbol{y} \in \mathbb{R}^{m}$ we can use the matrix exponential, defined as

$$e^{At} = I + \frac{At}{1!} + \frac{A^2t^2}{2!} + \dots \tag{3.77}$$

where $A \in \mathbb{R}^{n\times n}$, I is the $n \times n$ identity matrix. The matrix exponential satisfies $\dot{\boldsymbol{z}}(t) = A\boldsymbol{z}(t)$ and $\boldsymbol{z}(0) = I$, leading to $\boldsymbol{z}(t) = e^{At}$. We can verify by checking:

1. The initial condition: $\boldsymbol{z}(t) = e^{At}\big|_{\boldsymbol{t=0}} = \left(I + \frac{A\cdot 0}{1!} + \frac{A^2\cdot 0^2}{2!} + \dots\right) = I.$
2. The differential equation: $\dot{\boldsymbol{z}}(t) = \frac{de^{At}}{dt} = \left(A + \frac{A^2\cdot t}{1!} + \dots\right) = Ae^{At} = A\boldsymbol{z}(t).$

The series expansion is efficient for sparse A, but typically e^{At} is computed using the inverse Laplace transform

$$e^{At} = \mathcal{L}^{-1}((sI - A)^{-1}). \tag{3.78}$$

The solution to $\dot{\boldsymbol{x}}(t) = A\boldsymbol{x}(t) + B\boldsymbol{u}(t)$ is

$$\boldsymbol{x}(t) = e^{At}\boldsymbol{x}_0 + \int_0^t e^{A(t-\tau)}B\boldsymbol{u}(\tau)d\tau \tag{3.79}$$

and for $\mathbf{y}(t) = C\mathbf{x}(t) + D\mathbf{u}(t)$ is

$$\mathbf{y}(t) = Ce^{At}\mathbf{x}_0 + \int_0^t e^{A(t-\tau)}CB\mathbf{u}(\tau)d\tau + D\mathbf{u} \tag{3.80}$$

Although not explicitly verified here, Eqs (3.79) and (3.80) satisfy both the initial condition and differential equation.

3.5.2.1 Computing e^{At} and Modal Interpretation

Computing e^{At} is often time consuming or even impossible for a general A. To simplify, we can diagonalize A using a transformation matrix. The eigenvalues of A are the roots of the characteristic equation

$$det(\lambda I - A) = \mathbf{0}. \tag{3.81}$$

An eigenvalue λ exists if there is a nonzero eigenvector, $\mathbf{v}$, such that

$$A\mathbf{v} = \lambda\mathbf{v}. \tag{3.82}$$

If A has n eigenvectors that are linearly independent, they can form the columns of matrix transformation, T, such that

$$AT = T\Lambda \tag{3.83}$$

where $\Lambda = \text{diag}\,(\lambda_1, \ldots, \lambda_n)$. Multiplying both sides by T^{-1}, we obtain

$$T^{-1}AT = \Lambda \tag{3.84}$$

which can also be expressed as $A = T\Lambda T^{-1}$. Then, the matrix exponential e^{At} can be computed as

$$e^{At} = Te^{\Lambda t}T^{-1} \tag{3.85}$$

where $e^{\Lambda t} = \begin{bmatrix} e^{\lambda_1 t} & & \\ & \ddots & \\ & & e^{\lambda_n t} \end{bmatrix}$, $T = \begin{bmatrix} | & & | \\ \mathbf{v}_1 & \cdots & \mathbf{v}_n \\ | & & | \end{bmatrix}$, and $T^{-1} = \begin{bmatrix} - & \mathbf{w}_1^T & - \\ & \vdots & \\ - & \mathbf{w}_n^T & - \end{bmatrix}$ is a column of row vectors.

Expanding e^{At} as a summation yield

$$e^{At} = \sum_{i=1}^{n} e^{\lambda_i t}\mathbf{v}_i\mathbf{w}_i^T. \tag{3.86}$$

If A is diagonalizable, then $\dot{\mathbf{x}}(t) = A\mathbf{x}(t)$ with initial condition $\mathbf{x}(0)$, and has the solution

$$\mathbf{x}(t) = e^{At}\mathbf{x}(0) = Te^{\Lambda t}T^{-1}\mathbf{x}(0) = \sum_{i=1}^{n} e^{\lambda_i t}\mathbf{v}_i\mathbf{w}_i^T\mathbf{x}(0). \tag{3.87}$$

The state solution is a linear combination of the system modes, $\mathbf{v}_i e^{\lambda_i t}$, where

- $e^{\lambda_i t}$ determines the nature of the time response,
- $\mathbf{v}_i$ determines how each state contributes to that mode,
- $\mathbf{w}_i^T\mathbf{x}(0)$ determines the extent to which the initial condition excites the mode.

For example, for $\lambda = -5$ and corresponding eigenvector $\mathbf{v} = [1 \;\; -5 \;\; 20]^T$ the mode e^{-5t} contributes 20 times more to state x_3 than x_1.

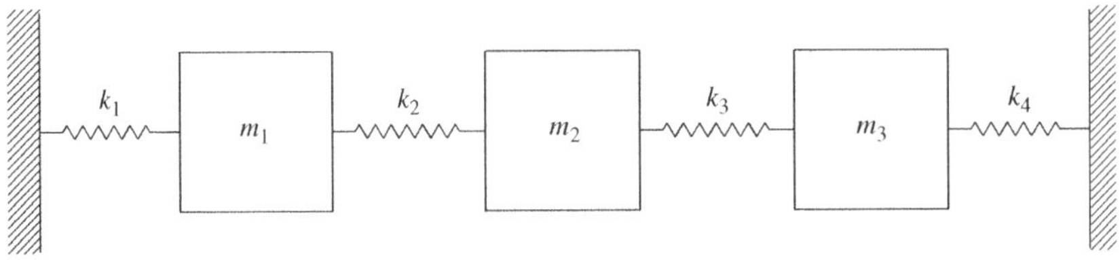

Figure 3.66 Mass-spring system with three masses.

Example: Mass-spring system.

Figure 3.66 illustrates a mass-spring system with three masses that are connected to walls on both sides.

The equations of motion for this system are

$$m_1\ddot{x}_1 = -(k_1 + k_2)x_1 + k_2x_2$$
$$m_2\ddot{x}_2 = -(k_2 + k_3)x_2 + k_2x_1 + k_3x_2$$
$$m_3\ddot{x}_3 = -(k_3 + k_4)x_3 + k_3x_2.$$

We can write this more compactly in state-space form as

$$\dot{\boldsymbol{x}} = A\boldsymbol{x}$$

where

$$\boldsymbol{x} = [x_1 \quad x_2 \quad x_3\dot{x}_1 \quad \dot{x}_2 \quad \dot{x}_3]^T, A = \begin{bmatrix} 0 & I \\ -M^{-1}K & 0 \end{bmatrix}, M = \begin{bmatrix} m_1 & 0 & 0 \\ 0 & m_2 & 0 \\ 0 & 0 & m_3 \end{bmatrix}, \text{ and}$$

$$K = \begin{bmatrix} k_1 + k_2 & -k_2 & 0 \\ -k_2 & k_2 + k_3 & -k_3 \\ 0 & -k_3 & k_3 + k_4 \end{bmatrix}.$$

The eigenvalues and eigenvectors of this undamped system are

$$\lambda_1 = \pm 1.85j \text{ and } \mathbf{v}_1 = \begin{bmatrix} 1.00 \\ -1.41 \\ 1.00 \\ \pm 1.85j \\ \mp 2.61j \\ \pm 1.85j \end{bmatrix}; \lambda_2 = \pm 1.41j \text{ and } \mathbf{v}_2 = \begin{bmatrix} 1.00 \\ 0.00 \\ -1.00 \\ \pm 1.41j \\ 0.00 \\ \pm 1.41j \end{bmatrix}; \lambda_3 = \pm 0.76j \text{ and } \mathbf{v}_3 = \begin{bmatrix} 1.00 \\ 1.41 \\ 1.00 \\ \pm 0.77j \\ \pm 1.08j \\ \pm 0.77j \end{bmatrix}.$$

We have normalized each eigenvector, $\mathbf{v}_i$, by its first component to simplify the interpretation of the relative displacements of the masses and how each state contributes to the corresponding mode.

Let us simulate this system and examine how the 3 modes of this system behave for different initial conditions. Since we are only concerned with the displacement of each mass, we only consider the real part of the initial conditions such that $\boldsymbol{x}_i(0) = \mathbb{R}(\mathbf{v}_i)$. The plot for mode 1, $\lambda_1 = \pm 1.85j$, is shown in Figure 3.67.

The highest frequency mode has masses m_1 and m_3 moving in the same direction while m_2 moves in the opposite direction. The plot for mode 2, $\lambda_2 = \pm 1.41j$, is shown in Figure 3.68.

The middle-frequency mode keeps m_2 stationary, while the two other masses move in opposition. The plot for mode 3, $\lambda_3 = \pm 0.77j$ is shown in Figure 3.69.

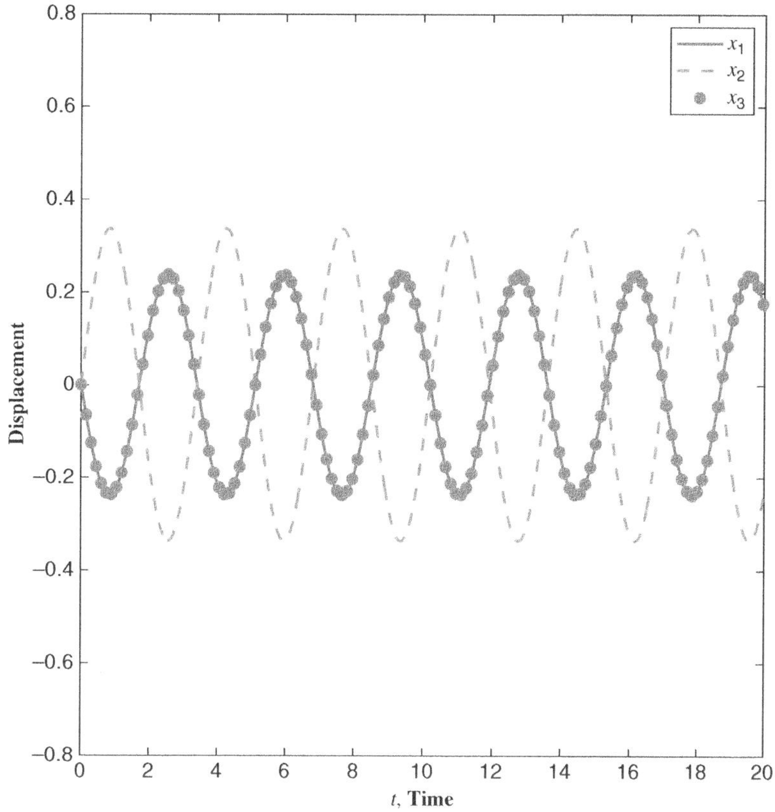

Figure 3.67 Displacements for mode with $\lambda_1 = \pm 1.85j$.

The lowest frequency mode has all masses moving in phase, with m_2. The middle mass, m_2, exhibiting higher amplitude motion. Finally, for a random initial condition combining all modes, as shown in Figure 3.70, the displacement reflects contributions from each mode.

3.5.3 State-space Stability

The state equation, $\dot{\boldsymbol{x}}(t) = A\boldsymbol{x}(t)$, is called internally stable if the state $\boldsymbol{x}(t)$, $t > 0$, remains bounded for any bounded initial state, $\boldsymbol{x}_0(t)$. This system is also called asymptotically internally stable if $\boldsymbol{x}(t)$ decays to $\mathbf{0}$ as $t \to \infty$. Finally, the system is exponentially stable there exits $\alpha > 0$ and $m > 0$ such that $\|\boldsymbol{x}(t)\| \leq me^{-\alpha t} \cdot \|\boldsymbol{x}_0\|$, Every solution of $\dot{\boldsymbol{x}}(t) = A\boldsymbol{x}(t)$ is bounded by a decaying exponential in t defined by m, α, and $\|\boldsymbol{x}_0\|$. The system $\dot{\boldsymbol{x}}(t) = A\boldsymbol{x}(t)$ is exponentially stable if and only if all the eigenvalues of A are in the left-half plane [8].

Example: Stability check.
Check whether the following system is given by

$$\dot{\boldsymbol{x}} = A\boldsymbol{x} + B\boldsymbol{u}, \boldsymbol{y} = C\boldsymbol{x} + D\boldsymbol{u},$$

where $A = \begin{bmatrix} -1 & 0 & 0 \\ 0 & 3 & 0 \\ 0 & 0 & -6 \end{bmatrix}$, $B = \begin{bmatrix} 1/10 \\ -1/6 \\ 1/15 \end{bmatrix}$, $C = \begin{bmatrix} 1 & 1 & 1 \end{bmatrix}$ is stable? Clearly, this system is not internally stable since A has an eigenvalue at 3. The transfer function is

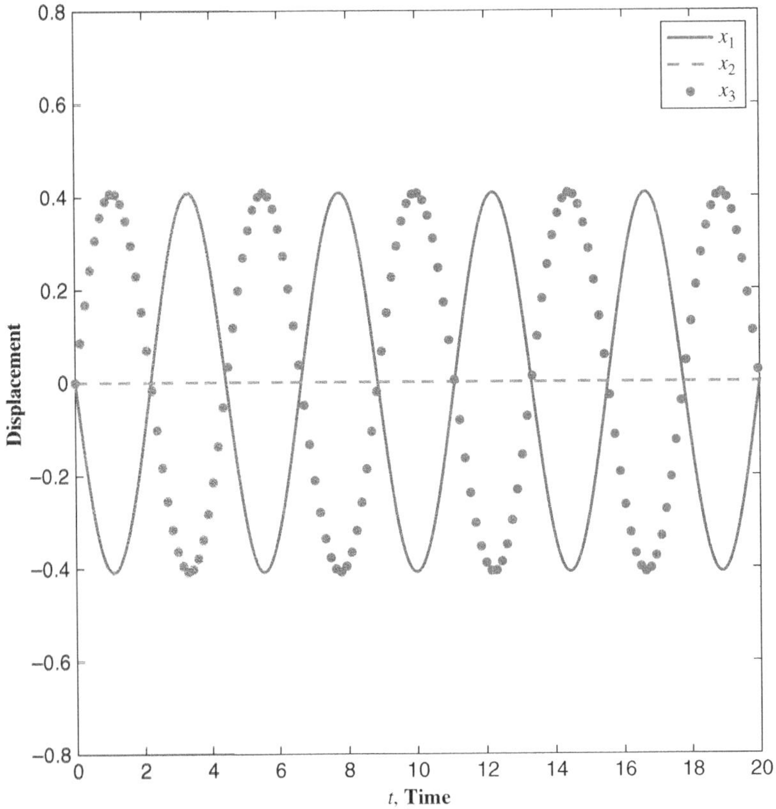

Figure 3.68 Displacements for mode with $\lambda_2 = \pm 1.41j$.

$$G(s) = C(sI - A)^{-1}B = \frac{1}{6}\frac{s+4}{(s+1)(s+6)}.$$

This system is BIBO stable but not internally stable. The mode at $\lambda = 3$ is uncontrollable from $\boldsymbol{u}$ and thus does not appear in the transfer function.

3.5.3.1 Canonical Forms

The input-output transfer function, $\frac{Y(s)}{U(s)}$, can be represented as a rational fraction of polynomials in s

$$\frac{Y(s)}{U(s)} = \frac{b_m s^m + b_{m-1}s^{m-1} + \cdots + b_0}{s^n + a_{n-1}s^{n-1} + \cdots + a_0}.$$

If we multiply both the numerator and denominator by $X(s)$, and applying the inverse Laplace transform the equations for the input and output become

$$u(t) = x^{(n)} + a_{n-1}x^{(n-1)} + \cdots + a_0 x$$

$$y(t) = b_m x^{(m)} + b_{m-1}x^{(m-1)} + \cdots + b_0 x.$$

Defining state variables $x_1 = x$, $x_2 = \dot{x}$, $\cdots$, $x_n = x^{(n-1)}$, the system can be written in state-space form as

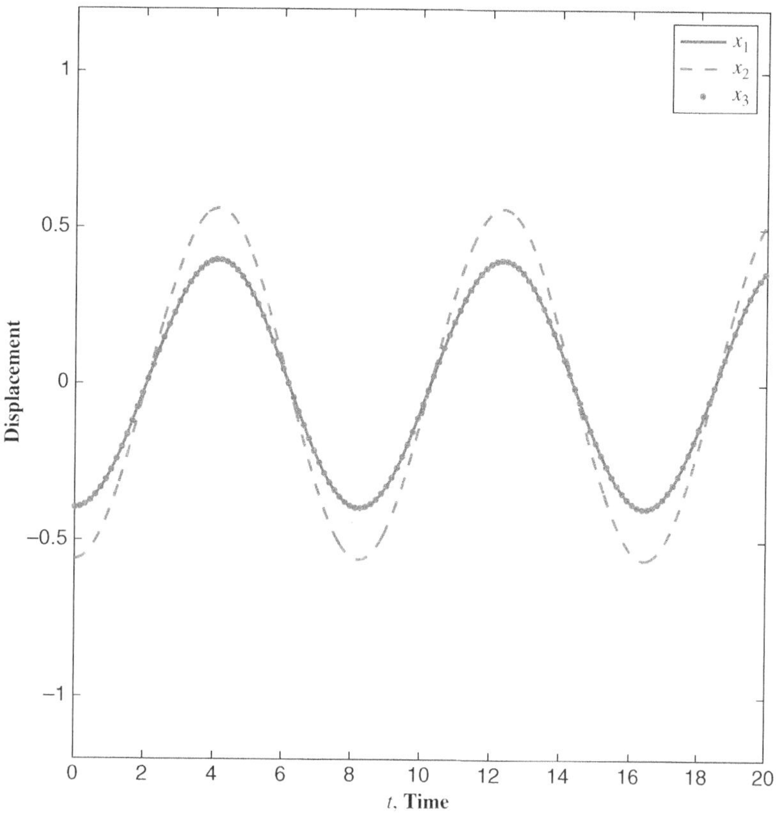

Figure 3.69 Displacements for mode with $\lambda_3 = \pm 0.77j$.

$$\dot{\boldsymbol{x}} = \begin{bmatrix} 0 & 1 & 0 & \cdots & 0 \\ 0 & 0 & 1 & \cdots & 0 \\ & & & \ddots & \\ 0 & 0 & \cdots & \cdots & 1 \\ -a_0 & -a_1 & \cdots & \cdots & -a_{n-1} \end{bmatrix} \boldsymbol{x} + \begin{bmatrix} 0 \\ 0 \\ \vdots \\ \vdots \\ 1 \end{bmatrix} \boldsymbol{u}, \tag{3.88}$$
$$\boldsymbol{y} = [b_0 \quad b_1 \quad\quad b_m \quad 0 \quad \cdots \quad 0]\,\boldsymbol{x}.$$

This is called controllable canonical form.

The transfer function, $\frac{Y(s)}{U(s)}$, can be expressed by cross-multiplying to obtain

$$s^n Y(s) + a_{n-1}s^{n-1}Y(s) + \cdots + a_0 Y(s) = b_{n-1}s^{n-1}U(s) + b_{n-2}s^{n-2}U(s) + \cdots + b_0 U(s).$$

Rewriting $Y(s)$ recursively in terms of $U(s)$ leads to

$$\begin{aligned} x_n(s) &= \frac{1}{s}(b_0 U(s) - a_0 Y(s)) \\ x_{n-1}(s) &= \frac{1}{s}(b_1 U(s) - a_1 Y(s) + x_n(s)) \\ &\vdots \\ x_1(s) &= \frac{1}{s}(b_{n-1}U(s) - a_{n-1}Y(s) + x_2(s)) \\ Y(s) &= x_1(s). \end{aligned}$$

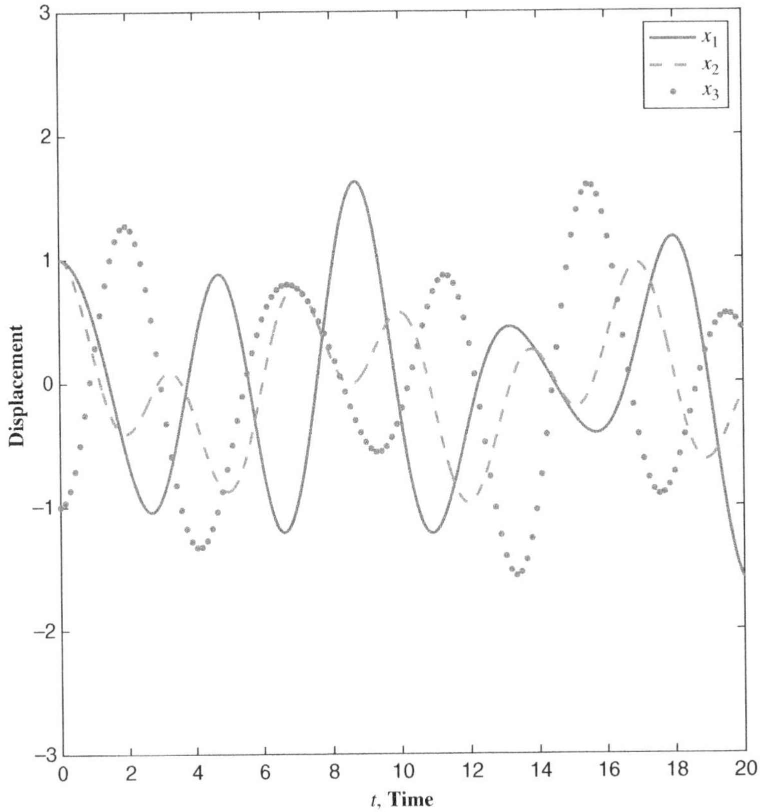

Figure 3.70 Displacements for a random input.

In state-space form

$$\dot{\boldsymbol{x}} = \begin{bmatrix} -a_{n-1} & 1 & 0 & \cdots & 0 \\ -a_{n-2} & 0 & 1 & \cdots & 0 \\ & & & \ddots & \\ 0 & 0 & \cdots & \cdots & 1 \\ -a_0 & 0 & \cdots & \cdots & 0 \end{bmatrix} \boldsymbol{x} + \begin{bmatrix} b_{n-1} \\ b_{n-2} \\ \vdots \\ \vdots \\ b_0 \end{bmatrix} \boldsymbol{u} \tag{3.89}$$

$$\boldsymbol{y} = [1 \quad 0 \quad \cdots \quad \cdots \quad \cdots \quad \cdots \quad 0]\boldsymbol{x}.$$

This is called observable canonical form [8].

3.5.4 State Feedback Design

Let us place the poles of the following system

$$\begin{bmatrix} \dot{x}_1 \\ \dot{x}_2 \end{bmatrix} = \begin{bmatrix} -1 & a \\ 3 & -2 \end{bmatrix} \begin{bmatrix} x_1 \\ x_2 \end{bmatrix} + \begin{bmatrix} 0 \\ 1 \end{bmatrix} u.$$

A block diagram of the full-state feedback system is shown in Figure 3.71.

The control law is $u = -K\boldsymbol{x}$ where $K = -[k_1 \quad k_2]$. The closed-loop system can be described by $\dot{\boldsymbol{x}} = (A - BK)\boldsymbol{x}$ where

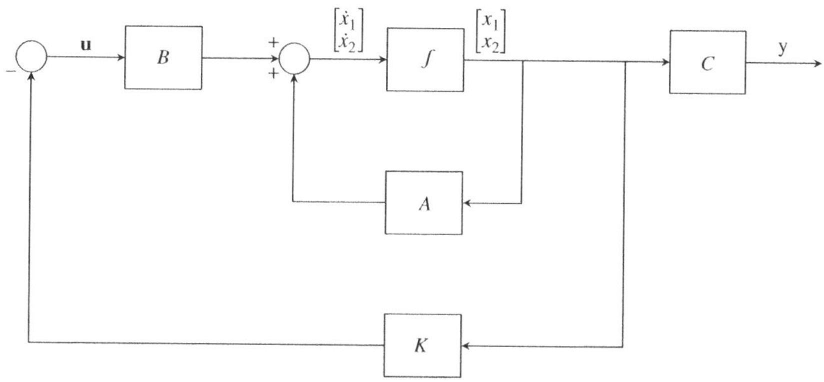

Figure 3.71 Block diagram of a system with full-state feedback.

$$(A - BK) = \begin{bmatrix} -1 & a \\ 3 - k_1 & -2 - k_2 \end{bmatrix}.$$

The poles are given by the roots of the closed-loop characteristic polynomial

$$\det(sI - A + BK) = s^2 + s(k_2 + 3) + 2 + k_2 + a(k_1 - 3) = 0.$$

Assume the desired closed-loop poles to be at -2 and -3. To design a feedback controller with these poles, we need to compare the closed-loop characteristic polynomial with the desired polynomial

$$(s + 2)(s + 3) = s^2 + 5s + 6.$$

Matching coefficients,

$$k_2 + 3 = 5 \Rightarrow k_2 = 2$$
$$2 + k_2 + a(k_1 - 3) = 6 \Rightarrow k_1 = \frac{2}{a} + 3.$$

Thus, the state feedback control law is

$$u = -\begin{bmatrix} \frac{2}{a} + 3 & 2 \end{bmatrix} \begin{bmatrix} x_1 \\ x_2 \end{bmatrix}.$$

If $a \neq 0$, the closed-loop system poles are -2 and -3. When $a = 0$, we are no longer able to place the closed-loop poles independently since k_1 does not influence the closed-loop characteristic polynomial.

$$s^2 + s(k_2 + 3) + 2 + k_2 = (s + 2 + k_2)(s + 1) = 0.$$

The mode at $s = -1$ cannot be changed since k_1 no longer influences the closed-loop poles.

3.5.5 Controllability

The system with the state equation $\dot{\boldsymbol{x}} = A\boldsymbol{x} + B\boldsymbol{u}$ and $A \in \mathbb{R}^{n \times n}$ is controllable if the rank$(\mathcal{C}) = n$ such that

$$\mathcal{C} = [B \;\; AB \;\; \cdots \;\; A^{n-1}B] \tag{3.90}$$

is called the controllability matrix [8].

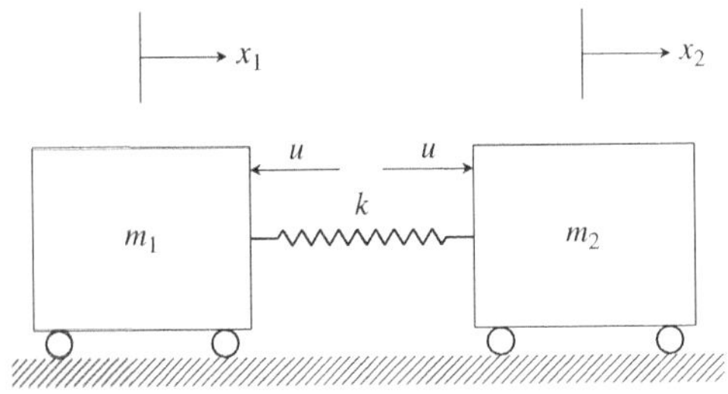

Figure 3.72 Two mass-spring systems with internal forces.

Example: Uncontrollable dynamics.
A two mass-spring system is shown in Figure 3.72.

The forces, u, are internal to the system. Because of Newton's Third Law of Motion, the location of the center of mass of a closed system cannot be changed by forces internal to the system. Show why this system is uncontrollable.

Let

$$x_3 = \dot{x}_1,$$

$$x_4 = \dot{x}_2,$$

Then

$$\dot{x}_1 = x_3$$
$$\dot{x}_2 = x_4$$
$$\dot{x}_3 = \frac{-k}{m_1}(x_1 - x_2) - \frac{u}{m_1}$$
$$\dot{x}_4 = \frac{-k}{m_2}(x_2 - x_1) - \frac{u}{m_2}.$$

The state-space matrices are

$$A = \begin{bmatrix} 0 & 0 & 1 & 0 \\ 0 & 0 & 0 & 1 \\ -k/m_1 & k/m_1 & 0 & 0 \\ k/m_2 & -k/m_2 & 0 & 0 \end{bmatrix} \text{ and } B = \begin{bmatrix} 0 \\ 0 \\ -1/m_1 \\ 1/m_2 \end{bmatrix}.$$

The controllability matrix is

$$\mathcal{C} = [B \quad AB \quad A^2B \quad A^3B]$$

$$= \begin{bmatrix} 0 & -1/m_1 & 0 & k/m_1^2 + k/(m_1m_2) \\ 0 & 1/m_2 & 0 & -k/m_2^2 - k/(m_1m_2) \\ -1/m_1 & 0 & k/m_1^2 + k/(m_1m_2) & 0 \\ 1/m_2 & 0 & -k/m_2^2 - k/(m_1m_2) & 0 \end{bmatrix}.$$

The rank of $\mathcal{C}$ is 2, which means the controllability matrix is not full rank and the system is not controllable. We can examine the dynamics of the system to see why this is the case. The center of mass of this system is

$$x_c = \frac{m_1x_1 + m_2x_2}{m_1 + m_2}$$

and consider a transformation of the state variables

$$\bar{x}_1 = x_c$$
$$\bar{x}_2 = x_1 - x_2$$
$$\bar{x}_3 = \dot{\bar{x}}_1$$
$$\bar{x}_4 = \dot{\bar{x}}_2.$$

This new set of coordinates produces the following state-space matrices

$$\bar{A} = \begin{bmatrix} 0 & 0 & 1 & 0 \\ 0 & 0 & 0 & 1 \\ 0 & 0 & 0 & 0 \\ 0 & 0 & 0 & -k/m_1 - k/m_2 \end{bmatrix} \text{ and } \bar{B} = \begin{bmatrix} 0 \\ 0 \\ 0 \\ 1/m_1 - 1/m_2 \end{bmatrix}.$$

There is only one nonzero element in $\bar{B}$. Therefore, the internal force u can cause a displacement between x_1 and x_2 but not the states x_1 and x_2 independently. To independently change the states, an external force is needed.

3.5.5.1 Pole-placement

For a system in controllable canonical form, with

$$K = [k_1 \;\; k_2 \;\; \cdots \;\; k_n]$$

the modified system matrix becomes

$$A - bK = \begin{bmatrix} 0 & 1 & 0 & \cdots & 0 \\ 0 & 0 & 1 & \cdots & 0 \\ & & & \ddots & \\ 0 & 0 & \cdots & \cdots & 1 \\ -a_0 - k_1 & -a_1 - k_2 & \cdots & \cdots & -a_{n-1} - k_n \end{bmatrix}. \tag{3.91}$$

The eigenvalues of $A - bK$ are given by the roots of det $(sI - A + bK)$ which are

$$s^n + (a_{n-1} + k_n)s^{n-1} + (a_{n-2} + k_{n-1})s^{n-2} + \cdots + a_0 + k_1 = 0. \tag{3.92}$$

Compare this with the desired closed-loop polynomial

$$s^n + c_1 s^{n-1} + c_2 s^{n-2} + \cdots + c_{n-1}s + c_n = 0 \tag{3.93}$$

and we can solve the feedback control law gains

$$\begin{aligned} k_1 &= c_n - a_0 \\ k_2 &= c_{n-1} - a_1 \\ &\vdots \\ k_n &= c_1 - a_{n-1}. \end{aligned} \tag{3.94}$$

This procedure is called pole-placement and can be easily computed using computational software like Matlab. It is not necessary to transform the system $\dot{\boldsymbol{x}} = A\boldsymbol{x} + B\boldsymbol{u}$ to controllable canonical form to perform the pole placement; we can do pole-placement directly on (A, B) by computing the characteristic equation $(A - BK, \; B)$ directly [8].

Example: Pole-placement.
Consider the system

$$\dot{\boldsymbol{x}}(t) = \begin{bmatrix} 1 & 1 \\ 1 & 2 \end{bmatrix} \boldsymbol{x}(t) + \begin{bmatrix} 1 \\ 0 \end{bmatrix} u.$$

The open-loop characteristic equation is

$$\det(sI - A) = (s-1)(s-2) - 1 = s^2 - 3s + 1 = 0$$

and the system is unstable. Let us try to place poles such that they are at $s = -5, -6$.

For the closed-loop system

$$A - BK = \begin{bmatrix} 1 & 1 \\ 1 & 2 \end{bmatrix} - \begin{bmatrix} 1 \\ 0 \end{bmatrix} [k_1 \quad k_2].$$

The closed-loop characteristic equation is

$$\det(sI - A - BK) = s^2 + (k_1 - 3)s + (1 - 2k_1 + k_2) = 0.$$

The desired characteristic equation is

$$\alpha_c(s) = (s+5)(s+6) = s^2 + 11s + 30.$$

Matching coefficients

$$k_1 - 3 = 11 \Rightarrow k_1 = 14$$
$$1 - 2k_1 + k_2 = 30 \Rightarrow k_2 = 57.$$

Thus, the feedback gain is $K = [14 \quad 57]$.

The procedure shown in the examples so far shows how to design a full-state feedback controller for second-order systems by hand. We can extend this to higher-order systems too but it is often tedious. Ackermann's formula provides a systematic way to compute the state-feedback gains,

$$K = [0 \quad \cdots \quad 0 \quad 1]\, \mathcal{C}^{-1} \alpha_c(A) \tag{3.95}$$

where $\mathcal{C} = [B \quad AB \quad A^2B \quad \cdots \quad A^{n-1}B]$ is the controllability matrix, $\alpha_c(s)$ is the desired characteristic equation for the closed-loop poles, which we evaluate for $s = A$. The system must be controllable because we are inverting the controllability matrix.

Example: Pole-placement using Ackermann's formula.
Let us use Ackermann's formula to calculate the full-state controller for the previous example. Recall the desired characteristic equation, $\alpha_c(s)$, for the closed-loop poles is

$$\alpha_c(s) = (s+5)(s+6) = s^2 + 11s + 30.$$

Let us calculate the controllability matrix

$$\mathcal{C} = [B \quad AB] = \left[\begin{bmatrix} 1 \\ 0 \end{bmatrix} \quad \begin{bmatrix} 1 & 1 \\ 1 & 2 \end{bmatrix}\begin{bmatrix} 1 \\ 0 \end{bmatrix}\right] = \begin{bmatrix} 1 & 1 \\ 0 & 1 \end{bmatrix}.$$

Using Ackerman's formula, the control gains are

$$K = [0 \quad 1]\mathcal{C}^{-1}(A^2 + 11A + 30I)$$

which we calculate to get

$$K = [14 \quad 57].$$

Using Ackermann's formula can also be tedious and error prone, so it is better to rely on software tools like Matlab.

So far, we have analyzed how to select K to stabilize A such that $\lambda_i(A - BK)$ achieves the state dynamics we desire. But where should we put the closed-loop poles? A common approach is to use time-domain specifications to locate dominant poles as roots of

$$s^2 + 2\zeta\omega_n s + \omega_n^2 = 0. \tag{3.96}$$

Specifications for second-order responses like rise time, settling time, time to peak, and peak overshoot can help guide the placement of closed-loop poles. The remaining poles can be placed much faster than the second-order behavior, maintaining the same damped frequency ω_d and adjusting the real parts to be 2–3 times faster than $\zeta\omega_n$. However, poles should not be placed too far left, as this leads to avoid excessive control effort. Alternatively, linear quadratic regulation (LQR) can optimize a cost function to determine pole placement. LQR will be discussed later.

3.5.6 Reference Input Design in State-space

For an output $y(t)$ to track reference input $r(t)$ as $t \to \infty$, we require that $y(t) \approx r(t)$. In the frequency domain, $\lim_{t\to\infty} y(t) = \lim_{s\to 0} sY(s) \Longrightarrow \left.\frac{Y(s)}{R(s)}\right|_{s=0} = 1$. The transfer function from $R(s)$ to $Y(s)$ should be 1 or 0 dB at DC steady state. If $r(t)$ is a constant over time, then in steady-state ($\dot{\boldsymbol{x}} = \boldsymbol{0}$) we want our state-space model to be

$$\begin{matrix} \boldsymbol{0} = A\boldsymbol{x}_{ss} + B\boldsymbol{u}_{ss} \\ r = C\boldsymbol{x}_{ss} + D\boldsymbol{u}_{ss} \end{matrix} \Rightarrow \begin{bmatrix} \boldsymbol{0} \\ r \end{bmatrix} = \begin{bmatrix} A & B \\ C & D \end{bmatrix} \begin{bmatrix} \boldsymbol{x}_{ss} \\ \boldsymbol{u}_{ss} \end{bmatrix}. \tag{3.97}$$

If the inverse exists, then

$$\begin{bmatrix} \boldsymbol{x}_{ss} \\ \boldsymbol{u}_{ss} \end{bmatrix} = \begin{bmatrix} A & B \\ C & D \end{bmatrix}^{-1} \begin{bmatrix} \boldsymbol{0} \\ r \end{bmatrix} = \begin{bmatrix} N_x r \\ N_u r \end{bmatrix}. \tag{3.98}$$

We can compute the required reference input to achieve a desired output by setting $\boldsymbol{u} = \boldsymbol{u}_{ss} - K(\boldsymbol{x} - \boldsymbol{x}_{ss})$ and then when $\boldsymbol{x} = \boldsymbol{x}_{ss}$ and $\boldsymbol{u} = \boldsymbol{u}_{ss}$ then we get Eq (3.97). We have effectively inverted the plant description to achieve this. Then, plugging Eq (3.98) into $\boldsymbol{u} = \boldsymbol{u}_{ss} - K(\boldsymbol{x} - \boldsymbol{x}_{ss})$ we get that

$$\boldsymbol{u} = -K\boldsymbol{x} + (KN_x + N_u)r = -K\boldsymbol{x} + \bar{N}r \tag{3.99}$$

where $\bar{N} = [KN_x \ \ N_u]^T$. The block diagram of the full-state feedback system with a reference input is shown in Figure 3.73 [8].

For full-state feedback, the closed-loop dynamics are

$$\dot{\boldsymbol{x}}(t) = (A - BK)\boldsymbol{x}(t) + B\bar{N}r(t), \boldsymbol{y}(t) = C\boldsymbol{x}(t) \tag{3.100}$$

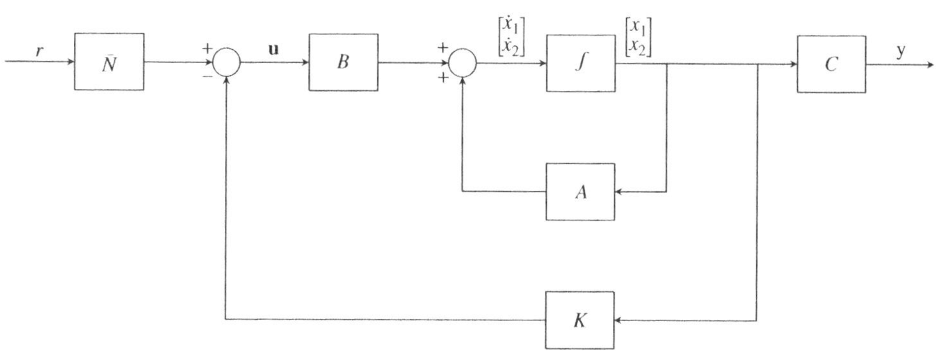

Figure 3.73 Block diagram of a system with full-state feedback and reference input, $r(t)$.

with the transfer function

$$G(s) = \frac{Y(s)}{R(s)} = C(sI - (A - BK))^{-1}B\overline{N} \tag{3.101}$$

and we can compute $\overline{N}$ by setting $\left.\frac{Y(s)}{R(s)}\right|_{s=0} = 1 = -C(A - BK)^{-1}B\overline{N}$ to get

$$\overline{N} = -(C(A - BK)^{-1}B)^{-1}. \tag{3.102}$$

3.5.7 Estimator Design in State-space

In state feedback, it is often assumed that the state vector is directly measurable, but this is often not the case. The observer is a signal reconstruction device which provides an estimate of inaccessible states [8]. Suppose a plant whose dynamical behavior is governed by the state-space equations

$$\dot{\boldsymbol{x}}(t) = A\boldsymbol{x}(t) + B\boldsymbol{u}(t), \boldsymbol{y}(t) = C\boldsymbol{x}(t) + D\boldsymbol{u}(t),$$

has both its input $\boldsymbol{u}$ and output $\boldsymbol{y}$ connected as shown in Figure 3.74.

If we denote the state estimate of $\boldsymbol{x}(t)$ as $\hat{\boldsymbol{x}}(t)$, then an inspection of the block diagram in Figure 3.22 shows that its state-space equation is

$$\dot{\hat{\boldsymbol{x}}}(t) = (A - LC)\hat{\boldsymbol{x}}(t) + B\boldsymbol{u}(t) + L\boldsymbol{y}(t). \tag{3.103}$$

and the estimation error is given by

$$\dot{\tilde{\boldsymbol{x}}}(t) = (A - LC)\tilde{\boldsymbol{x}}(t) \tag{3.104}$$

where $\tilde{\boldsymbol{x}}(t) = \hat{\boldsymbol{x}}(t) - \boldsymbol{x}(t)$. The solution of this system is $\tilde{\boldsymbol{x}}$

$$\tilde{\boldsymbol{x}}(t) = e^{(A-LC)t}\tilde{\boldsymbol{x}}(0). \tag{3.105}$$

Choosing the gain matrix, L, such that $(A - LC)$ has eigenvalues with negative real parts ensures $\hat{\boldsymbol{x}}(t) \rightarrow \boldsymbol{x}(t)$ as $t \rightarrow \infty$ (the asymptotic estimate), regardless of the input, $\boldsymbol{u}(t)$. In other words, we can select gain L to improve the convergence of the estimation error (and speed it up).

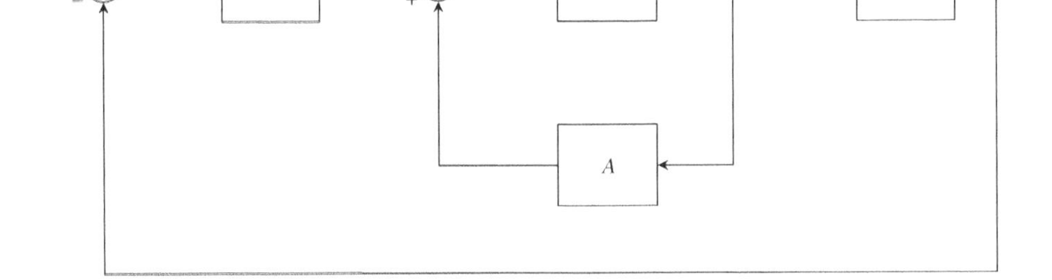

Figure 3.74 Block diagram of the full-order observer.

3.5.8 Observability

The system with state and output equations, $\dot{\boldsymbol{x}}(t) = A\boldsymbol{x}(t) + B\boldsymbol{u}(t)$ and $\boldsymbol{y}(t) = C\boldsymbol{x}(t)$, is observable if the rank$(\mathcal{O}) = n$, where

$$\mathcal{O} = \begin{bmatrix} C \\ CA \\ \vdots \\ CA^{n-1} \end{bmatrix} \tag{3.106}$$

is called the observability matrix [8].

The poles of the closed-loop system may be placed by selecting L such that $\alpha_e(s) = \det(sI - A + LC)$, provided that $(A,\ C)$ is observable. The procedure for selecting L is very similar to that used for selecting K used for full-state feedback. The standard observer configuration is the dual of the state feedback configuration [8]. Similarly, we can apply Ackermann's formula such that the estimator equivalent is

$$L = \alpha_e(A)\,\mathcal{O}^{-1}\begin{bmatrix} 0 \\ \vdots \\ 0 \\ 1 \end{bmatrix}. \tag{3.107}$$

Example: Estimator design.
Consider a system with matrices

$$A = \begin{bmatrix} -1 & 1.5 \\ 1 & -2 \end{bmatrix},\ B = \begin{bmatrix} 1 \\ 0 \end{bmatrix},\ C = [1 \quad 0],\ D = 0.$$

The initial state $\boldsymbol{x}(0) = \begin{bmatrix} -0.5 \\ -1 \end{bmatrix}$ is not precisely known. While the system is stable, the dominant eigenvalue, $\lambda_{max}(A) = -0.18$, is close to zero and nearly critically stable. The observability matrix is

$$\mathcal{O} = \begin{bmatrix} C \\ CA \end{bmatrix} = \begin{bmatrix} 1 & 0 \\ -1 & 1.5 \end{bmatrix}.$$

The rank of the observability matrix is rank $(\mathcal{O}) = 2$, so the system is observable. The observer gain, L, is designed to place the closed-loop poles at -3 and -4. Since the initial conditions are not well known, the state estimate starts at $\hat{\boldsymbol{x}}(0) = [0 \quad 0]^T$.

The open-loop dynamics and estimator are

$$\dot{\boldsymbol{x}}(t) = A\boldsymbol{x}(t) + B\boldsymbol{u}(t),\ \boldsymbol{y}(t) = C\boldsymbol{x}(t),$$

$$\dot{\hat{\boldsymbol{x}}}(t) = A\hat{\boldsymbol{x}}(t) + B\boldsymbol{u}(t),\ \hat{\boldsymbol{y}}(t) = C\hat{\boldsymbol{x}}(t).$$

In state-space form, the open-loop model is

$$\begin{bmatrix} \dot{\boldsymbol{x}}(t) \\ \dot{\hat{\boldsymbol{x}}}(t) \end{bmatrix} = \begin{bmatrix} A & 0 \\ 0 & A \end{bmatrix}\begin{bmatrix} \boldsymbol{x}(t) \\ \hat{\boldsymbol{x}}(t) \end{bmatrix} + \begin{bmatrix} B \\ B \end{bmatrix}\boldsymbol{u}(t)$$

with initial conditions $[\boldsymbol{x}(0) \quad \hat{\boldsymbol{x}}(0)]^T = [-0.5 \quad -1 \quad 0 \quad 0]^T$. The closed-loop dynamics and estimator are

$$\dot{\boldsymbol{x}}(t) = A\boldsymbol{x}(t) + B\boldsymbol{u}(t)$$
$$\dot{\hat{\boldsymbol{x}}}(t) = (A - LC)\hat{\boldsymbol{x}}(t) + B\boldsymbol{u}(t) + LC\boldsymbol{x}(t).$$

In state-space form, the closed-loop model is

$$\begin{bmatrix} \dot{x}(t) \\ \dot{\hat{x}}(t) \end{bmatrix} = \begin{bmatrix} A & 0 \\ LC & A - LC \end{bmatrix} \begin{bmatrix} x(t) \\ \hat{x}(t) \end{bmatrix} + \begin{bmatrix} B \\ B \end{bmatrix} \boldsymbol{u}(t). \tag{3.108}$$

The input $\boldsymbol{u}(t)$ is designed with varying amplitudes to stimulate the system and observe its dynamics. Given both these models let us simulate both the open-loop estimator and the closed-loop estimator states as well as the estimation error.

The open-loop estimator error shown in Figure 3.75 will eventually converge to zero but very slowly.

Figure 3.76 shows that the closed-loop estimator error converges more rapidly and has better performance. We have demonstrated how to use the plant's input and output to estimate the state vector asymptotically.

Suppose the state estimate is fed back to the plant input, as shown in Figure 3.77.

The equations describing the entire system (plant + observer + controller) are

$$\begin{aligned} \dot{\hat{\boldsymbol{x}}}(t) &= (A - BK - LC)\hat{\boldsymbol{x}}(t) + LC\boldsymbol{x}(t) \\ \dot{\boldsymbol{x}}(t) &= A\boldsymbol{x}(t) - BK\hat{\boldsymbol{x}}(t) \\ u(t) &= -K\hat{\boldsymbol{x}}(t). \end{aligned} \tag{3.109}$$

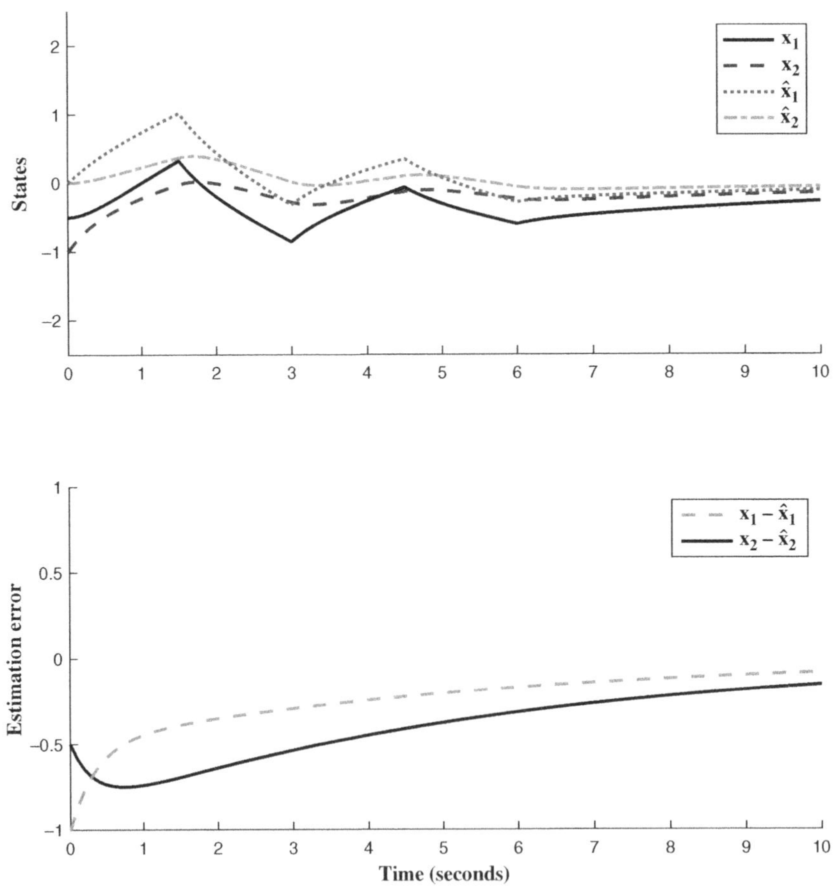

Figure 3.75 Open-loop response.

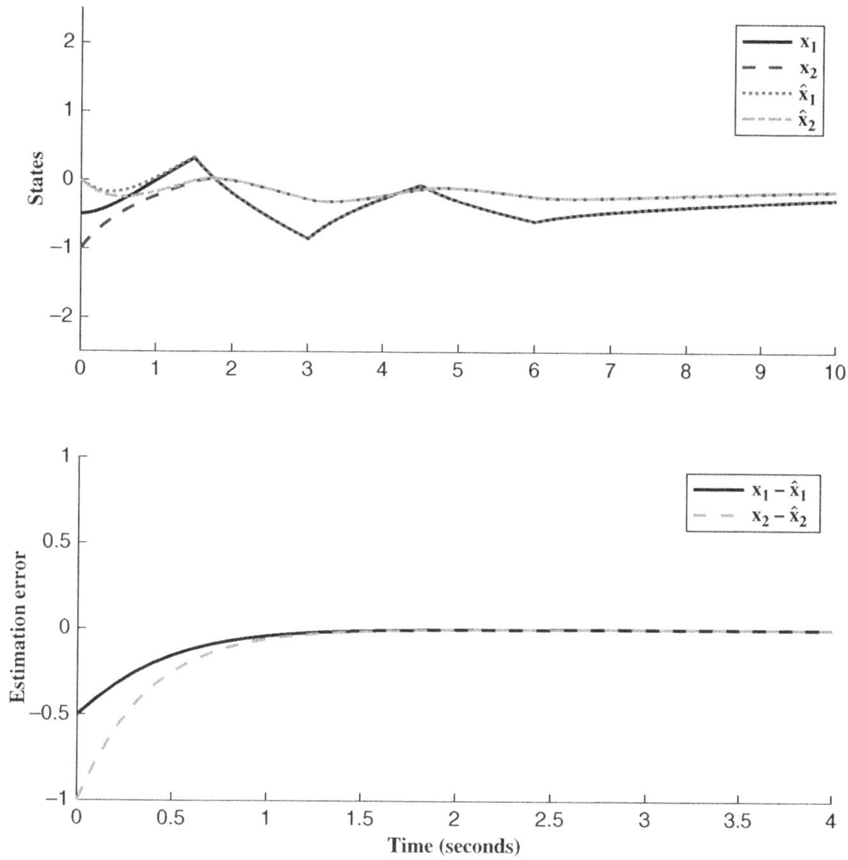

Figure 3.76 Closed-loop response.

The estimation error $\tilde{\boldsymbol{x}}(t) = \hat{\boldsymbol{x}}(t) - \boldsymbol{x}(t)$ propagates as

$$\dot{\tilde{\boldsymbol{x}}}(t) = (A - LC)\tilde{\boldsymbol{x}}(t).$$

Rewriting the state equations in terms of $\boldsymbol{x}(t)$ and $\tilde{\boldsymbol{x}}(t)$

$$\dot{\boldsymbol{x}}(t) = (A - BK)\boldsymbol{x}(t) - BK\tilde{\boldsymbol{x}}(t),$$

then the state space equations are

$$\begin{bmatrix} \dot{\boldsymbol{x}}(t) \\ \dot{\tilde{\boldsymbol{x}}}(t) \end{bmatrix} = \begin{bmatrix} A - BK & -BK \\ 0 & A - LC \end{bmatrix} \begin{bmatrix} \boldsymbol{x}(t) \\ \tilde{\boldsymbol{x}}(t) \end{bmatrix}. \tag{3.110}$$

Since the eigenvalues of a block diagonal matrix are the combined eigenvalues of the diagonal blocks, the eigenvalues of the composite system are those of $(A - BK)$ and $(A - LC)$.

3.5.9 The Separation Theorem

The problem of arbitrarily assigning the closed-loop poles of a system using feedback can be "separated" into two parts [8]

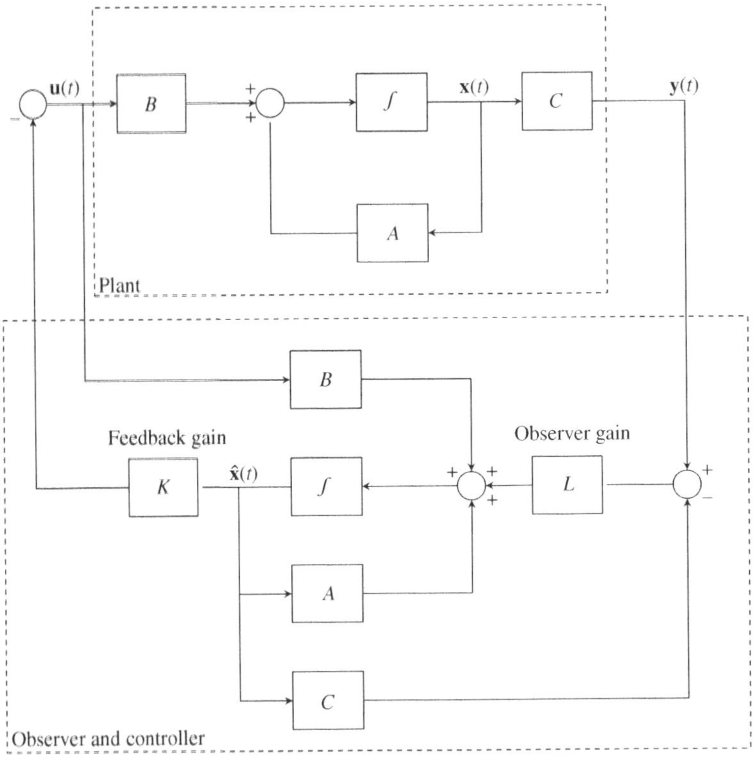

Figure 3.77 Block diagram of the observer in feedback configuration.

1. designing an observer to provide a set of asymptotically accurate state estimates,
2. designing a pole-assigned state feedback matrix as though the true states were available for direct measurement.

Example: Controller and estimator design.

Let $G(s) = \dfrac{1}{s^2+s+1}$, with the state-space model given by

$$A = \begin{bmatrix} 0 & 1 \\ -1 & -1 \end{bmatrix}, B = \begin{bmatrix} 0 \\ 1 \end{bmatrix}, C = [1 \quad 0], \text{ and } D = 0.$$

Design the regulator to place the poles at $s = -4 \pm 4j$

$$\det(sI - (A - BK)) \rightarrow K = [31 \quad 7].$$

The regulator pole time constant is $\tau_c = 1/\zeta\omega_n \approx 1/4 = 0.25$ seconds. The estimator pole time constant should be faster than the regulator time constant. Let us pick $\tau_c = 1/10 = 0.1$ seconds and use real poles for our desired characteristic equation $\alpha_e(s) = (s+10)^2$. Using Ackermann's formula

$$L = \alpha_e(A) \begin{bmatrix} C \\ CA \end{bmatrix}^{-1} \begin{bmatrix} 0 \\ 1 \end{bmatrix} = \begin{bmatrix} 99 & 19 \\ 19 & 80 \end{bmatrix} \begin{bmatrix} 0 \\ 1 \end{bmatrix} = \begin{bmatrix} 19 \\ 80 \end{bmatrix}.$$

The compensator is

$$A - BK - LC = \begin{bmatrix} 0 & 1 \\ -1 & -1 \end{bmatrix} - \begin{bmatrix} 0 \\ 1 \end{bmatrix} [31 \quad 7] - \begin{bmatrix} 19 \\ 80 \end{bmatrix} [1 \quad 0] = \begin{bmatrix} -19 & 1 \\ -112 & -8 \end{bmatrix}.$$

Therefore, in state-space form

$$\dot{\hat{\boldsymbol{x}}}(t) = (A - BK - LC)\hat{\boldsymbol{x}}(t) + L\boldsymbol{y}(t) = \begin{bmatrix} -19 & 1 \\ -112 & -8 \end{bmatrix}\hat{\boldsymbol{x}}(t) + \begin{bmatrix} 19 \\ 80 \end{bmatrix}\boldsymbol{y}(t)$$

$$\boldsymbol{u}(t) = -K\hat{\boldsymbol{x}}(t) = -[31 \quad 7]\hat{\boldsymbol{x}}(t).$$

The compensator transfer function is

$$G_c(s) = \frac{U(s)}{E(s)} = K(sI - (A - BK - LC))^{-1}L$$

$$= 1149 \cdot \frac{s + 2.553}{s^2 + 27s + 264}.$$

3.5.9.1 Closed-loop Poles

When you design a feedback controller using $\boldsymbol{u} = -K\boldsymbol{x}$ the control input $\boldsymbol{u}$ depends on the state feedback gain matrix K such that:

- if the poles of the closed-loop system (compared to the open-loop system) are moved farther into the left-half plane, the larger the "control" input.
- if the closed-loop poles remain closer to the original open-loop poles, less control input is required, which means a smaller gain matrix.

Estimator poles:

Estimator poles are typically placed faster than controller poles to ensure quicker convergence of estimation errors relative to the system dynamics but:

- placing estimator poles too far to the left increases the bandwidth of the estimator, causing more sensor noise to pass on to the control actuator.
- a common heuristic places observer poles **about 2–5 times farther [8]** to the left than the slowest closed-loop pole. This ensures that state estimation converges much faster than the dynamics of the controlled system.

3.6 Introduction to Discrete-time Systems

In a discrete-time control system, continuous signals are sampled at fixed intervals defined by the sampling period, h. The analog-to-digital (A/D) conversion process maps a continuous signal $\boldsymbol{y}(t)$ to a discrete sequence $\boldsymbol{y}_d(k)$ at each sampling instant as shown in Figure 3.78 [6].

Following computation, the digital signal is converted back to a continuous signal through digital-to-analog (D/A) conversion using a zero-order hold (ZOH) mechanism, which holds the signal constant over each sampling interval. Given a linear-time invariant (LTI) continuous-time system

$$\dot{\boldsymbol{x}}(t) = A\boldsymbol{x}(t) + B\boldsymbol{u}(t)$$
$$\boldsymbol{y}(t) = C\boldsymbol{x}(t) + D\boldsymbol{u}(t)$$
$$\boldsymbol{x}(0) = \boldsymbol{x}_0,$$

the continuous-time solution to this system is

$$\boldsymbol{x}(t) = e^{A(t-t_0)}\boldsymbol{x}(t_0) + \int_{t_0}^{t} e^{A(t-\tau)}B\boldsymbol{u}(\tau)\, d\tau.$$

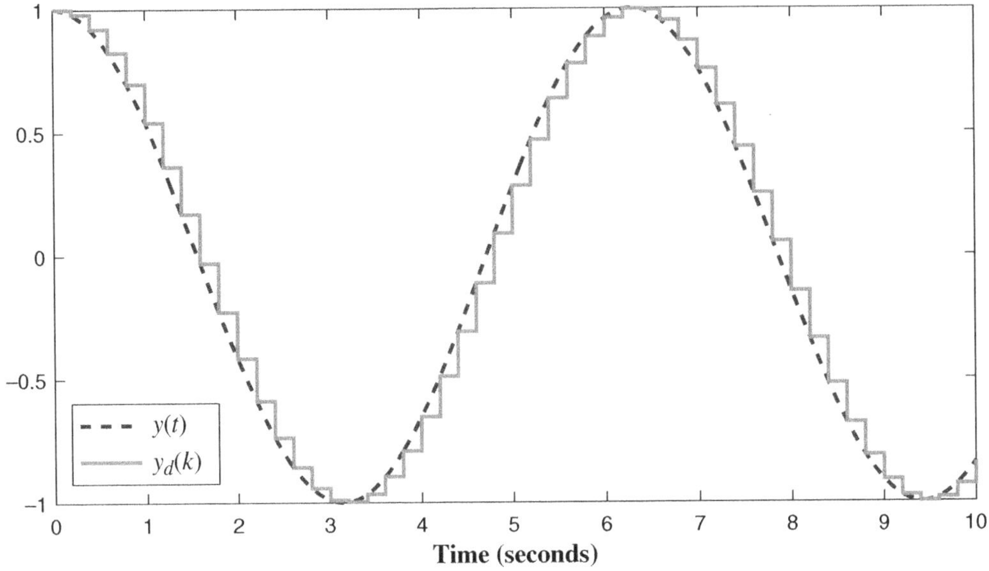

Figure 3.78 Sampling a continuous signal, $h = 200$ ms.

We will sample $\boldsymbol{x}(t)$ at sample times 0, h, $2h$, $3h$, etc. The k'th sample is $\boldsymbol{x}_d(k) = \boldsymbol{x}(kh)$. Let $t_0 = kh$ and $t = kh + h$ then

$$\boldsymbol{x}_d(k+1) = \boldsymbol{x}(kh+h) = e^{AT}\boldsymbol{x}(kh) + \int_{kh}^{kh+h} e^{A(kh+h-\tau)} B\boldsymbol{u}(\tau)\, d\tau.$$

Also, if we let $\alpha = \tau - kT$ then

$$\boldsymbol{x}_d(k+1) = e^{Ah}\boldsymbol{x}_d(k) + \int_0^h e^{A(h-\alpha)} B\boldsymbol{u}(kh+\alpha)\, d\alpha.$$

Using the zero-order hold (ZOH) assumption, the input signal $\boldsymbol{u}$ is constant between intervals $[kh, (k+1)h)$ [6] then

$$\boldsymbol{u}(t) = \boldsymbol{u}_d(k) \text{ for } kh \le t < (k+1)h, \tag{3.111}$$

and so, we have

$$\boldsymbol{x}_d(k+1) = e^{Ah}\boldsymbol{x}_d(k) + \left(\int_0^h e^{A\alpha} B\, d\alpha\right)\boldsymbol{u}_d(k). \tag{3.112}$$

Therefore, we have the discretized system

$$\begin{aligned} \boldsymbol{x}_d(k+1) &= A_d\boldsymbol{x}_d(k) + B_d\boldsymbol{u}_d(k) \\ \boldsymbol{y}_d(k) &= C_d\boldsymbol{x}_d(k) + D_d\boldsymbol{u}_d(k) \\ \boldsymbol{x}_d(0) &= \boldsymbol{x}_0 \end{aligned} \tag{3.113}$$

where $A_d = e^{Ah}$, $B_d = \int_0^h e^{A\alpha} B\, d\alpha$, $C_d = C$, and $D_d = D$. If $\boldsymbol{u}(t)$ is piecewise constant, A_d, B_d, C_d, and D_d provide the exact evolution of state and output samples at discrete times kh.

Example: Discretization of continuous-time system.
Given the following linear dynamical system

$$\dot{\boldsymbol{x}}(t) = \begin{bmatrix} 0 & 1 \\ 0 & 0 \end{bmatrix} \boldsymbol{x}(t) + \begin{bmatrix} 0 \\ 1 \end{bmatrix} \boldsymbol{u}(t), \boldsymbol{y}(t) = [1 \quad 0] \boldsymbol{x}(t)$$

compute the discretization.

Solution:

$$A_d = e^{Ah} = I + Ah + \frac{1}{2}A^2h^2 + \cdots = \begin{bmatrix} 1 & h \\ 0 & 1 \end{bmatrix}.$$

Also,

$$B_d = \int_0^h e^{A\alpha} B \, d\alpha = \begin{bmatrix} \frac{1}{2}h^2 \\ h \end{bmatrix}.$$

The discretization is

$$\boldsymbol{x}_d(k+1) = \begin{bmatrix} 1 & h \\ 0 & 1 \end{bmatrix} \boldsymbol{x}_d(k) + \begin{bmatrix} \frac{1}{2}h^2 \\ h \end{bmatrix} \boldsymbol{u}_d(k)$$

$$\boldsymbol{y}_d(k) = [1 \quad 0] \boldsymbol{x}_d(k).$$

We can also compute the discretization using the fact $\frac{d}{dt}e^{Ah} = A$ so that

$$B_d = \int_0^h e^{A\alpha} B d\alpha = A^{-1}\left[e^{At}B\right]_0^h = A^{-1}\left(e^{Ah} - I\right)B$$

and so, if A is invertible then

$$B_d = A^{-1}\left(e^{Ah} - I\right)B.$$

3.6.1 Discrete-time Linear Dynamical Systems

A discrete-time linear dynamical system (LDS) has the form

$$\begin{aligned} \boldsymbol{x}(t+1) &= A\boldsymbol{x}(t) + B\boldsymbol{u}(t) \\ \boldsymbol{y}(t) &= C\boldsymbol{x}(t) + D\boldsymbol{u}(t) \end{aligned} \tag{3.114}$$

and an autonomous discrete-time LDS is

$$\boldsymbol{x}(t+1) = A\boldsymbol{x}(t)$$

where t is the time variable. We can also express the discrete-time LDS using the equivalent subscript notation

$$\begin{aligned} \boldsymbol{x}_{t+1} &= A\boldsymbol{x}_t + B\boldsymbol{u}_t \\ \boldsymbol{y}_t &= C\boldsymbol{x}_t + D\boldsymbol{u}_t. \end{aligned} \tag{3.115}$$

When the state is a scalar such that $x(t+1) = ax(t)$, then the solution is

$$x(t) = a^t x(t). \tag{3.116}$$

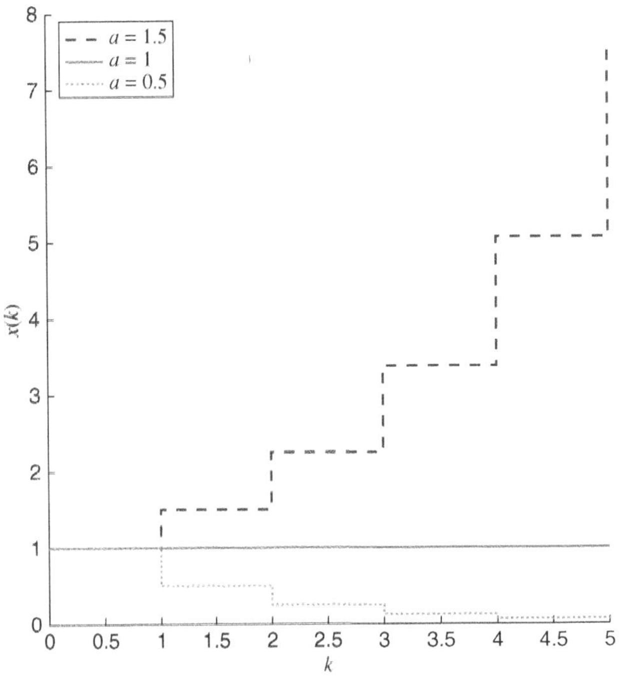

Figure 3.79 Discrete-time solution for $|a| < 1$, $|a| = 1$, and $|a| > 1$.

As shown in Figure 3.79, the discrete-time LDS is stable if $|a| < 1$ and unstable if $|a| > 1$. We can discretize the continuous-time LDS, $\dot{x}(t) = \sigma x(t)$

$$x_d(t+1) = e^{\sigma h} x_d(t) \tag{3.117}$$

so if $\sigma < 0$ then $|e^{\sigma h}| < 1$ and the LDS is still stable for any discretization, $h > 0$.

For linear discrete-time systems, the behavior of the system is determined by the eigenvectors and eigenvalues. For $A\mathbf{v} = \lambda \mathbf{v}$, the solution with $\boldsymbol{x}(0) = \mathbf{v}$ is

$$\boldsymbol{x}(t) = A^t \mathbf{v} = \lambda^t \mathbf{v}. \tag{3.118}$$

The solution $\boldsymbol{x}(t) = \lambda^t \mathbf{v}$ is known as a mode of the system, which we discussed for continuous-time systems. The behavior of the mode is determined by the magnitude of the eigenvalue [6]

- if $|\lambda| < 1$, then $\|\boldsymbol{x}(t)\|$ decreases as t increases, indicating the mode is stable.
- if $|\lambda| > 1$, then $\|\boldsymbol{x}(t)\|$ grows with increasing t, so the mode is unstable.

3.6.1.1 Solution of Discrete-time Linear Dynamical Systems

The solution of discrete-time LDS is determined through linear equations of the form $\boldsymbol{x}(t+1) = A\boldsymbol{x}(t) + B\boldsymbol{u}(t)$. After iterating this relationship over t steps, we can express $\boldsymbol{x}(t)$ as

$$\boldsymbol{x}(t) = A^t \boldsymbol{x}(0) + \begin{bmatrix} A^{t-1}B & A^{t-2}B & \cdots & AB & B \end{bmatrix} \begin{bmatrix} \boldsymbol{u}(0) \\ \boldsymbol{u}(1) \\ \vdots \\ \boldsymbol{u}(t-2) \\ \boldsymbol{u}(t-1) \end{bmatrix}.$$

This time evolution can be expressed in a more compact form as

$$x(t) = \underbrace{A^t x(0)}_{\text{zero input response}} + \underbrace{\textstyle\sum_{\tau=0}^{t-1} A^{t-1-\tau} B u(\tau)}_{\text{zero state response}}. \tag{3.119}$$

The zero-input response reflects state evolution without input, while the zero-state response sums contributions from past inputs [6].

3.6.2 Block Toeplitz Matrices in Discrete-time Systems

The output sequence $y(t)$ over time can be represented in terms of block Toeplitz matrices as follows

$$\begin{bmatrix} y(0) \\ y(1) \\ y(2) \\ \vdots \\ y(t) \end{bmatrix} = \begin{bmatrix} D & & & & \\ CB & D & & & \\ CAB & CB & D & & \\ \vdots & & & \ddots & \\ CA^{t-1}B & CA^{t-2}B & \cdots & CB & D \end{bmatrix} \begin{bmatrix} u(0) \\ u(1) \\ u(2) \\ \vdots \\ u(t) \end{bmatrix} + \begin{bmatrix} C \\ CA \\ CA^2 \\ \vdots \\ CA^t \end{bmatrix} x(0) \tag{3.120}$$

The block Toeplitz matrix structure provides a powerful representation of the output sequence $y(0), y(1), \ldots$ in terms of the input sequence $u(0), u(1), \ldots$ and the initial state $x(0)$. A block Toeplitz matrix has constant diagonal blocks from the top-left to the bottom-right and can used to find controllers and estimators [6].

3.7 Introduction to Optimal Control

In classical control problems (i.e. transfer function, root locus, Bode plots), it is assumed that the system model is known, often expressed through (linear) differential equations. The control objective in these cases is to design the input $u(t)$ as a function of the output $y(t)$ so that the system exhibits desirable behavior. This section introduces optimal control, extending the state-space methods discussed earlier. Unlike classical methods, optimal control addresses practical challenges such as

- MIMO systems (multiple-inputs multiple-outputs),
- Measurement noise and imprecision,
- Constraints (e.g. temperature limits or physical boundaries).

By framing control problems through an optimization perspective, this approach offers a modern and versatile alternative to traditional techniques [7].

3.7.1 Linear Algebra Review

This section provides a basic review of linear algebra concepts relevant to advanced control problems, emphasizing the importance of matrix operations and other properties in understanding estimation and control.

3.7.1.1 Vector Operations

Vectors are treated as columns by default. Consider two vectors $\boldsymbol{x} = [x_1 \quad \cdots \quad x_m]^T \in \mathbb{R}^m$ and $\boldsymbol{y} = [y_1 \quad \cdots \quad y_n]^T \in \mathbb{R}^n$. When $m = n$, their inner product is

$$\langle \boldsymbol{x}, \boldsymbol{y} \rangle = \boldsymbol{x}^T \boldsymbol{y} = \sum_{k=1}^{n} x_i y_i. \tag{3.121}$$

The inner product is a scalar and satisfies commutativit such that $\langle \boldsymbol{x}, \boldsymbol{y} \rangle = \langle \boldsymbol{y}, \boldsymbol{x} \rangle$. The norm of a vector measures its "size," with the most used being the Euclidean norm or 2-norm,

$$\|\boldsymbol{x}\| = \left(\sum_{i=1}^{n} x_i^2 \right)^{1/2}. \tag{3.122}$$

For all $\boldsymbol{x}, \boldsymbol{y} \in \mathbb{R}^n$ and $\alpha \in \mathbb{R}$, a vector norm satisfies

1. Positivity: $\|\boldsymbol{x}\| \geq \boldsymbol{0}$ for all $\boldsymbol{x}$.
2. Definiteness: If $\|\boldsymbol{x}\| = 0$ then $\boldsymbol{x} = \boldsymbol{0}$.
3. Triangle inequality: $\|\boldsymbol{x} + \boldsymbol{y}\| \leq \|\boldsymbol{x}\| + \|\boldsymbol{y}\|$.

We can define the angle between the two vectors $\boldsymbol{x}, \boldsymbol{y} \in \mathbb{R}^n$ as the angle $\theta \in [0, \pi]$ such that

$$\cos\theta = \frac{\langle \boldsymbol{x}, \boldsymbol{y} \rangle}{\|\boldsymbol{x}\| \, \|\boldsymbol{y}\|}. \tag{3.123}$$

When $\langle \boldsymbol{x}, \boldsymbol{y} \rangle = 0$, the vectors are orthogonal and the angle between them is $\frac{\pi}{2}$.

3.7.1.2 Matrix Multiplication

Matrix multiplication is a core operation in control problems. For matrices $A \in \mathbb{R}^{m \times p}$ and $B \in \mathbb{R}^{p \times n}$, their product $C = AB \in \mathbb{R}^{m \times n}$, is defined as

$$c_{ij} = \sum_{k=1}^{p} a_{ik} b_{kj}. \tag{3.124}$$

Key properties of matrix multiplication include

- Noncommutativity: $AB \neq BA$, in general.
- Distributivity: $A(B + C) = AB + AC$ whenever the dimensions make sense.
- Associativity: $(AB)C = A(BC)$.

3.7.1.3 Subspaces

A subspace, $S \subseteq \mathbb{R}^n$ is a set of vectors that satisfies three key properties

1. Contains the zero vector: $\boldsymbol{0} \in S$.
2. Closed under addition: If $\boldsymbol{x}, \boldsymbol{y} \in S$, then $\boldsymbol{x} + \boldsymbol{y} \in S$.
3. Closed under scalar multiplication: If $x \in S$ and $\alpha \in \mathbb{R}$, then $\alpha \boldsymbol{x} \in S$.

A set of vectors $\{\boldsymbol{v}_1, \ldots, \boldsymbol{v}_k\}$ spans S if every vector in S can be written as a linear combination of these vectors, written as $S = \text{span}(\boldsymbol{v}_1, \ldots, \boldsymbol{v}_k)$. If the spanning set is linearly independent, it forms a basis for S. The number of vectors in any basis of S is the dimension, denoted $\dim(S)$. Every vector in S can be uniquely represented as a linear combination of its basis vectors.

3.7.1.4 Range, Null Space, and Rank

Given a matrix $A \in \mathbb{R}^{m \times n}$, two fundamental subspaces associated with it are the range and null space. The range of A is the set of all vectors that can be reached by multiplying A with some vector $\boldsymbol{x} \in \mathbb{R}^n$ such that

$$\text{range}(A) = \{A\boldsymbol{x} \mid \boldsymbol{x} \in \mathbb{R}^n\} \tag{3.125}$$

where the range is a subspace of $\mathbb{R}^m$. The null space of A is the set of vectors that map to zero through A

$$\text{null}(A) = \{\boldsymbol{x} \in \mathbb{R}^n \mid A\boldsymbol{x} = \boldsymbol{0}\} \tag{3.126}$$

where the null space is a subspace of $\mathbb{R}^n$. The column rank of A, denoted rank(A), is the dimension of the range of A (the span of its columns). Similarly, the row rank is the dimension of the span of its rows. A key result is that the row rank and column rank are always equal, so we refer to both simply as the rank of A. We can enumerate these facts as

- $\text{rank}(A) = \dim(\text{range}(A))$
- If $\text{rank}(A) = m$, the matrix A has full row rank (all rows are linearly independent)
- If $\text{rank}(A) = n$, the matrix A has full column rank (all columns are linearly independent)

In general, $\text{rank}(A) \leq \min(m, n)$. If the matrix has either full row or column rank, we say it has full rank. Moreover, $\text{rank}(A) = \text{rank}(A^T)$, which provides useful symmetry in many control problems.

Example: Range, null space, and rank of matrices.

Find the range, null space, and rank of the following matrix $A = \begin{bmatrix} 1 & -1 \\ -1 & 1 \\ 0 & 0 \end{bmatrix}$.

Solution:

The range or column space of a matrix is the set of all linear combinations of its columns. The column vectors of A are $\mathbf{c}_1 = \begin{bmatrix} 1 \\ -1 \\ 0 \end{bmatrix}$ and $\mathbf{c}_2 = \begin{bmatrix} -1 \\ 1 \\ 0 \end{bmatrix}$. We see that $\mathbf{c}_1 = -\mathbf{c}_2$ so the columns are linearly dependent, and the range of A is spanned by just one vector

$$\text{range}(A) = \text{span}\left(\begin{bmatrix} 1 \\ -1 \\ 0 \end{bmatrix} \right) \subseteq \mathbb{R}^3.$$

Therefore, the dimension of the range (or column space) is $\dim(\text{range}(A)) = 1$. The null space corresponds to all solutions $\boldsymbol{x} = \begin{bmatrix} x_1 \\ x_2 \end{bmatrix}$ such that

$$A \begin{bmatrix} x_1 \\ x_2 \end{bmatrix} = \begin{bmatrix} 0 \\ 0 \\ 0 \end{bmatrix}.$$

This gives the system of equations

$$\begin{aligned} x_1 - x_2 &= 0 \Rightarrow x_1 = x_2 \\ -x_1 + x_2 &= 0 \Rightarrow x_1 = x_2. \end{aligned}$$

Thus, the null space of A is

$$\text{null}(A) = \text{span}\left(\begin{bmatrix} 1 \\ 1 \end{bmatrix} \right) \subseteq \mathbb{R}^2.$$

The dimension of the null space is $\dim(\text{null}(A)) = 1$. The rank of A is the same as the dimension of the range (column space)

$$\text{rank}(A) = \dim(\text{range}(A)) = 1.$$

3.7.1.5 Orthonormal Basis and Orthogonal Complement

Every subspace, $S \subseteq \mathbb{R}^n$ has an orthonormal basis, which is a set of vectors $\boldsymbol{u}_1, \ldots, \boldsymbol{u}_k$ that have the following properties

- Norm: Each vector $\boldsymbol{u}_i$ has unit length, $\|\boldsymbol{u}_i\| = 1$.
- Orthogonality: Each pair of distinct vectors is orthogonal, meaning

$$\langle \boldsymbol{u}_i, \boldsymbol{u}_j \rangle = \begin{cases} 1 \text{ if } i = j \\ 0 \text{ if } i \neq j. \end{cases} \tag{3.127}$$

- Spanning: $\text{span}(\boldsymbol{u}_1, \ldots, \boldsymbol{u}_k) = S$.

Given any spanning set $\boldsymbol{v}_1, \ldots, \boldsymbol{v}_k$ for a subspace S, we can use the Gram–Schmidt process to construct an orthonormal basis. A familiar example of an orthonormal basis for $\mathbb{R}^n$ is the standard basis

$$\boldsymbol{e}_1 = [1 \quad 0 \quad \cdots \quad 0]^T, \boldsymbol{e}_2 = [0 \quad 1 \quad \cdots \quad 0]^T, \boldsymbol{e}_3 = [0 \quad 0 \quad \cdots \quad 1]^T.$$

3.7.1.6 Properties of Orthonormal Sets and Semi-orthogonal Matrices

Given an orthonormal set of vectors $\boldsymbol{u}_1, \ldots, \boldsymbol{u}_k$ a semi-orthogonal matrix $U \subseteq \mathbb{R}^{n \times k}$ can be formed by stacking the vectors as columns

$$U = [\boldsymbol{u}_1 \quad \cdots \quad \boldsymbol{u}_k].$$

This matrix satisfies

$$U^T U = I_k,$$

and preserves inner products and norms

$$\langle U\boldsymbol{x}, U\boldsymbol{y} \rangle = \langle \boldsymbol{x}, \boldsymbol{y} \rangle, \ \|U\boldsymbol{x}\| = \boldsymbol{x}. \tag{3.128}$$

If U is square ($k = n$), it is an orthogonal matrix, and it satisfies

$$U^T U = UU^T = I_n, \ U^{-1} = U^T. \tag{3.129}$$

3.7.1.7 Orthogonal Complement

The orthogonal complement of a subspace $S \subseteq \mathbb{R}^n$, denoted $S^\perp$, is the set of all vectors orthogonal to every vector in S defined as

$$S^\perp = \{\boldsymbol{x} \in \mathbb{R}^n \mid \langle \boldsymbol{x}, \boldsymbol{s} \rangle = 0 \text{ for all } s \in S\}. \tag{3.130}$$

Key properties include

- $S^\perp$ is a subspace
- Any vector $\boldsymbol{v} \in \mathbb{R}^n$ can be uniquely decomposed as $\boldsymbol{v} = \boldsymbol{x} + \boldsymbol{y}$, where $\boldsymbol{x} \in S$ and $\boldsymbol{y} \in S^\perp$.
- The dimensions of S and $S^\perp$ satisfy $\dim(S) + \dim(S^\perp) = n$.

3.7.2 Linear Equations

A set of m linear equations in variables $x_1, \ldots, x_n$ takes the forms

$$\begin{matrix} a_{11}x_1 + a_{12}x_2 + \cdots + a_{1n}x_n = b_1 \\ \vdots \\ a_{m1}x_1 + a_{m2}x_2 + \cdots + a_{mn}x_n = b_m \end{matrix} \rightarrow \begin{bmatrix} a_{11} & \cdots & a_{1n} \\ \vdots & \ddots & \vdots \\ a_{m1} & \cdots & a_{mn} \end{bmatrix} \begin{bmatrix} x_1 \\ \vdots \\ x_n \end{bmatrix} = \begin{bmatrix} b_1 \\ \vdots \\ b_m \end{bmatrix} \Rightarrow A\boldsymbol{x} = \boldsymbol{b}. \tag{3.131}$$

There are three cases that can occur when solving $A\boldsymbol{x} = \boldsymbol{b}$.

1. There are no solutions. For example, $x_1 + x_2 = 2$ and $x_1 + x_2 = 5$.
2. There is exactly one solution. For example, $x_1 = 2$ and $x_2 = 1$.
3. There are infinitely many solutions. For example, $x_1 + x_2 = 4$.

A solution can exist if and only if $\boldsymbol{b} \in \text{range}(A)$. If a solution $\hat{\boldsymbol{x}}$ does exist and there is a nonzero $\boldsymbol{v} \in \text{null}(A)$, then $\hat{\boldsymbol{x}} + \alpha\boldsymbol{v}$ is also a solution for any $\alpha \in \mathbb{R}$, so there are infinitely many solutions.

3.7.3 Symmetric Matrices, Positive Definiteness, and Ellipsoids

A matrix $A \in \mathbb{R}^{n\times n}$ is symmetric if $A = A^T$. Symmetric matrices have real distinct eigenvalues and corresponding orthogonal eigenvectors. For a symmetric A

- The quadratic function $f(\boldsymbol{x}) = \boldsymbol{x}^T A\boldsymbol{x}$ which maps $\mathbb{R}^n$ to $\mathbb{R}$ is called a quadratic form.
- Is positive definite if $\boldsymbol{x}^T A\boldsymbol{x} > 0$ for all nonzero $\boldsymbol{x} \in \mathbb{R}^n$.
- Is positive semidefinite if $\boldsymbol{x}^T A\boldsymbol{x} \geq 0$ for all nonzero $\boldsymbol{x} \in \mathbb{R}^n$.

Therefore, if A is symmetric and with eigenvalues $\lambda_1 \geq \cdots \geq \lambda_n$, then $\lambda_1\|\boldsymbol{x}\|^2 \geq \boldsymbol{x}^T A\boldsymbol{x} \geq \lambda_n\|\boldsymbol{x}\|^2$ for all $\boldsymbol{x} \in \mathbb{R}^n$. This gives the following test for positive definiteness

$$A > 0 \Leftrightarrow \lambda_{\min}(A) > 0. \tag{3.132}$$

Therefore, the positive definiteness of a matrix can also be checked by verifying that all its eigenvalues are positive [6].

3.7.3.1 Ellipsoids

Positive definite matrices also define ellipsoids. Given a positive definite matrix B, the corresponding ellipsoid is defined as

$$E = \{\boldsymbol{x} \in \mathbb{R}^n \mid \boldsymbol{x}^T B^{-1}\boldsymbol{x} \leq 1\}.$$

Ellipsoids offer a geometric interpretation of sensitivity in estimation and control. Figure 3.80 illustrates the lengths of the principal axes of the ellipsoid are given by the square roots of the eigenvalues of B, and the directions of these axes correspond to the eigenvectors.

3.7.4 Estimation and Control Problems

Many engineering problems can be framed as solving the equation $A\boldsymbol{x} = \boldsymbol{b}$. Understanding how to approach such equations is fundamental to both estimation and control contexts.

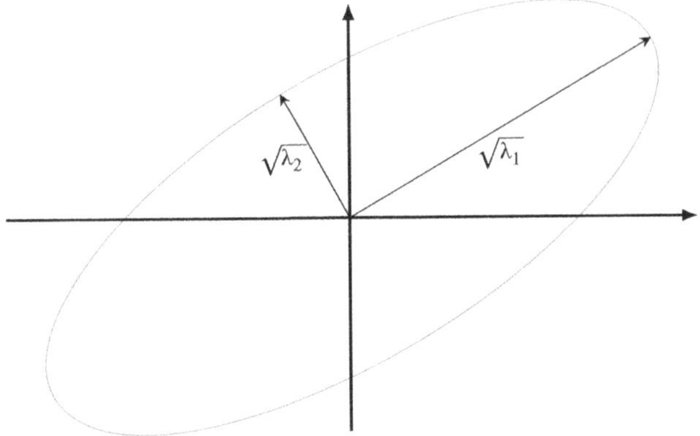

Figure 3.80 Ellipsoid with shorter axes indicates higher sensitivity, while longer axes indicate lower sensitivity.

In estimation problems, the matrix $A \in \mathbb{R}^{m \times n}$, is tall ($m > n$), meaning there are more equations than unknowns. This is also called an overdetermined system, where an exact solution may not exist [7]. Instead, the goal is to find an approximate solution $\hat{\boldsymbol{x}}$ such that $A\hat{\boldsymbol{x}} \approx \boldsymbol{b}$. One way to interpret this is by decomposing A into its rows

$$\boldsymbol{b} = A\boldsymbol{x} = \begin{bmatrix} - & \tilde{\boldsymbol{a}}_1^T & - \\ & \vdots & \\ - & \tilde{\boldsymbol{a}}_m^T & - \end{bmatrix} \boldsymbol{x} = \begin{bmatrix} \tilde{\boldsymbol{a}}_1^T \boldsymbol{x} \\ \vdots \\ \tilde{\boldsymbol{a}}_m^T \boldsymbol{x} \end{bmatrix} = \begin{bmatrix} \langle \tilde{\boldsymbol{a}}_1, \boldsymbol{x} \rangle \\ \vdots \\ \langle \tilde{\boldsymbol{a}}_m, \boldsymbol{x} \rangle \end{bmatrix}. \tag{3.133}$$

Here, each component $\boldsymbol{b}$ is the inner product of the corresponding row of A with $\boldsymbol{x}$. The result of the inner products is $b_1, \ldots, b_m$, which can be thought of as m linear measurements of $\boldsymbol{x}$, the quantity we are trying to estimate.

In control problems, the matrix $A \in \mathbb{R}^{m \times n}$, is wide ($m < n$), meaning there are more variables than equations. This makes the system underdetermined, with typically many solutions for $A\boldsymbol{x} = \boldsymbol{b}$ [7]. The challenge is to design a solution $\boldsymbol{x}$ such that $\boldsymbol{b}$ achieves specific performance goals or constraints. A common way to interpret this is by decomposing A into its columns

$$\boldsymbol{y} = A\boldsymbol{x} = \begin{bmatrix} | & & | \\ \boldsymbol{a}_1 & \cdots & \boldsymbol{a}_n \\ | & & | \end{bmatrix} \begin{bmatrix} x_1 \\ \vdots \\ x_n \end{bmatrix} = \boldsymbol{a}_1 x_1 + \cdots + \boldsymbol{a}_n x_n \tag{3.134}$$

Here, each column $\boldsymbol{a}_i$ represents an actuator. The task is to find the best combination of these n actuators to achieve the desired $\boldsymbol{b}$.

3.7.5 Singular Value Decomposition

Suppose $A \in \mathbb{R}^{m \times n}$, and it is tall and full rank, then there exists a factorization called the economy Singular Value Decomposition (economy SVD)

$$A = U_1 \Sigma_1 V_1^T \tag{3.135}$$

where, $U_1 \in \mathbb{R}^{m \times r}$ and $V_1 \in \mathbb{R}^{n \times r}$ both have orthonormal columns, and $\Sigma_1 \in \mathbb{R}^{r \times r}$ is a diagonal matrix diag$(\sigma_1, \ldots, \sigma_r)$ with $\sigma_1 \geq \sigma_2 \geq \ldots \geq \sigma_r > 0$. Figure 3.81 shows the matrix multiplication in the economy SVD, with rectangles indicating each matrix and its dimensions.

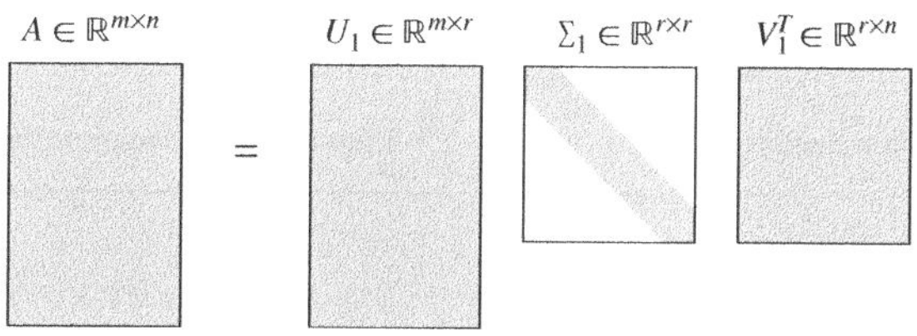

Figure 3.81 The economy SVD for an $m \times n$ matrix.

Each σ_i is called a singular value of A. The columns of U_1 are the left singular vectors, and the columns of V_1 are the right singular vectors. The SVD can also be expressed as rank-1 matrices

$$A = \sum_{i=1}^{r} \sigma_i \boldsymbol{u}_i \boldsymbol{v}_i^T. \tag{3.136}$$

The singular values σ_i are unique for a given matrix A. However, the singular vectors are not unique. We can extend U_1 and V_1 to form full orthogonal matrices $U \in \mathbb{R}^{m \times m}$ and $V \in \mathbb{R}^{n \times n}$, and express the full singular value decomposition (SVD) as

$$A = U\Sigma V^T = \begin{bmatrix} U_1 & U_2 \end{bmatrix} \begin{bmatrix} \Sigma_1 & 0 \\ 0 & 0 \end{bmatrix} \begin{bmatrix} V_1^T \\ V_2^T \end{bmatrix}. \tag{3.137}$$

Here, $U \in \mathbb{R}^{m \times m}$ and $V \in \mathbb{R}^{n \times n}$ are orthogonal matrices, and $\Sigma \in \mathbb{R}^{m \times n}$ matches the dimensions of A. The following properties relate the singular vectors to the matrix A

$$A\boldsymbol{v}_i = \sigma_i \boldsymbol{u}_i, \quad A^T \boldsymbol{u}_i = \sigma_i \boldsymbol{v}_i, \text{ for } i = 1, \ldots, r \tag{3.138}$$

and

$$A\boldsymbol{v}_i = 0, \quad A^T \boldsymbol{u}_i = 0, \text{ for } i > r. \tag{3.139}$$

These relations imply that the left singular vectors $\boldsymbol{u}_i$ span the range of A, while the remaining left singular vectors span the orthogonal complement of the range. The SVD also tells us the rank of a matrix such that if A has rank r, then A has r nonzero singular values [6]. Some key properties for the SVD can be summarized as

- $\text{range}(U_1) = \text{range}(A)$ and $\text{range}(U_2) = \text{range}(A)^\perp$,
- $\text{range}(V_1) = \text{null}(A)^\perp$ and $\text{range}(V_2) = \text{null}(A)$,
- $\text{rank}(A) = r$ then A has r nonzero singular values

Also, the singular values of A are the square roots of the nonzero eigenvalues of A^TA or AA^T because

$$A^TA = (U\Sigma V^T)^T(U\Sigma V^T) = V\Sigma^T\Sigma V^{-1}$$

where $\Sigma^T\Sigma = \text{diag}(\sigma_1^2, \sigma_2^2, \ldots, \sigma_r^2, 0, \ldots, 0) = \text{diag}(\lambda_1, \lambda_2, \ldots, \lambda_r, 0, \ldots, 0) \in \mathbb{R}^{n \times n}$. Therefore, this is the same as finding the eigenvalue decomposition of A^TA. A comparison of eigenvalues and singular values is shown in Table 3.5.

3.7.5.1 Control Ellipsoid

We can also define a control ellipsoid for the matrix $A \in \mathbb{R}^{m \times n}$, which is typically wide (more columns than rows) and of full rank, the control ellipsoid E in the output space $\mathbb{R}^m$ is defined as

$$E = \{\boldsymbol{y} \in \mathbb{R}^m | \boldsymbol{y}^T(AA^T)^{-1}\boldsymbol{y} \leq 1\}. \tag{3.140}$$

Table 3.5 Comparison of eigenvalues and singular values.

Property	Eigenvalue decomposition	Singular value decomposition
Matrix Type	Must be square	Can be any size
Values	Eigenvalues can be real or complex	Singular values are always real and positive
Vectors	Eigenvalues can be real or complex	Singular vectors are always real
Definition	$Ax = \lambda x$	$Av = \sigma u$ and $A^T u = \sigma v$
Matrix Form	$AX = X\Lambda$	$AV = U\Sigma$
Invertibility	X may not be invertible (nondiagonalizable A)	U and V are always orthogonal

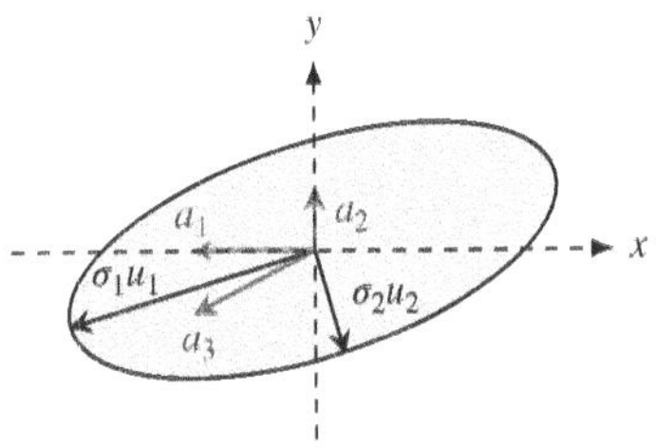

Figure 3.82 Control ellipsoid for hovercraft using the SVD [6].

The right singular vectors $\boldsymbol{v}_i$ of A represent directions in the input space $\mathbb{R}^n$, while the left singular vectors $\boldsymbol{u}_i$ correspond to directions in the output space $\mathbb{R}^m$. Each input direction $\boldsymbol{v}_i$ is mapped to an output direction $\boldsymbol{u}_i$ and scaled by the corresponding singular value σ_i. The singular values σ_i quantify the "actuator authority" in the direction of $\boldsymbol{u}_i$. Larger singular values indicate directions where control inputs have a strong influence. Consider a hovercraft where forces are applied via three thrusters $\boldsymbol{a}_1$, $\boldsymbol{a}_2$, and $\boldsymbol{a}_3$. The total force is given $\boldsymbol{y} = A\boldsymbol{x}$ where $\boldsymbol{x}$ represents the power supplied to each [6], and

$$A = [\boldsymbol{a}_1 \quad \boldsymbol{a}_2 \quad \boldsymbol{a}_3] = \begin{bmatrix} -1 & 0 & -1 \\ 0 & 0.5 & -0.5 \end{bmatrix}$$

represents the thruster directions. The singular values $\sigma_1 = 1.4668$ and $\sigma_2 = 0.5904$ define the ellipsoid axes, indicating the effectiveness of the control input in each principal direction of the ellipsoid in Figure 3.82. The longest axis corresponds to the direction where thrust is most efficiently applied.

3.7.5.2 Condition Number

The condition number, $\kappa(A)$, measures how sensitive the solution of a linear system is to small changes in the data. It is defined as

$$\kappa(A) = \|A\| \cdot \|A^{-1}\|, \text{ if } A \text{ is invertible,}$$

and ∞ otherwise. Here, the matrix norm, $\|A\|$, is defined as

$$\|A\| = \max\left\{ \frac{\|A\boldsymbol{x}\|}{\|\boldsymbol{x}\|} \mid \boldsymbol{x} \in \mathbb{R}^n, \boldsymbol{x} \neq \boldsymbol{0} \right\}, \tag{3.141}$$

where this norm, $\|A\| = \sigma_1(A)$, corresponds to the maximum singular value. Similarly, $\|A^{-1}\|$ is equal to the inverse of the smallest singular value of A, so $\|A^{-1}\| = 1/\sigma_n$. Therefore, for A with singular values $\sigma_1 \geq \sigma_2 \geq \ldots \geq \sigma_n > 0$,

$$\kappa(A) = \sigma_1/\sigma_n. \tag{3.142}$$

For a perturbed linear system $A(\boldsymbol{x} + \delta\boldsymbol{x}) = (\boldsymbol{b} + \delta\boldsymbol{b})$, the relative error in $\boldsymbol{x}$ based on the relative error in $\boldsymbol{b}$ satisfies

$$\frac{\|\delta x\| / \|x\|}{\|\delta b\| / \|b\|} \leq \|A\| \cdot \|A^{-1}\| = \kappa(A). \tag{3.143}$$

This inequality shows that the condition number $\kappa(A)$ sets an upper limit on the ratio of relative errors in $\boldsymbol{x}$ and $\boldsymbol{b}$. A larger $\kappa(A)$ implies that even small errors in $\boldsymbol{b}$ may lead to significant errors in the solution $\boldsymbol{x}$ [7].

3.7.6 Least Squares Estimation

Consider the equation $A\boldsymbol{x} = \boldsymbol{b}$, where $A \in \mathbb{R}^{m \times n}$ where $m > n$, meaning A is a tall matrix [7]. The system is overdetermined, and so this is the estimation setup. There is no exact solution $\boldsymbol{x}$ such that $A\boldsymbol{x} = \boldsymbol{b}$. Instead, we aim to find an approximate solution that minimizes the Euclidean norm of the residual error, $\boldsymbol{r} = A\boldsymbol{x} - \boldsymbol{b}$. Therefore, the least squares problem can be formulated as

$$\|\hat{\boldsymbol{r}}\|^2 = \min_{\boldsymbol{x} \in \mathbb{R}^n} \|A\boldsymbol{x} - \boldsymbol{b}\|^2. \tag{3.144}$$

Squaring the norm removes the square root, simplifying the optimization. Conceptually, the objective is to find $\hat{\boldsymbol{b}} \in \text{range}(A)$ closest to $\boldsymbol{b}$. The optimal $\hat{\boldsymbol{x}}$, which results in the smallest residual error $\hat{\boldsymbol{r}} = A\hat{\boldsymbol{x}} - \boldsymbol{b}$, is such that $\hat{\boldsymbol{r}}$ lies in the orthogonal complement of the range, $\text{range}(A)^{\perp}$. Figure 3.83 provides a visualization of this concept.

3.7.6.1 Derivation of Least Squares Solution

While the least squares problem can be interpreted geometrically, it is often derived analytically using Calculus. For a continuously differentiable function $f : \mathbb{R}^n \to \mathbb{R}$, the necessary condition for a minimum is the gradient must be zero. Applying this to the squared norm of the residual with respect to $\boldsymbol{x}$

$$\nabla_{\boldsymbol{x}} \|A\boldsymbol{x} - \boldsymbol{b}\|^2 = \nabla_{\boldsymbol{x}} (A\boldsymbol{x} - \boldsymbol{b})^T (A\boldsymbol{x} - \boldsymbol{b}) = 2(A^T A)\boldsymbol{x} - 2A^T \boldsymbol{b}.$$

At the optimal point $\boldsymbol{x} = \hat{\boldsymbol{x}}$, the gradient is zero, leading to the normal equations

$$A^T A\hat{\boldsymbol{x}} = A^T \boldsymbol{b}. \tag{3.145}$$

If $A \in \mathbb{R}^{m \times n}$ has full column rank, then $A^T A$ is invertible, and the least squares problem has a unique solution

$$\hat{\boldsymbol{x}} = (A^T A)^{-1} A^T \boldsymbol{b}. \tag{3.146}$$

In this case, $A^{\dagger} = (A^T A)^{-1} A^T$ is called the pseudoinverse of A [6]. The pseudoinverse is a generalization of the inverse, and it has these properties.

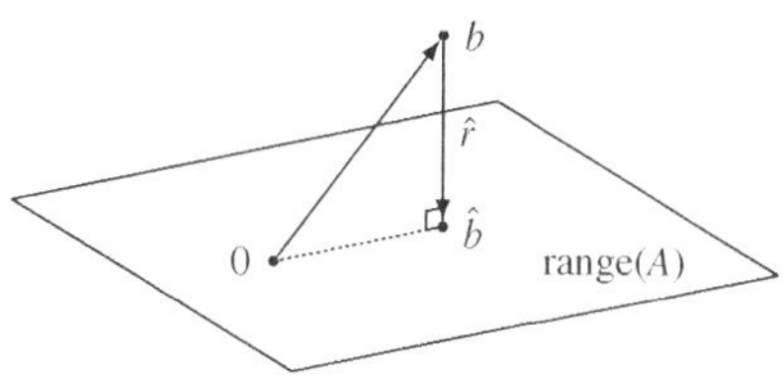

Figure 3.83 Geometrically, find the point in the range of A that is closest to $\boldsymbol{b}$. This occurs when the residual $\hat{r}$ is orthogonal to the range of A.

- $A^{\dagger}A = I_n$ ($n \times n$ identity matrix), meaning $A^{\dagger}$ acts as a left-inverse of A.
- $AA^{\dagger}$ is the projection matrix that maps any vector onto the range(A).
- If A is square and invertible, the pseudoinverse reduces to the standard inverse $A^{\dagger} = A^{-1}$.

3.7.6.2 Linear Regression

The most popular example of least squares is called linear regression. Suppose our model uses a linear combination of functions

$$f(t) = x_1 f_1(t) + x_2 f_2(t) + \cdots + x_n f_n(t)$$

and we collect m data samples such that $y_i = f(t_i)$ where $i = 1, \ldots, m$. In matrix form, we have

$$\begin{bmatrix} y_1 \\ y_2 \\ \vdots \\ y_m \end{bmatrix} = \begin{bmatrix} f_1(t_1) & \cdots & f_n(t_1) \\ f_1(t_2) & & f_n(t_2) \\ \vdots & & \vdots \\ f_1(t_m) & \cdots & f_n(t_m) \end{bmatrix} \begin{bmatrix} x_1 \\ x_2 \\ \vdots \\ x_n \end{bmatrix}$$

and can find a least-squares estimate for $\boldsymbol{x}$, $\boldsymbol{x}_{\text{est}} = A^{\dagger}\boldsymbol{y}$.

In engineering, polynomial models are common

$$f(t) = x_1 + x_2 t + x_3 t^2 + \cdots + x_n t^{n-1}.$$

With data samples $y_i = f(t_i) = \sum_{j=1}^{n} x_j t_i^{j-1}$, this becomes $\boldsymbol{y} = A\boldsymbol{x}$ where

$$A = \begin{bmatrix} 1 & t_1 & t_1^2 & \cdots & t_1^2 \\ 1 & t_2 & t_2^2 & \cdots & t_2^{n-1} \\ \vdots & & & & \vdots \\ 1 & t_m & t_1^2 & \cdots & t_m^{n-1} \end{bmatrix} \tag{3.147}$$

is called the Vandermonde matrix [6]. Therefore, if $m \geq n$ and A has full rank we can use linear regression. Figure 3.84 illustrates seventh-order and eleventh-order polynomial fits. A seventh-order fit captures the underlying trend, whereas the eleventh-order fit matches the data exactly but risks overfitting the data [6]. Overfitting occurs when the higher-order fit captures noise in the data rather than the underlying trend, reducing its ability to generalize and perform well on new, unseen data.

3.7.7 Minimum-norm Optimization

In control problems, the goal is often to find an input vector $\boldsymbol{x}$ such that $A\boldsymbol{x} = \boldsymbol{b}$, where $A \in \mathbb{R}^{m\times n}$ is a wide matrix ($m < n$) [7]. Since there are more potential inputs than outputs, the system is underdetermined, resulting in infinitely many solutions. Among all possible solutions, the solution with the smallest Euclidean norm, known as the minimum-norm solution, minimizes $\|\boldsymbol{x}\|$, formulated as

$$\min_{\boldsymbol{x}\in\mathbb{R}^n} \|\boldsymbol{x}\|^2 \text{ subject to } A\boldsymbol{x} = \boldsymbol{b}. \tag{3.148}$$

Here, $A\boldsymbol{x} = \boldsymbol{b}$ is a constraint, and the objective function is $\|\boldsymbol{x}\|$. Conceptually, the set of all solutions to $A\boldsymbol{x} = \boldsymbol{b}$, lies in the affine space, $\boldsymbol{x}_p + \text{null}(A)$. The affine space is a shifted subspace that passes through the point $\boldsymbol{x}_p$ instead of the origin. Among all points in the solution set $\boldsymbol{x}_p + \text{null}(A)$, there exists

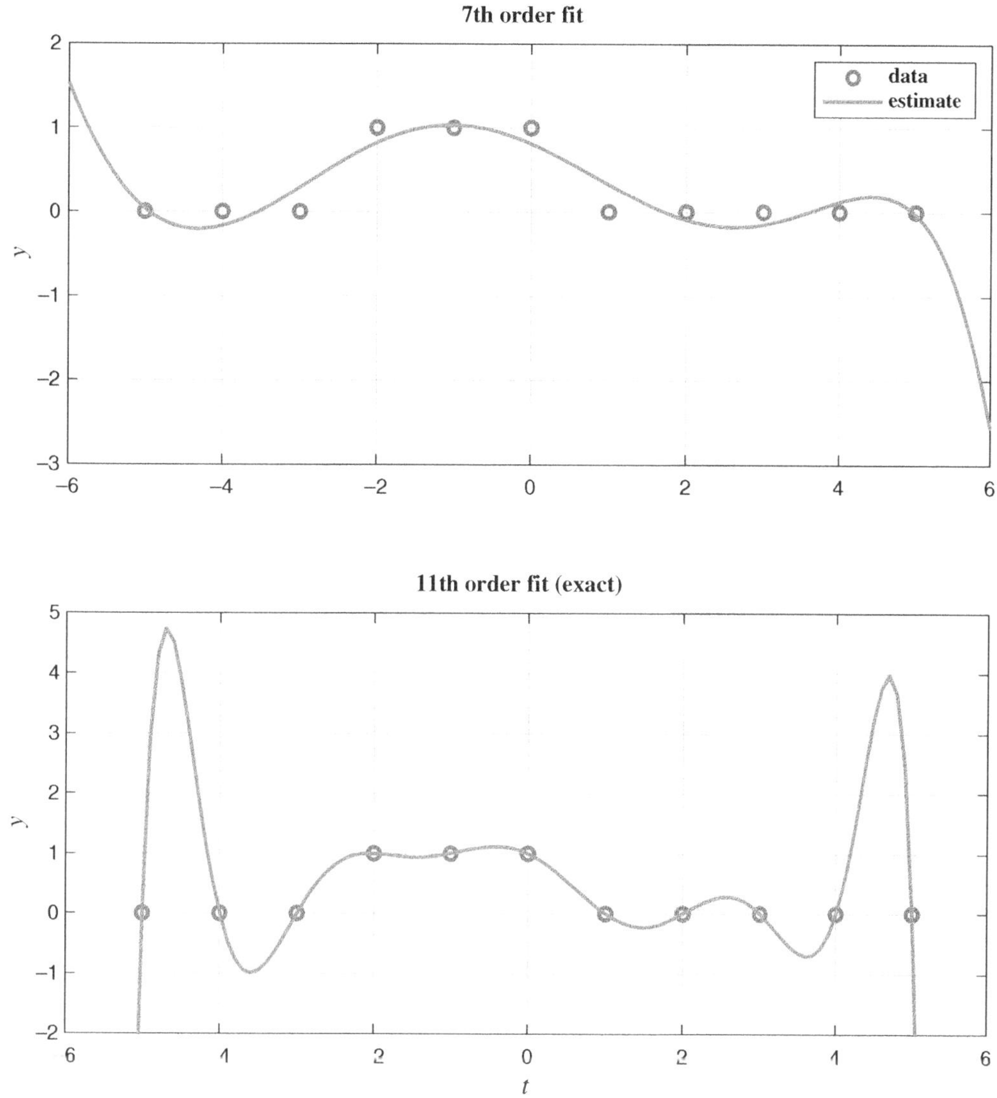

Figure 3.84 Seventh and eleventh-order polynomials fit to the data [6].

a point $\hat{\boldsymbol{x}}$ with the minimum norm, meaning it lies closest to the origin. This implies that $\hat{\boldsymbol{x}}$ must be orthogonal to every vector in null(A), such that $\hat{\boldsymbol{x}} \in \text{null}(A)^{\perp}$. Figure 3.85 shows a visualization of this concept.

Having established the geometric intuition behind the minimum-norm solution, we now turn to an analytical approach using Lagrange multipliers. The Lagrangian function is defined as

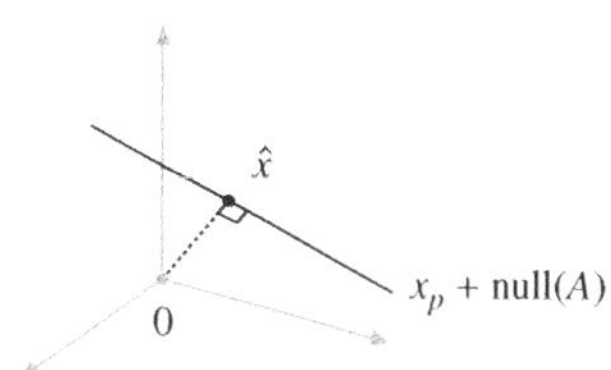

Figure 3.85 Geometric depiction of the input space $\mathbb{R}^n$ where want to minimize $\|\boldsymbol{x}\|$ subject to $A\boldsymbol{x} = \boldsymbol{b}$. We are looking for the point in $\hat{\boldsymbol{x}}$ + null(A) that is closest to $\mathbf{0}$. This happens when $\hat{\boldsymbol{x}} \in \text{null}(A)^{\perp}$.

$$\mathcal{L}(\boldsymbol{x}, \lambda) = \|\boldsymbol{x}\|^2 + \lambda^T(A\boldsymbol{x} - \boldsymbol{b}) \tag{3.149}$$

where λ is the vector of Lagrange multipliers. Taking the derivatives with respect to both $\boldsymbol{x}$ and λ and setting them to zero gives the optimality conditions

$$\frac{\partial \mathcal{L}}{\partial \hat{\boldsymbol{x}}} = 2\hat{\boldsymbol{x}}^T + A^T\lambda = \boldsymbol{0},$$
$$\frac{\partial \mathcal{L}}{\partial \lambda} = (A\hat{\boldsymbol{x}} - \boldsymbol{b})^T = \boldsymbol{0}.$$

From the first condition

$$\hat{\boldsymbol{x}} = -\frac{1}{2}A^T\lambda.$$

Substituting into the second equation yields

$$AA^T\lambda = -2\boldsymbol{b} \Rightarrow \lambda = A^T(AA^T)^{-1}\boldsymbol{b}.$$

Thus, the minimum-norm solution is

$$\hat{\boldsymbol{x}} = A^T(AA^T)^{-1}\boldsymbol{b}. \tag{3.150}$$

If A has full row rank, then there exists a unique solution to the minimum-norm problem given by $\hat{\boldsymbol{x}} = A^T(AA^T)^{-1}\boldsymbol{b}$. The matrix $A^\dagger = A^T(AA^T)^{-1}$ is (also) called the pseudoinverse of A [6]. It has the following properties

- $AA^\dagger = I_n$ meaning $A^\dagger$ acts as a right inverse of A.
- If A is square and full rank (invertible), then the left and right inverses coincide or $A^\dagger = A^T(AA^T)^{-1} = (A^TA)^{-1}A^T = A^{-1}$.

Suppose the matrix $A \in \mathbb{R}^{m\times n}$ is not full rank. Then we have the following options

- Estimation setup or A is tall, which will typically have many least squares solutions.
- Control setup or A is wide, which will have no $\boldsymbol{x}$ that satisfies $A\boldsymbol{x} = \boldsymbol{b}$.

Example: Controlling a hovercraft [6].
Consider the hovercraft shown in Figure 3.86, with its vertices located at $(\pm 1, \pm 1)$. The corners are numbered 1 through 4 with the top left corner 1, the top right 2, the bottom left 3, and the bottom right 4. At corner 1 there is a thruster with 80% efficiency pointing in the NE direction. At corner 2, there is a thruster with 20% efficiency pointing in the S direction. At corner 3, there is a thruster with 20% efficiency pointing in the SE direction. At corner 4, there is a thruster with 90% efficiency pointing in the SW direction. Let x_i be the power input to the thruster at corner i.

a) Let $\boldsymbol{y} \in \mathbb{R}^2$ be the vector of the total force acting on the body passing through the center of mass. Find the relationship between $\boldsymbol{y}$ and $\boldsymbol{x}$.

b) In which direction can the hovercraft move most efficiently? And least efficiently?

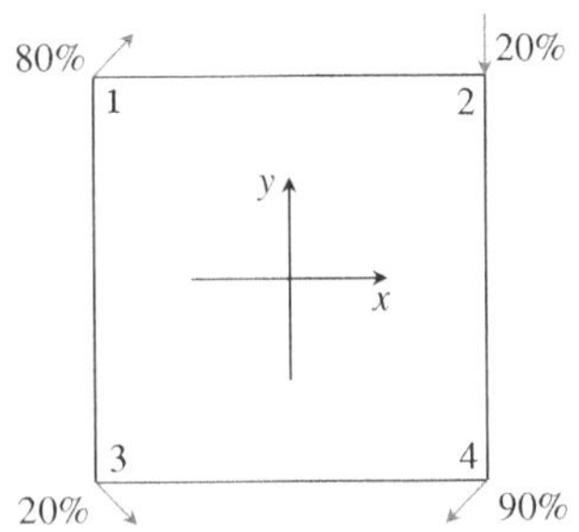

Figure 3.86 Hovercraft thruster pointing NE at corner 1, N at corner 2, SE at corner 3, and SW at corner 4.

c) Find a controller for the vehicle, which given $\boldsymbol{y}_{\text{des}}$ returns the most efficient vector of inputs $\boldsymbol{x}_{\text{control}}$ to apply to the thrusters.

d) Suppose now we want to apply forces in such a way that there is no resulting torque on the vehicle about the center of mass. Again, find a controller, which given a desired force returns the most efficient vector of inputs $\boldsymbol{x}_{\text{control}}$ to apply to the thrusters, which will not cause the vehicle to rotate.

Solution:

a) The unit vector directions of the thrusters are

Thruster 1: $\begin{bmatrix} 1/\sqrt{2} \\ 1/\sqrt{2} \end{bmatrix}$; Thruster 2: $\begin{bmatrix} 0 \\ -1 \end{bmatrix}$; Thruster 3: $\begin{bmatrix} 1/\sqrt{2} \\ -1/\sqrt{2} \end{bmatrix}$; Thruster 4: $\begin{bmatrix} -1/\sqrt{2} \\ -1/\sqrt{2} \end{bmatrix}$.

Considering the directions with the efficiencies we get

$$\boldsymbol{y} = A\boldsymbol{x} \Rightarrow \begin{bmatrix} y_1 \\ y_2 \end{bmatrix} = \underbrace{\begin{bmatrix} 1/\sqrt{2} & 0 & 1/\sqrt{2} & -1/\sqrt{2} \\ 1/\sqrt{2} & -1 & -1/\sqrt{2} & -1/\sqrt{2} \end{bmatrix} \begin{bmatrix} 0.8 & 0 & 0 & 0 \\ 0 & 0.2 & 0 & 0 \\ 0 & 0 & 0.2 & 0 \\ 0 & 0 & 0 & 0.9 \end{bmatrix}}_{A} \begin{bmatrix} x_1 \\ x_2 \\ x_3 \\ x_4 \end{bmatrix}$$

$$\Rightarrow A = \begin{bmatrix} 4/5\sqrt{2} & 0 & 1/5\sqrt{2} & -9/10\sqrt{2} \\ 4/5\sqrt{2} & -0.2 & -1/5\sqrt{2} & -9/10\sqrt{2} \end{bmatrix}.$$

b) The hovercraft is most efficient in the $\boldsymbol{u}_1$ direction (first input direction) and least efficient in the $\boldsymbol{u}_2$ direction (second input direction) where

$$\boldsymbol{u}_1 = \begin{bmatrix} -0.697 \\ -0.717 \end{bmatrix} \text{ and } \boldsymbol{u}_2 = \begin{bmatrix} -0.717 \\ 0.697 \end{bmatrix}.$$

c) We want the minimum norm solution of $\boldsymbol{y} = A\boldsymbol{x}$

$$\boldsymbol{x}_{\text{control}} = A^T(AA^T)^{-1}\boldsymbol{y} \Rightarrow A^\dagger = A^T(AA^T)^{-1} = \begin{bmatrix} 0.5154 & 0.2577 \\ 1.6059 & -1.6970 \\ 2.3999 & -2.3355 \\ -0.5799 & -0.2899 \end{bmatrix}.$$

d) The torque on the hovercraft is $\boldsymbol{T}_i = \boldsymbol{r}_i \times \boldsymbol{F}_i$, where $\boldsymbol{r}_i$ is the vector from the center of mass to the ith thruster and $\boldsymbol{F}_i$ is the thruster force. The total torque on the hovercraft is

$$\boldsymbol{T} = \sum_{i=1}^{4} \boldsymbol{r}_i \times \boldsymbol{F}_i = \sum_{i=1}^{4} (\boldsymbol{r}_1\boldsymbol{F}_2 - \boldsymbol{r}_2\boldsymbol{F}_1).$$

Assuming that the center of mass is at $(\mathbf{0}, \mathbf{0})$ then

$$\boldsymbol{r} = \begin{bmatrix} -\mathbf{1} & \mathbf{1} & -\mathbf{1} & \mathbf{1} \\ \mathbf{1} & \mathbf{1} & -\mathbf{1} & -\mathbf{1} \end{bmatrix}.$$

From before

$$\boldsymbol{F} = \boldsymbol{y} = \begin{bmatrix} 4/5\sqrt{2} & 0 & 1/5\sqrt{2} & -9/10\sqrt{2} \\ 4/5\sqrt{2} & -0.2 & -1/5\sqrt{2} & -9/10\sqrt{2} \end{bmatrix}\boldsymbol{x}$$

and so the torque is

$$\boldsymbol{T} = [-8/5\sqrt{2} \quad -1/5 \quad 2/5\sqrt{2} \quad -9/5\sqrt{2}]\boldsymbol{x}.$$

We can build a new matrix A_{TF} which stacks the forces and torques

$$\begin{bmatrix} \boldsymbol{F} \\ \boldsymbol{T} \end{bmatrix} = A_{TF}\boldsymbol{x}$$

where

$$A_{TF} = \begin{bmatrix} 4/5\sqrt{2} & 0 & 1/5\sqrt{2} & -9/10\sqrt{2} \\ 4/5\sqrt{2} & 1/5 & -1/5\sqrt{2} & -9/10\sqrt{2} \\ -8/5\sqrt{2} & -1/5 & 2/5\sqrt{2} & -9/5\sqrt{2} \end{bmatrix}.$$

The minimum norm input is given by

$$\boldsymbol{x}_{\mathbf{min}} = \underbrace{A_{TF}{}^{T}(A_{TF}A_{TF}{}^{T})^{-1}}_{A_{TF}^{\dagger}} \begin{bmatrix} \boldsymbol{F_{des}} \\ \mathbf{0} \end{bmatrix}$$

$$\Rightarrow A_{TF}^{\dagger} = \begin{bmatrix} 0.7381 & 0.1518 & -0.4344 \\ 1.649 & -1.7176 & -0.0842 \\ 2.369 & -2.3210 & 0.0592 \\ -0.3887 & -0.3808 & -0.3730 \end{bmatrix}.$$

3.7.8 Multiobjective Least Squares

Least squares minimizes error, while the minimum-norm problem seeks to find the smallest control effort. These two objectives can be defined as

$$J_1(\boldsymbol{x}) = \|A\boldsymbol{x} - \boldsymbol{b}\|^2, \; J_2(\boldsymbol{x}) = \|\boldsymbol{x}\|^2,$$

where $A \in \mathbb{R}^{m \times n}$ and $\boldsymbol{b} \in \mathbb{R}^n$. We want to simultaneously minimize both the error and control effort. In multiobjective optimization, we usually find a set of solutions known as the Pareto-optimal front that are feasible points where an improvement in one objective results in the detriment of the other objective [6]. To balance these competing objectives, we can combine J_1 and J_2 into a single-cost function

$$\min_{\boldsymbol{x} \in \mathbb{R}^n} J_1(\boldsymbol{x}) + \lambda J_2(\boldsymbol{x}) = \min_{\boldsymbol{x} \in \mathbb{R}^n} \|A\boldsymbol{x} - \boldsymbol{b}\|^2 + \lambda \|\boldsymbol{x}\|^2. \tag{3.151}$$

where $\lambda > 0$ is a weighting parameter controlling the trade-off between error (J_1) and control effort (J_2). To solve this, we can reformulate the problem using block matrices

$$\min_{\boldsymbol{x} \in \mathbb{R}^n} \left\| \begin{bmatrix} A \\ \sqrt{\lambda} I \end{bmatrix} \boldsymbol{x} - \begin{bmatrix} \boldsymbol{b} \\ \mathbf{0} \end{bmatrix} \right\|^2, \tag{3.152}$$

which leads to the normal equations

$$(A^T A + \lambda I)\boldsymbol{x} = A^T \boldsymbol{b}.$$

The solution is

$$\boldsymbol{x} = (A^T A + \lambda I)^{-1} A^T \boldsymbol{b}. \tag{3.153}$$

The parameter λ effectively trades off both objectives, affecting the slope of the Pareto-optimal front. As $\lambda \to 0$, the solution prioritizes the least-squares objective, minimizing the error at the expense of a larger control effort. Conversely, as $\lambda \to \infty$, the solution prioritizes minimizing the control norm. This trade-off is represented by the Pareto-optimal front in Figure 3.87, where varying λ shifts the solution along the curve [7].

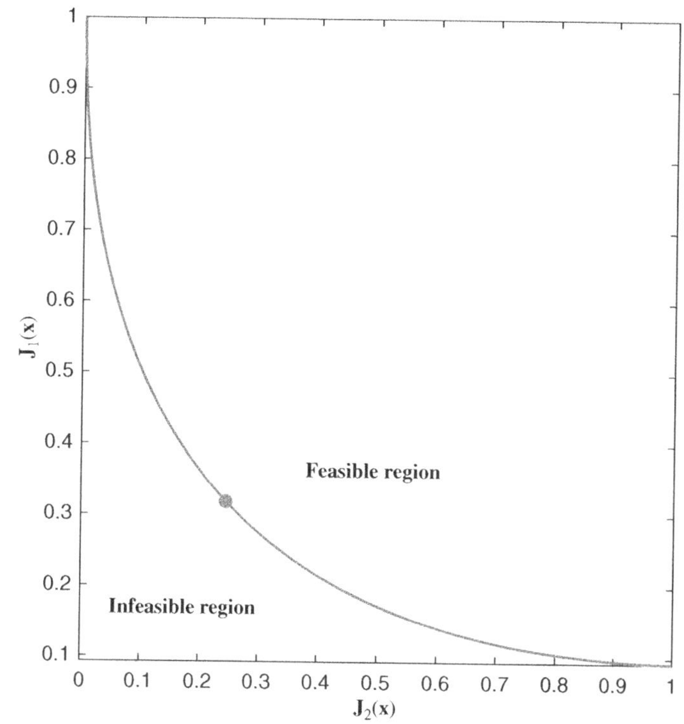

Figure 3.87 The Pareto-optimal front. The red point on the curve represents the best design for balancing between error and control input.

3.7.9 Completing the Square

Completing the square for a quadratic expression $ax^2 + 2bxy + dy^2 = 0$ isolates the x dependent term and can be rewritten as

$$a\left(x + \frac{by}{a}\right)^2 + \left(d - \frac{b^2}{a}\right)y^2.$$

This minimum value occurs at $x_{opt} = -\frac{by}{a}$, resulting in $\left(d - \frac{b^2}{a}\right)y^2$ for $a > 0$. For the matrix case, given symmetric matrices $A \in \mathbb{R}^{n \times n}$, $B \in \mathbb{R}^{m \times m}$, and $D \in \mathbb{R}^{m \times m}$. The quadratic form is

$$\min_{\boldsymbol{x}} \begin{bmatrix} \boldsymbol{x} \\ \boldsymbol{y} \end{bmatrix}^T \begin{bmatrix} A & B \\ B^T & D \end{bmatrix} \begin{bmatrix} \boldsymbol{x} \\ \boldsymbol{y} \end{bmatrix} = \min_{\boldsymbol{x}} (\boldsymbol{x}^T A \boldsymbol{x} + 2\boldsymbol{x}^T B \boldsymbol{y} + \boldsymbol{y}^T D \boldsymbol{y}). \tag{3.154}$$

Assuming A is invertible, we can rewrite this expression as

$$\min_{\mathbf{x}} (\boldsymbol{x} + A^{-1}B\boldsymbol{y})^T A(\boldsymbol{x} + A^{-1}B\boldsymbol{y}) + \boldsymbol{y}^T(D - B^T A^{-1} B)\boldsymbol{y}. \tag{3.155}$$

The minimum value is achieved when $\boldsymbol{x} = -A^{-1} B\boldsymbol{y}$, and the resulting quadratic form in $\boldsymbol{y}$ demonstrates how the optimal solution for $\boldsymbol{x}$ depends linearly on $\boldsymbol{y}$.

3.7.10 Block Matrix Inversion

If A^{-1} or D^{-1} exists then we can use either the LDU decomposition (Lower-diagonal-upper) or the UDL decomposition (Upper-diagonal-lower) on the following block matrices

$$\begin{bmatrix} A & B \\ C & D \end{bmatrix} = \begin{bmatrix} I & 0 \\ CA^{-1} & I \end{bmatrix} \begin{bmatrix} A & 0 \\ 0 & D - CA^{-1}B \end{bmatrix} \begin{bmatrix} I & A^{-1}B \\ 0 & I \end{bmatrix}$$
$$\begin{bmatrix} A & B \\ C & D \end{bmatrix} = \begin{bmatrix} I & BD^{-1} \\ 0 & I \end{bmatrix} \begin{bmatrix} A - BD^{-1}C & 0 \\ 0 & D \end{bmatrix} \begin{bmatrix} I & 0 \\ D^{-1}C & I \end{bmatrix}. \tag{3.156}$$

The corresponding block inverses for the LDU and UDL factorizations are

$$\begin{bmatrix} A & B \\ C & D \end{bmatrix}^{-1} = \begin{bmatrix} I & -A^{-1}B \\ 0 & I \end{bmatrix} \begin{bmatrix} A^{-1} & 0 \\ 0 & (D - CA^{-1}B)^{-1} \end{bmatrix} \begin{bmatrix} I & 0 \\ -CA^{-1} & I \end{bmatrix} \tag{3.157}$$

$$\begin{bmatrix} A & B \\ C & D \end{bmatrix}^{-1} = \begin{bmatrix} I & 0 \\ D^{-1}C & I \end{bmatrix} \begin{bmatrix} (A - BD^{-1}C)^{-1} & 0 \\ 0 & D^{-1} \end{bmatrix} \begin{bmatrix} I & -BD^{-1} \\ 0 & I \end{bmatrix}. \tag{3.158}$$

Efficient computation of matrix inverses is a key aspect of control algorithms. The Sherman–Morrison Formula provides a way to compute the inverse of a modified matrix efficiently [6]

$$(A - BD^{-1}C)^{-1} = A^{-1} + A^{-1}B(D - CA^{-1}B)^{-1}CA^{-1}. \tag{3.159}$$

3.7.11 The Linear Quadratic Regulator

The linear quadratic regulator (LQR) is a method for determining the optimal control input for a discrete-time linear system

$$\boldsymbol{x}_{t+1} = A\boldsymbol{x}_t + B\boldsymbol{u}_t, \; \boldsymbol{x}_0 = \text{initial state},$$

where $\boldsymbol{x}_t \in \mathbb{R}^n$, represents the system's state at time t, $\boldsymbol{u}_t \in \mathbb{R}^m$ is the control input at time t, and $A \in \mathbb{R}^{m\times n}$ and $B \in \mathbb{R}^{n\times m}$, are system matrices. Our goal is to design a control sequence $\boldsymbol{u}_0$, $\boldsymbol{u}_1$, … that achieves effective regulation while minimizing control effort. We seek to minimize a cost function, J, that balances state regulation and control effort over a specified time horizon, N. The LQR is cost function is given by

$$J(\boldsymbol{u}) = \sum_{t=0}^{N-1} (\boldsymbol{x}_t^T Q \boldsymbol{x}_t + \boldsymbol{u}_t^T R \boldsymbol{u}_t) + \boldsymbol{x}_N^T Q_f \boldsymbol{x}_N \tag{3.160}$$

where

- $Q \geq 0$ and $Q_f \geq 0$ are the state cost and final state cost matrices, respectively.
- $R > 0$ is the input cost matrix.
- The first term $\boldsymbol{x}_t^T Q \boldsymbol{x}_t$ penalizes state deviations, encouraging small state values.
- The second term $\boldsymbol{u}_t^T R \boldsymbol{u}_t$ penalizes the magnitude of the control input, promoting small input effort.
- The final term $\boldsymbol{x}_N^T Q_f \boldsymbol{x}_N$ measures the state deviation at the final time N.

By carefully selecting the weighting matrices Q, Q_f, and R, we can prioritize the importance of state regulation versus input minimization [6].

3.7.11.1 Dynamic Programming Approach

Practically, the LQR problem is usually solved using dynamic programming to recursively minimize the value function, $V_t(\boldsymbol{x}_t)$, which represents the minimum cost from the state $\boldsymbol{x}_t$ at time t to the final state over the time interval

$$V_t(\boldsymbol{x}_t) = \min_{\boldsymbol{u}_t,\ldots,\boldsymbol{u}_{N-1}} \sum_{\tau=t}^{N-1} (\boldsymbol{x}_\tau^T Q \boldsymbol{x}_\tau + \boldsymbol{u}_\tau^T R \boldsymbol{u}_\tau) + \boldsymbol{x}_N^T Q_f \boldsymbol{x}_N \tag{3.161}$$

subject to constraints $\boldsymbol{x}_{\tau+1} = A\boldsymbol{x}_{\tau+1} + B\boldsymbol{u}_\tau$ for $\tau = t, \ldots, N-1$ [6]. The function $V_t(\boldsymbol{x}_t)$ is also called the minimum cost-to-go for the LQR problem, starting from the state $\boldsymbol{x}_t$ at time t [6]. The cost-to-go has a quadratic form

$$V_t(\boldsymbol{x}_t) = \boldsymbol{x}_t^T P_t \boldsymbol{x}_t,$$

where $P_t = P_t^T \geq 0$. We can find P_t recursively, working backward from $t = N$. The optimal $\boldsymbol{u}$ can be expressed in terms of P_t. The cost-to-go with no time left is the final state cost is $V_N(\boldsymbol{x}_N) = \boldsymbol{x}_N^T Q_f \boldsymbol{x}_N$ and so $P_N = Q_f$.

The minimization can be performed in any order such that we can pull out the first stage cost out of the sum

$$V_t(\boldsymbol{x}_t) = \min_{\boldsymbol{u}_t} \left(\boldsymbol{x}_t^T Q \boldsymbol{x}_t + \boldsymbol{u}_t^T R \boldsymbol{u}_t + \min_{\boldsymbol{u}_{t+1},\ldots,\boldsymbol{u}_{N-1}} \sum_{\tau=t+1}^{N-1} (\boldsymbol{x}_\tau^T Q \boldsymbol{x}_\tau + \boldsymbol{u}_\tau^T R \boldsymbol{u}_\tau) + \boldsymbol{x}_N^T Q_f \boldsymbol{x}_N \right),$$

and let $\boldsymbol{z} = \boldsymbol{x}_t$ and $\boldsymbol{w} = \boldsymbol{u}_t$ so we can optimize over $\boldsymbol{u}_t$ separately and get

$$V_t(\boldsymbol{z}) = \min_{\boldsymbol{w}} \left(\boldsymbol{z}^T Q \boldsymbol{z} + \boldsymbol{w}^T R \boldsymbol{w} + \min_{\boldsymbol{u}_{t+1},\ldots,\boldsymbol{u}_{N-1}} \sum_{\tau=t+1}^{N-1} (\boldsymbol{x}_\tau^T Q \boldsymbol{x}_\tau + \boldsymbol{u}_\tau^T R \boldsymbol{u}_\tau) + \boldsymbol{x}_N^T Q_f \boldsymbol{x}_N \right),$$

which can be expressed as the following recursive Bellman equation

$$V_t(\boldsymbol{z}) = \min_{\boldsymbol{w}} (\boldsymbol{z}^T Q \boldsymbol{z} + \boldsymbol{w}^T R \boldsymbol{w} + V_{t+1}(A\boldsymbol{z} + B\boldsymbol{w})). \tag{3.162}$$

The dynamic programming principle roughly states that if the total cost for the path from **A** to **E** is the minimum cost, then the cost along any portion of the path, such as from **A** to **C** is also the minimum cost. This principle is captured by the Bellman equation since

- $\boldsymbol{z}^T Q \boldsymbol{z} + \boldsymbol{w}^T R \boldsymbol{w}$ is the cost incurred at time t when the control input $\boldsymbol{u}_t = \boldsymbol{w}$ is applied.
- $V_{t+1}(A\boldsymbol{z} + B\boldsymbol{w})$ represents the minimum cost-to-go starting from the state $\boldsymbol{x}_t = A\boldsymbol{z} + B\boldsymbol{w}$.

This Bellman equation provides a recursive relationship for V_t in terms of V_{t+1}. Since $V_N(\boldsymbol{z}) = \boldsymbol{z}^T P_N \boldsymbol{z}$ where $P_N = Q_f$ then we want to show that $V_{N-1}(\boldsymbol{z})$ has the same quadratic form [6]. By the dynamic programming principle, $V_{N-1}(\boldsymbol{z})$ is expressed as

$$\begin{aligned} V_{N-1}(\boldsymbol{z}) &= \min_{\boldsymbol{w}} (\boldsymbol{z}^T Q \boldsymbol{z} + \boldsymbol{w}^T R \boldsymbol{w} + (A\boldsymbol{z} + B\boldsymbol{w})^T P_N (A\boldsymbol{z} + B\boldsymbol{w})) \\ &= \min_{\boldsymbol{w}} \left(\begin{bmatrix} \boldsymbol{z} \\ \boldsymbol{w} \end{bmatrix}^T \begin{bmatrix} A^T P_N A + Q & A^T P_N B \\ B^T P_N A & B^T P_N B + R \end{bmatrix} \begin{bmatrix} \boldsymbol{z} \\ \boldsymbol{w} \end{bmatrix} \right) \end{aligned}$$

where the optimal control input $\boldsymbol{w}^*$ is found by setting the derivative of $V_{N-1}(\boldsymbol{z})$ with respect to $\boldsymbol{w}$ to zero and solving for $\boldsymbol{w}^*$

$$\boldsymbol{w}^* = -(R + B^T P_N B)^{-1} B^T P_N A \boldsymbol{z}. \tag{3.163}$$

Substituting the optimal input $\boldsymbol{w}^*$ into the value function $V_{N-1}(\boldsymbol{z})$, we obtain

$$V_{N-1}(\boldsymbol{z}) = \boldsymbol{z}^T \underbrace{(Q + A^T P_N A - A^T P_N B (R + B^T P_N B)^{-1} B^T P_N A)}_{P_{N-1}} \boldsymbol{z}.$$

We conclude that V_{N-1} is quadratic since $V_{N-1}(z) = z^T P_{N-1} z$. Since the function $V_t(z)$ is quadratic, such that $V_t(z) = z^T P_t z$ then $V_{t-1}(z)$ is also quadratic. This backward iterative method computes the cost-to-go matrices P_t for all time steps t using the Riccati equation. Specifically, P_{t-1} is given by

$$P_{t-1} = Q + A^T P_t A - A^T P_t B(R + B^T P_t B)^{-1} B^T P_t A. \tag{3.164}$$

Since the terminal cost, $P_N = Q_f$, is known then we can compute P_{N-1}, P_{N-2}, …, P_0 recursively using the Riccati equation. This backward recursion in time is also called the Riccati recursion for P_t. The optimal control input, $\boldsymbol{u}_t^{\mathrm{LQR}}$, is computed at each time step as

$$\boldsymbol{u}_t^{\mathrm{LQR}} = -(R + B^T P_{t+1} B)^{-1} B^T P_{t+1} A \boldsymbol{x}_t. \tag{3.165}$$

3.7.11.2 Summary of LQR Solution via Dynamic Programming

The LQR problem is solved using dynamic programming as follows [6].

1. Initialization: set $P_N = Q_f$, the terminal cost matrix.
2. Riccati Recursion for P_t: for $t = N, \ldots, 1$, compute

$$P_{t-1} = Q + A^T P_t A - A^T P_t B(R + B^T P_t B)^{-1} B^T P_t A.$$

3. Feedback Gain Computation: for $t = 0, \ldots, N-1$, compute the state feedback gain

$$K_t = -(R + B^T P_{t+1} B)^{-1} B^T P_{t+1} A. \tag{3.166}$$

4. Optimal Control Law: for $t = 0, \ldots, N-1$, the optimal control input is

$$\boldsymbol{u}_t^{\mathrm{LQR}} = K_t \boldsymbol{x}_t = -(R + B^T P_{t+1} B)^{-1} B^T P_{t+1} A \boldsymbol{x}_t.$$

The optimal control input, $\boldsymbol{u}_t$, is expressed as a linear function of the state, referred to as linear state feedback. This method does not require knowledge of the initial state $\boldsymbol{x}_0$, and we can compute $K_0, \ldots, K_{N-1}$ in advance. At runtime, the control input $\boldsymbol{u}_t$ depends only on the current state, $\boldsymbol{x}_t$, and not past states.

3.7.11.3 Steady-state Regulator

The cost-to-go matrices P_t converges to steady-state value P_{ss} satisfying the discrete-time Algebraic Riccati Equation (ARE)

$$P_{\mathrm{ss}} = Q + A^T P_{\mathrm{ss}} A - A^T P_{\mathrm{ss}} B(R + B^T P_{\mathrm{ss}} B)^{-1} B^T P_{\mathrm{ss}} A. \tag{3.167}$$

For t far from the time horizon, N, the LQR optimal control law is approximately a linear, constant state feedback of the form

$$u(t) = K_{\mathrm{ss}} x(t), \text{ where } K_{\mathrm{ss}} = -(R + B^T P_{\mathrm{ss}} B)^{-1} B^T P_{\mathrm{ss}} A. \tag{3.168}$$

This steady-state regulator form is widely used in practical applications.

Example: Spring-mass-damper [6].

We consider a mass-spring-damper system in Figure 3.88 with three masses, using parameters $k_i = 0.2$, $b_i = 0.6$, and $m_i = 1$. The six states represent the positions and velocities of the masses.

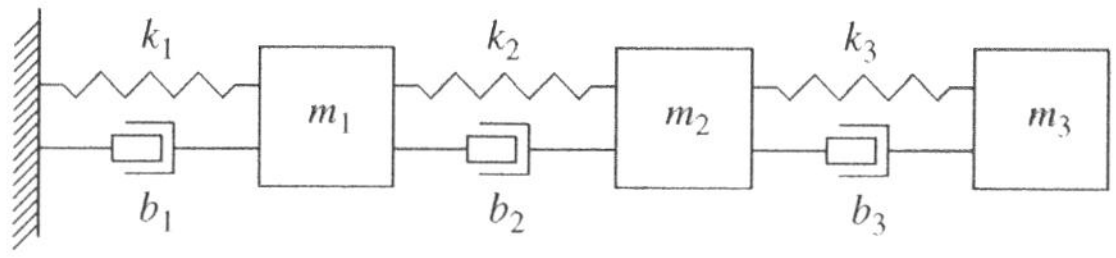

Figure 3.88 Spring-mass-damper system.

The equations of motion (EOM) for this system are $M\ddot{\boldsymbol{q}}(t) + D\dot{\boldsymbol{q}}(t) + K\boldsymbol{q}(\boldsymbol{t}) = \boldsymbol{F}(\boldsymbol{t})$.

a) Setup a 1 output, 3 input discrete-time system with a force input applied to each mass and the displacement of the first mass as the output. Use the sample period, $h = 0.25$, simulate the system from 0 to T, where $T = 100$. The initial state is $\boldsymbol{x}_0 = [1 \quad 2 \quad 3 \quad 0 \quad 0 \quad 0]^T$ and zero input force. Plot the state trajectories. Compute the energy of the output signal $J_{\text{out}} = \sum_{k=0}^{T} \|\boldsymbol{y}(k)\|^2$.

b) Pick a multiobjective weight $\mu = 0.01$ and use the pseudoinverse to find the input $\boldsymbol{u}_0, \ldots, \boldsymbol{u}_{T-1}$, which minimizes $J = J_{\text{out}} + \mu J_{in}$. Plot the state trajectories and compute J_{out} that is achieved by the optimal input.

c) Set this problem up for an equivalent LQR solution. Simulate the system with the LQR controller. Plot the state trajectories.

d) Verify that the input sequence is the same as input from part b). Compute J_{out}.

Solution:

a) Let us convert the EOM into a state-space system

$$\ddot{\boldsymbol{q}}(t) = -M^{-1}D\dot{\boldsymbol{q}}(t) - M^{-1}K\boldsymbol{q}(t) + M^{-1}\boldsymbol{F}(t)$$

where $M, D, K \in \mathbb{R}^{3\times3}$. Choose the state vector $\boldsymbol{x} = [q_1 \quad q_2 \quad q_3 \quad \dot{q}_1 \quad \dot{q}_2 \quad \dot{q}_3]^T$ such that

$$\begin{bmatrix} \dot{q}_1 \\ \dot{q}_2 \\ \dot{q}_3 \\ \ddot{q}_1 \\ \ddot{q}_2 \\ \ddot{q}_3 \end{bmatrix} = \underbrace{\left[\begin{array}{c|c} 0_{3\times3} & I_{3\times3} \\ \hline -M^{-1}K & -M^{-1}D \end{array}\right]}_{A_c} \begin{bmatrix} q_1 \\ q_2 \\ q_3 \\ \dot{q}_1 \\ \dot{q}_2 \\ \dot{q}_3 \end{bmatrix} + \underbrace{\begin{bmatrix} 0_{3\times3} \\ \hline I_{3\times3} \end{bmatrix}}_{B_c} \begin{bmatrix} \boldsymbol{F}_1 \\ \boldsymbol{F}_2 \\ \boldsymbol{F}_3 \end{bmatrix}$$

and

$$\boldsymbol{y} = \underbrace{[1 \quad 0 \quad 0 \quad 0 \quad 0 \quad 0]}_{C_c} [q_1 \quad q_2 \quad q_3 \quad \dot{q}_1 \quad \dot{q}_2 \quad \dot{q}_3]^T.$$

The discrete-time system is

$$A_d = e^{hA_c},\ B_d = A^{-1}\left(e^{hA_c} - I\right) B_c, \text{ and } C_d = C_c.$$

Now, the system propagates as follows

$$\boldsymbol{x}_{t+1} = A_d\boldsymbol{x}_t + B_d\boldsymbol{u}_t$$

and the state trajectories are shown in Figure 3.89.

The resulting output energy is $J_{\text{out}} = 27.86$.

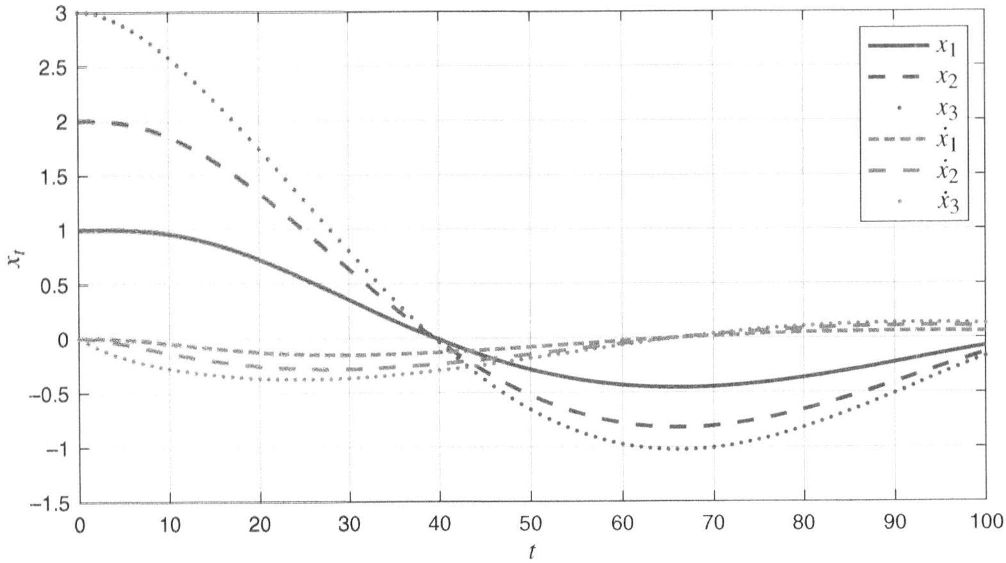

Figure 3.89 State trajectories starting at $\boldsymbol{x}_0$.

b) The least-squares problem is

$$\begin{bmatrix} \boldsymbol{y}_0 \\ \boldsymbol{y}_1 \\ \vdots \\ \boldsymbol{y}_{T-1} \end{bmatrix} = \begin{bmatrix} 0 & \cdots & \cdots & 0 \\ CB & 0 & \cdots & 0 \\ \vdots & \vdots & \ddots & \vdots \\ CA^{T-2}B & \cdots & CB & 0 \end{bmatrix} \begin{bmatrix} \boldsymbol{u}_0 \\ \boldsymbol{u}_1 \\ \vdots \\ \boldsymbol{u}_{T-1} \end{bmatrix} + \begin{bmatrix} C \\ CA \\ \vdots \\ CA^{T-1} \end{bmatrix} \boldsymbol{x}_0$$

and more compactly is

$$\bar{\boldsymbol{y}} = P_T\bar{\boldsymbol{u}} + J_T\boldsymbol{x}_0 \Rightarrow \bar{\boldsymbol{y}} - J_T\boldsymbol{x}_0 = P_T\bar{\boldsymbol{u}}.$$

The multiobjective problem is to minimize

$$\left\| \begin{bmatrix} P_T \\ \sqrt{\mu}I \end{bmatrix} \bar{\boldsymbol{u}} - \begin{bmatrix} -J_T\boldsymbol{x}_0 \\ 0 \end{bmatrix} \right\|^2$$

for which the solution is $\bar{\boldsymbol{u}}_{\mathbf{opt}} = (P_T{}^TP_T + \mu I)^{-1}(-P_T{}^TJ_T\boldsymbol{x}_0)$.

The state trajectories are shown in Figure 3.90.

The output energy is $J_{\text{out}} = 1.919$.

c) We need to pick Q, R, and Q_f that correspond to the least-squares solution. The input cost is $R = \mu I = 0.01I$, the state cost is $Q = C^TC$, and $Q_f = C^TC$.

State trajectories using state feedback using LQR gains are shown in Figure 3.91.

d) Figure 3.92 illustrates that the input sequence is the same for the least squares and LQR solutions.

The output energy is $J_{\text{out}} = 1.919$. This agrees with the least-squares solution.

3.7.12 Probability Review

This section introduces key probabilistic concepts for deriving the Kalman filter, including random variables, Gaussian distributions, random vectors, marginal, and conditional distributions.

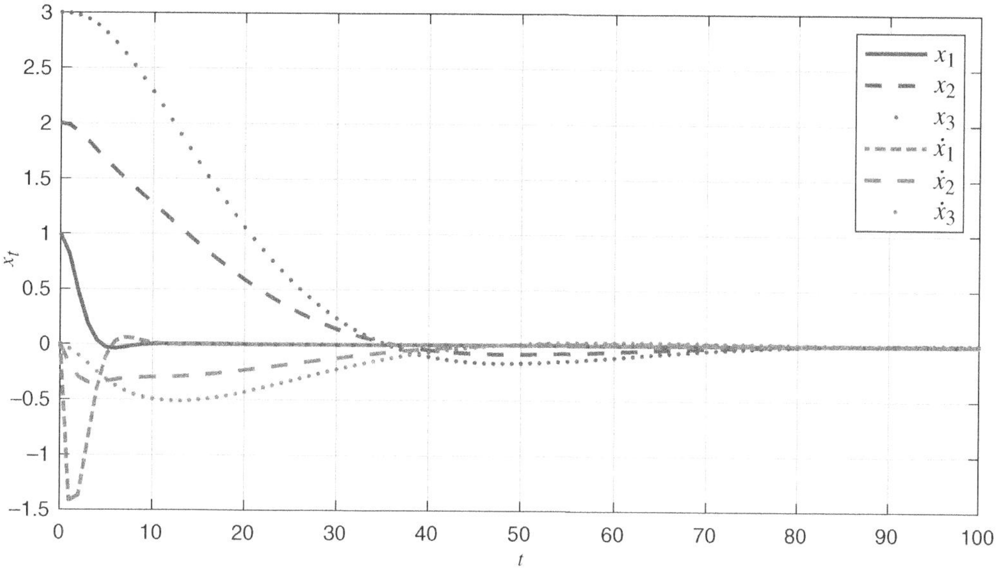

Figure 3.90 State trajectories for the minimum-norm solution.

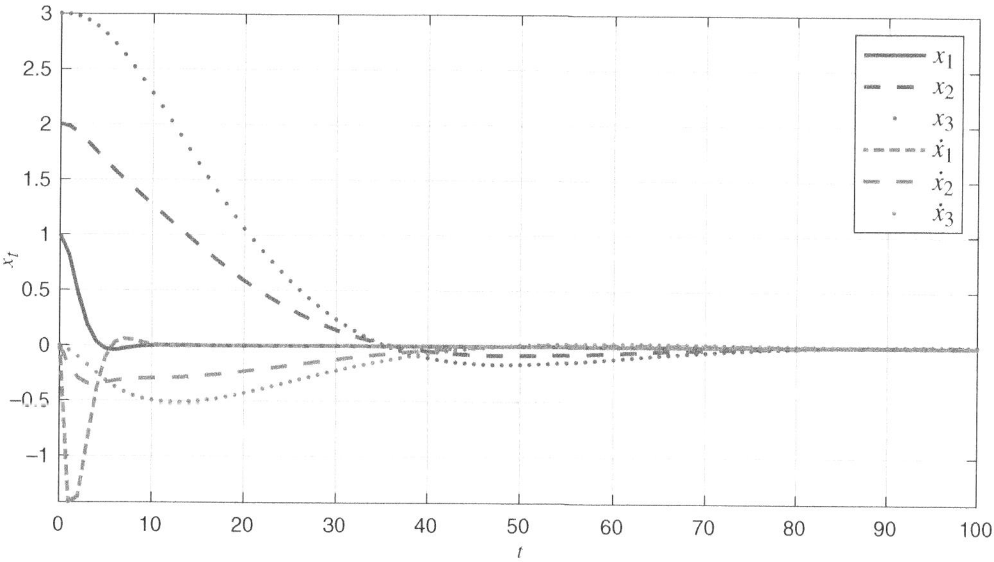

Figure 3.91 State trajectories using state feedback using LQR gains.

3.7.12.1 Random Variables

A random variable, $x \in \mathbb{R}$, may take on a variety of values associated with its probability density function (PDF), $f(x)$. The PDF, as depicted in Figure 3.93, specifies the relative likelihood of various outcomes. The probability that x lies within an interval $[a, b]$ is given by

$$P(a \leq X \leq b) = \int_{a}^{b} f(x)\, dx. \tag{3.169}$$

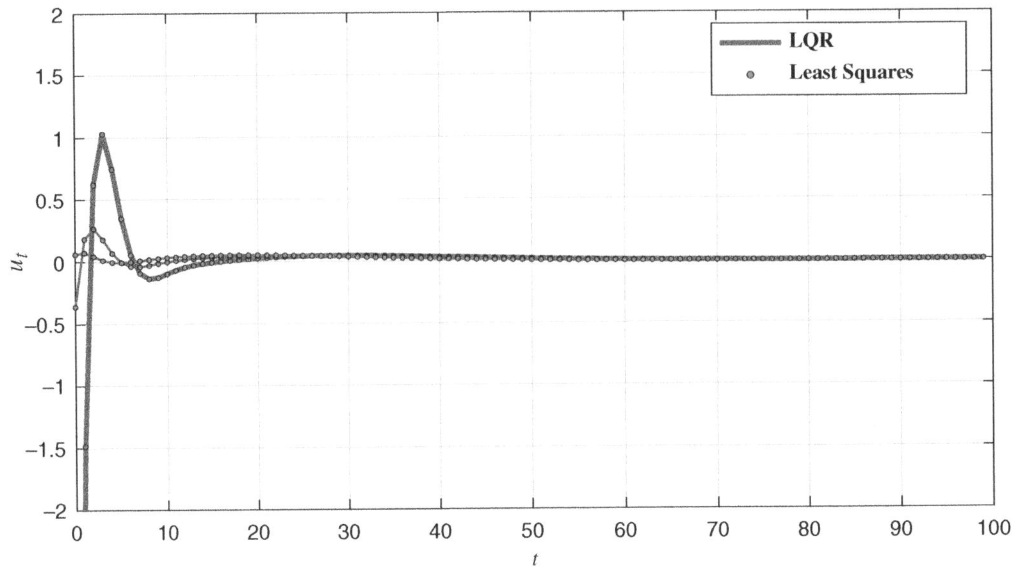

Figure 3.92 Input sequence for LQR versus least squares.

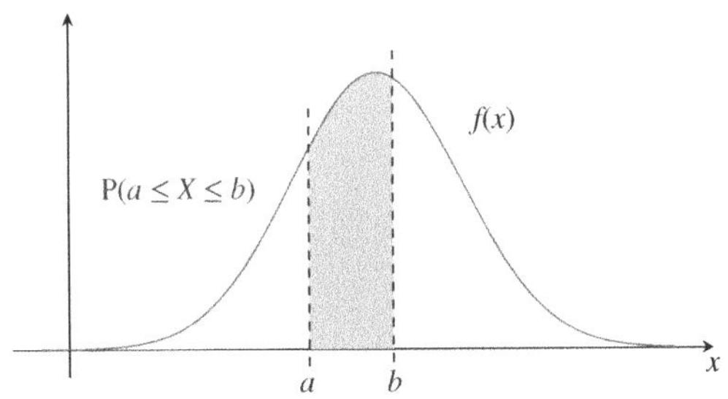

Figure 3.93 Probability density function.

The pdf must satisfy two conditions: (1) it is nonnegative, $f(x) \geq 0$ for all x and (2) it integrates to one $\int_{-\infty}^{\infty} f(x)\,dx = 1$. The cumulative density function (CDF), denoted $F(x)$, represents the probability that x will take a value less than or equal to some value

$$F(x) = \int_{-\infty}^{x} f(t)\,dt. \tag{3.170}$$

The CDF is a nondecreasing function where $\lim_{x\to\infty} F(x) = 1$. Important summary statistics of a random variable are

- Mean (or expected value): μ, is the average value or expected value of x,

$$\mu = \mathbf{E}(x) = \int_{-\infty}^{\infty} xf(x)\,dx. \tag{3.171}$$

- Variance: σ^2, is a measure of the spread of x around the mean,

$$\sigma^2 = \mathbf{Var}(x) = \mathbf{E}\left[(x - \mathbf{E}(x))^2\right] = \mathbf{E}(x^2) - \mathbf{E}(x)^2. \tag{3.172}$$

- Standard deviation: σ, is the square root of the variance, providing a direct measure of the average deviation of x from its mean [3].

3.7.12.1.1 Normal Distributions

One of the most fundamental distributions in probability theory is the normal (Gaussian) distribution. A random variable x that is drawn from a normally distributed with mean μ and variance σ^2 is denoted $x \sim N(\mu, \sigma^2)$. The PDF of a normal distribution is given by

$$f(x) = \frac{1}{\sqrt{2\pi\sigma^2}} e^{-\frac{(x-\mu)^2}{2\sigma^2}}. \qquad (3.173)$$

One can show that this integrates out to 1 by computing $\int_{-\infty}^{\infty} f(x)\, dx$. The bell curve shape of the normal distribution centers around the mean μ, with its spread determined by σ.

As shown in Figure 3.94, 68.3% of values lie within one standard deviation of the mean, 95.4% within two, and 99.7% within three. For any normally distributed variable $x \sim N(\mu, \sigma^2)$, we can standardize it using the transformation $z = \frac{x-\mu}{\sigma}$, resulting in a standard normal variable, $z \sim N(0,1)$, with mean 0 and variance 1 [7].

3.7.12.2 Random Vectors

In control systems, we often encounter random variables that are sets of variables that may be correlated. A random vector, $\boldsymbol{x} \in \mathbb{R}^n$, consists of multiple random variables that can exhibit correlations. The expected value, $\boldsymbol{\mu} \in \mathbb{R}^n$, of a random vector $\boldsymbol{x}$, is calculated elementwise, representing the average values of each component

$$\boldsymbol{\mu} = \mathbf{E}(\boldsymbol{x}) = \begin{bmatrix} \mathbf{E}(x_1) \\ \vdots \\ \mathbf{E}(x_n) \end{bmatrix} \in \mathbb{R}^n. \qquad (3.174)$$

The covariance matrix of a random vector $\boldsymbol{x}$, $\Sigma \in \mathbb{R}^{n \times n}$, captures the variances of each component along its diagonal and the covariances between pairs of components off the matrix diagonal. It can be written in the following ways [3]:

$$\mathbf{Cov}(\boldsymbol{x}) = \mathbf{E}\left[(\boldsymbol{x} - \mathbf{E}(\boldsymbol{x}))\,(\boldsymbol{x} - \mathbf{E}(\boldsymbol{x}))^T\right] = \begin{bmatrix} \mathbf{Cov}(x_1, x_1) & \mathbf{Cov}(x_1, x_2) & \cdots & \mathbf{Cov}(x_1, x_n) \\ \mathbf{Cov}(x_2, x_1) & \mathbf{Cov}(x_2, x_2) & \cdots & \mathbf{Cov}(x_2, x_n) \\ \vdots & \vdots & \ddots & \vdots \\ \mathbf{Cov}(x_n, x_1) & \mathbf{Cov}(x_n, x_2) & \cdots & \mathbf{Cov}(x_n, x_n) \end{bmatrix} = \Sigma. \qquad (3.175)$$

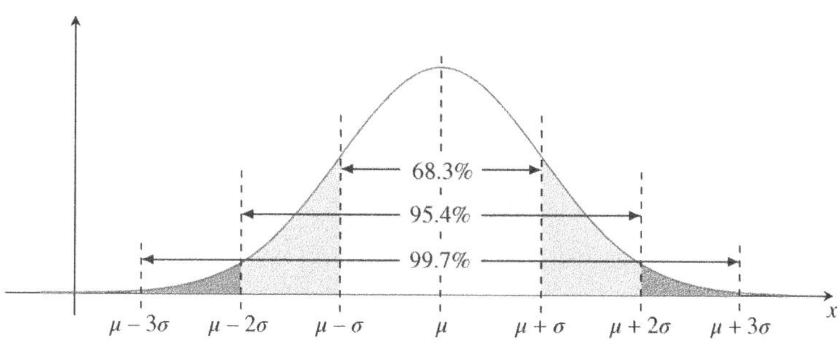

Figure 3.94 Normal distribution.

A positive covariance indicates that both variables increase together, while a negative covariance suggests that one variable tends to increase as the other decreases. The covariance between two random vectors $\boldsymbol{x} \in \mathbb{R}^n, \boldsymbol{y} \in \mathbb{R}^m$ is

$$\mathbf{Cov}(\boldsymbol{x}, \boldsymbol{y}) = \mathbf{E}(\boldsymbol{x}\boldsymbol{y}^T) - \mathbf{E}(\boldsymbol{x})\mathbf{E}(\boldsymbol{y})^T = \mathbf{E}(\boldsymbol{x}\boldsymbol{y}^T) - \boldsymbol{\mu}_x\boldsymbol{\mu}_y^T = \mathbf{Cov}(\boldsymbol{y}, \boldsymbol{x})^T \in \mathbb{R}^{m\times n}. \tag{3.176}$$

We may want to apply simple manipulations of random vectors like addition and subtraction. For two random vectors $\boldsymbol{x}, \boldsymbol{y} \in \mathbb{R}^n$, the following hold [3]

$$\begin{aligned} \mathbf{E}(\boldsymbol{x}+\boldsymbol{y}) &= \mathbf{E}(\boldsymbol{x}) + \mathbf{E}(\boldsymbol{y}) \\ \mathbf{E}(\mathbf{x}-\mathbf{y}) &= \mathbf{E}(\boldsymbol{x}) - \mathbf{E}(\boldsymbol{y}) \\ \mathbf{Cov}(\boldsymbol{x}+\boldsymbol{y}) &= \mathbf{Cov}(\boldsymbol{x}) + \mathbf{Cov}(\boldsymbol{y}) + \mathbf{Cov}(\boldsymbol{x}, \boldsymbol{y}) + \mathbf{Cov}(\boldsymbol{y}, \boldsymbol{x}) = \Sigma_x + \Sigma_y + \Sigma_{xy} + \Sigma_{yx} \\ \mathbf{Cov}(\boldsymbol{x}-\boldsymbol{y}) &= \mathbf{Cov}(\boldsymbol{x}) + \mathbf{Cov}(\boldsymbol{y}) - \mathbf{Cov}(\boldsymbol{x}, \boldsymbol{y}) - \mathbf{Cov}(\boldsymbol{y}, \boldsymbol{x}) = \Sigma_x + \Sigma_y - \Sigma_{xy} - \Sigma_{yx}. \end{aligned} \tag{3.177}$$

3.7.12.2.1 Multivariate Normal Distributions

A random vector $\boldsymbol{x} \in \mathbb{R}^n$ follows a multivariate normal distribution with mean $\boldsymbol{\mu}$ and covariance matrix Σ if its PDF is

$$f(\boldsymbol{x}) = \frac{1}{(2\pi)^{n/2}\sqrt{\det\Sigma}} \exp\left(-\frac{1}{2}(\boldsymbol{x}-\boldsymbol{\mu})^T\Sigma^{-1}(\boldsymbol{x}-\boldsymbol{\mu})\right). \tag{3.178}$$

A vector $\boldsymbol{x}$ that is normally distributed with mean $\boldsymbol{\mu}$ and covariance Σ can be denoted as $\boldsymbol{x} \sim N(\boldsymbol{\mu}, \Sigma)$.

Standardizing $\boldsymbol{x}$ with $z = \Sigma^{-1/2}(\boldsymbol{x}-\boldsymbol{\mu})$ transforms it into the standard normal distribution $\boldsymbol{z} \sim N(\mathbf{0}, I)$. We derived the probability that a normally distributed variable $x \sim N(\mu,\sigma)$ lies within a symmetric interval about its mean, for example, roughly 68.3% of values lie within one standard deviation of the mean, $P(\mu - \sigma \leq x \leq \mu + \sigma)$. Extending this idea to a multivariate Gaussian distribution leads us to the concept of a confidence ellipsoid where the contours of constant density form level sets defined as

$$S_\alpha = \{\boldsymbol{x} \in \mathbb{R}^n : (\boldsymbol{x}-\boldsymbol{\mu})^T\Sigma^{-1}(\boldsymbol{x}-\boldsymbol{\mu}) \leq \alpha\}. \tag{3.179}$$

To plot this ellipsoid, we can utilize the singular value decomposition or eigenvalue decomposition. The resulting confidence ellipsoid is centered at $\boldsymbol{\mu}$ and its axes align with the directions given by the eigenvectors $\boldsymbol{u}_i$, with corresponding axis lengths $\sqrt{\alpha\,\lambda_i}$, where λ_i are the eigenvalues. Figure 3.95 is an example of a 2D Gaussian with contours for a random vector $\boldsymbol{x} \sim N\left(\begin{bmatrix}\mu_x\\ \mu_y\end{bmatrix}, \begin{bmatrix}\Sigma_x & \Sigma_{xy}\\ \Sigma_{yx} & \Sigma_y\end{bmatrix}\right) = N\left(\begin{bmatrix}0\\0\end{bmatrix}, \begin{bmatrix}3 & -2\\ -2 & 2\end{bmatrix}\right)$. The ellipsoid corresponding to a given probability, p, is defined by $\{\boldsymbol{x} \in \mathbb{R}^n : (\boldsymbol{x}-\boldsymbol{\mu})^T\Sigma^{-1}(\boldsymbol{x}-\boldsymbol{\mu}) \leq \alpha\}$.

3.7.12.2.2 Linear Transformations of Random Vectors

When a Gaussian random vector $\boldsymbol{x} \sim N(\boldsymbol{\mu}_x, \Sigma_x)$ undergoes an affine transformation, $\boldsymbol{y} = A\boldsymbol{x} + \boldsymbol{b}$, where $A \in \mathbb{R}^{n\times n}$, $\boldsymbol{b} \in \mathbb{R}^n$ then the transformed vector $\boldsymbol{y}$ is also Gaussian. The mean and covariance of $\boldsymbol{y}$ are

$$\begin{aligned} \mathbf{E}(\boldsymbol{y}) &= \mathbf{E}(A\boldsymbol{x}+\boldsymbol{b}) = A\mathbf{E}(\boldsymbol{x}) + \boldsymbol{b} = A\boldsymbol{\mu}_x + \boldsymbol{b} \\ \mathbf{Cov}(\boldsymbol{y}) &= \mathbf{E}\left[(\boldsymbol{y}-\mathbf{E}(\boldsymbol{y}))\,(\boldsymbol{y}-\mathbf{E}(\boldsymbol{y}))^T\right] = A\mathbf{Cov}(\boldsymbol{x})A^T = A\Sigma_x A^T. \end{aligned} \tag{3.180}$$

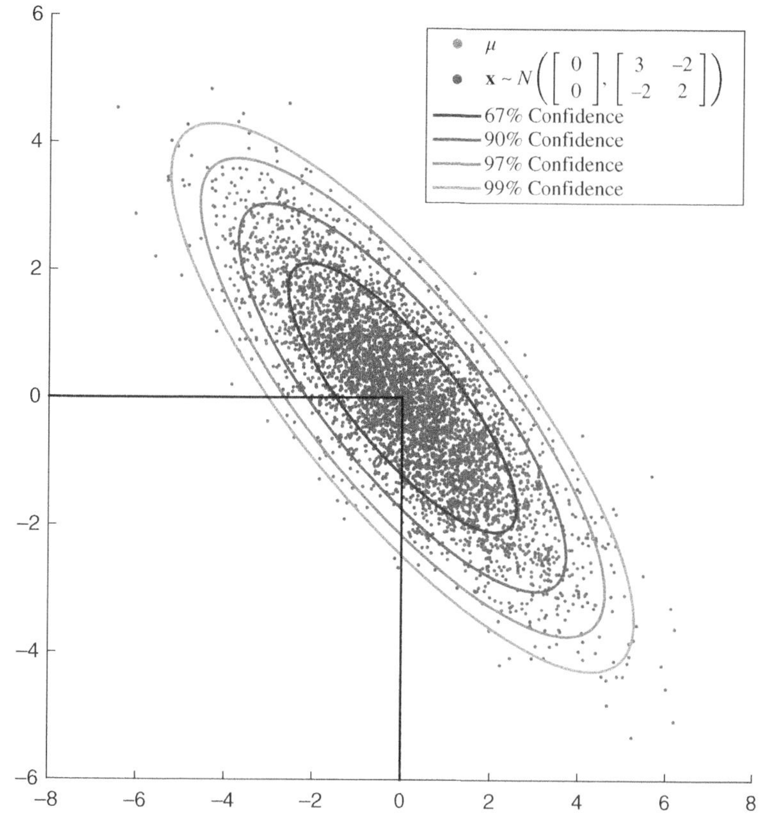

Figure 3.95 Confidence ellipsoids illustrate the proportion of points captured within each ellipse for different values of probability, *p*.

3.7.12.2.3 Joint Distribution of Random Vectors

In many estimation problems, we work with a larger vector that can be partitioned into sub-vectors $\boldsymbol{x}$ and $\boldsymbol{y}$, where $\boldsymbol{y}$ might represent measurements, while $\boldsymbol{x}$ is the state we seek to estimate. When the joint distribution is Gaussian, and $f(\boldsymbol{x}, \boldsymbol{y})$ is the density function, then

$$\begin{bmatrix} \boldsymbol{x} \\ \boldsymbol{y} \end{bmatrix} \sim N\left(\begin{bmatrix} \boldsymbol{\mu}_x \\ \boldsymbol{\mu}_y \end{bmatrix}, \begin{bmatrix} \Sigma_x & \Sigma_{xy} \\ \Sigma_{yx} & \Sigma_y \end{bmatrix}\right). \tag{3.181}$$

This encapsulates both the individual distributions of $\boldsymbol{x}$ and $\boldsymbol{y}$ and their dependence through the cross-covariance terms Σ_{xy} and Σ_{yx}.

3.7.12.3 Marginal Distributions

We often need to consider just the distribution of $\boldsymbol{x}$ while ignoring $\boldsymbol{y}$ for the joint pdf, $f(\boldsymbol{x}, \boldsymbol{y})$. This is called the marginal distribution of $\boldsymbol{x}$. Similarly, we can ignore $\boldsymbol{x}$ and consider only the distribution of $\boldsymbol{y}$. These marginal density functions as $f_x(\boldsymbol{x})$ and $f_y(\boldsymbol{y})$ are

$$\begin{aligned} f_x(\boldsymbol{x}) &= \int f(\boldsymbol{x}, \boldsymbol{y})\, d\boldsymbol{y} \\ f_y(\boldsymbol{x}) &= \int f(\boldsymbol{x}, \boldsymbol{y})\, d\boldsymbol{x}. \end{aligned} \tag{3.182}$$

3.7.12.3.1 Marginal Distribution of a Gaussian

Let us consider Gaussian random vectors $\boldsymbol{x} \in \mathbb{R}^r$ and $\boldsymbol{y} \in \mathbb{R}^{n-r}$ which are distributed together as

$$\begin{bmatrix} \boldsymbol{x} \\ \boldsymbol{y} \end{bmatrix} \sim N\left(\begin{bmatrix} \boldsymbol{\mu}_x \\ \boldsymbol{\mu}_y \end{bmatrix}, \begin{bmatrix} \Sigma_x & \Sigma_{xy} \\ \Sigma_{yx} & \Sigma_y \end{bmatrix} \right),$$

The joint PDF is

$$f(\boldsymbol{x}, \boldsymbol{y}) = c_1 \exp\left(-\frac{1}{2} \begin{bmatrix} \bar{\boldsymbol{x}} \\ \bar{\boldsymbol{y}} \end{bmatrix}^T \begin{bmatrix} \Sigma_x & \Sigma_{xy} \\ \Sigma_{yx} & \Sigma_y \end{bmatrix}^{-1} \begin{bmatrix} \bar{\boldsymbol{x}} \\ \bar{\boldsymbol{y}} \end{bmatrix} \right) \tag{3.183}$$

where $c_1 = \dfrac{1}{(2\pi)^{n/2}\sqrt{det(\Sigma)}}$, $\bar{\boldsymbol{x}} = \boldsymbol{x} - \boldsymbol{\mu}_x$, and $\bar{\boldsymbol{y}} = \boldsymbol{y} - \boldsymbol{\mu}_y$. Using the block-LDU factorization in Eq (3.67), we invert the covariance matrix and substitute it into the exponential term of the joint Gaussian

$$\begin{aligned} f(\boldsymbol{x}, \boldsymbol{y}) &= c_1 \exp\left(-\frac{1}{2} \begin{bmatrix} \bar{\boldsymbol{x}} \\ \bar{\boldsymbol{y}} - \Sigma_{yx}\Sigma_x^{-1}\bar{\boldsymbol{x}} \end{bmatrix}^T \begin{bmatrix} \Sigma_x^{-1} & 0 \\ 0 & (\Sigma_y - \Sigma_{yx}\Sigma_x^{-1}\Sigma_{xy})^{-1} \end{bmatrix} \begin{bmatrix} \bar{\boldsymbol{x}} \\ \bar{\boldsymbol{y}} - \Sigma_{yx}\Sigma_x^{-1}\bar{\boldsymbol{x}} \end{bmatrix} \right) \\ &= c_1 \exp\left(-\frac{1}{2}\bar{\boldsymbol{x}}^T\Sigma_x^{-1}\bar{\boldsymbol{x}} \right) \exp\left(-\frac{1}{2} (\bar{\boldsymbol{y}} - \Sigma_{yx}\Sigma_x^{-1}\bar{\boldsymbol{x}})^T (\Sigma_y - \Sigma_{yx}\Sigma_x^{-1}\Sigma_{xy})^{-1} (\bar{\boldsymbol{y}} - \Sigma_{yx}\Sigma_x^{-1}\bar{\boldsymbol{x}}) \right) \end{aligned}$$

Let us now compute the marginal distribution

$$\begin{aligned} f_x(\boldsymbol{x}) &= \int_{\mathbb{R}^{n-r}} f(\boldsymbol{x}, \boldsymbol{y}) d\boldsymbol{y} \\ &= c_1 \exp\left(-\frac{1}{2}\bar{\boldsymbol{x}}^T\Sigma_x^{-1}\bar{\boldsymbol{x}} \right) \int_{\mathbb{R}^{n-r}} \exp\left(-\frac{1}{2} (\bar{\boldsymbol{y}} - \Sigma_{xy}\Sigma_x^{-1}\bar{\boldsymbol{x}})^T (\Sigma_y - \Sigma_{yx}\Sigma_x^{-1}\Sigma_{xy})^{-1} (\bar{\boldsymbol{y}} - \Sigma_{xy}\Sigma_x^{-1}\bar{\boldsymbol{x}})^T \right) d\boldsymbol{y}. \end{aligned}$$

We can determine the normalization constants using the fact that $\int f(\boldsymbol{x}, \boldsymbol{y}) d\boldsymbol{y} = 1$. Evaluating the integral over the entire domain of $\boldsymbol{y}$ and simplifying, we arrive at the marginal distribution of $\boldsymbol{x}$

$$f_x(\boldsymbol{x}) = c_1 \exp\left(-\frac{1}{2}\bar{\boldsymbol{x}}^T\Sigma_x{}^{-1}\bar{\boldsymbol{x}} \right) = \frac{1}{(2\pi)^{r/2}\sqrt{\det(\Sigma_x)}} \exp\left(-\frac{1}{2}(\boldsymbol{x} - \boldsymbol{\mu}_x)^T\Sigma_x{}^{-1}(\boldsymbol{x} - \boldsymbol{\mu}_x) \right). \tag{3.184}$$

Therefore, if $\begin{bmatrix} \boldsymbol{x} \\ \boldsymbol{y} \end{bmatrix} \sim N(\boldsymbol{\mu}, \Sigma)$ then the marginal PDF of $\boldsymbol{x}$ is Gaussian and simplifies to $\boldsymbol{x} \sim N(\boldsymbol{\mu}_x, \Sigma_x)$.

3.7.12.4 Conditional Distributions

Conditional distributions allow us to update our estimate of $\boldsymbol{x}$ given an observation of $\boldsymbol{y}$. The conditional probability density function, $f_{x|y}(\boldsymbol{x}, \boldsymbol{y})$, is derived from the joint distribution but is a scaled since

$$f_{x|y}(\boldsymbol{x}, \boldsymbol{y}) = \frac{f(\boldsymbol{x}, \boldsymbol{y})}{f_y(\boldsymbol{y})} \tag{3.185}$$

where $f_y(\boldsymbol{y})$ is the marginal with respect to $\boldsymbol{y}$. Eq (3.185) is also known as Bayes Rule. Also, conditional distributions follow the standard properties of PDFs such that

$$\int f_{x|y}(\boldsymbol{x}, \boldsymbol{y}) d\boldsymbol{x} = \mathbf{1}. \tag{3.186}$$

3.7.12.4.1 Conditional Distribution of a Gaussian

When looking at the marginal we computed the joint PDF we can use the block-UDL factorization in Eq (3.158)

$$f(\boldsymbol{x}, \boldsymbol{y}) = c_1 \exp\left(-\frac{1}{2}\begin{bmatrix} \bar{\boldsymbol{x}} - \Sigma_{xy}\Sigma_y^{-1}\bar{\boldsymbol{y}} \\ \bar{\boldsymbol{y}} \end{bmatrix}^T \begin{bmatrix} (\Sigma_x - \Sigma_{xy}\Sigma_y^{-1}\Sigma_{yx})^{-1} & 0 \\ 0 & \Sigma_y^{-1} \end{bmatrix} \begin{bmatrix} \bar{\boldsymbol{x}} - \Sigma_{xy}\Sigma_y^{-1}\bar{\boldsymbol{y}} \\ \bar{\boldsymbol{y}} \end{bmatrix}\right)$$
$$= c_1 \exp\left(-\frac{1}{2}\bar{\boldsymbol{y}}^T\Sigma_y^{-1}\bar{\boldsymbol{y}}\right)\exp\left(-\frac{1}{2}(\bar{\boldsymbol{x}} - \Sigma_{xy}\Sigma_y^{-1}\bar{\boldsymbol{y}})^T(\Sigma_x - \Sigma_{xy}\Sigma_y^{-1}\Sigma_{yx})^{-1}(\bar{\boldsymbol{x}} - \Sigma_{xy}\Sigma_y^{-1}\bar{\boldsymbol{y}})\right)$$

Notice the first exponential term is the marginal distribution, $f_y(\boldsymbol{y})$. Therefore, the conditional pdf of $\boldsymbol{x}$ given $\boldsymbol{y}$ is

$$f_{x|y}(\boldsymbol{x}, \boldsymbol{y}) = \frac{f(\boldsymbol{x}, \boldsymbol{y})}{f_y(\boldsymbol{y})} = c_1 \exp\left(-\frac{1}{2}(\bar{\boldsymbol{x}} - \Sigma_{xy}\Sigma_y^{-1}\bar{\boldsymbol{y}})^T(\Sigma_x - \Sigma_{xy}\Sigma_y^{-1}\Sigma_{yx})^{-1}(\bar{\boldsymbol{x}} - \Sigma_{xy}\Sigma_y^{-1}\bar{\boldsymbol{y}})\right)$$
$$= c_1 \exp\left(-\frac{1}{2}\Big(\underbrace{\boldsymbol{x} - \boldsymbol{\mu}_x - \Sigma_{xy}\Sigma_y^{-1}(\boldsymbol{y} - \boldsymbol{\mu}_y)}_{-\mathbf{E}(\boldsymbol{x}|\boldsymbol{y})}\Big)^T \Big(\underbrace{\Sigma_x - \Sigma_{xy}\Sigma_y^{-1}\Sigma_{yx}}_{\mathbf{Cov}(\boldsymbol{x}|\boldsymbol{y})}\Big)^{-1} \Big(\underbrace{\boldsymbol{x} - \boldsymbol{\mu}_x - \Sigma_{xy}\Sigma_y^{-1}(\boldsymbol{y} - \boldsymbol{\mu}_y)}_{-\mathbf{E}(\boldsymbol{x}|\boldsymbol{y})}\Big)\right).$$

Therefore, if $\begin{bmatrix} \boldsymbol{x} \\ \boldsymbol{y} \end{bmatrix} \sim N(\boldsymbol{\mu}, \Sigma)$ then the conditional PDF of $\boldsymbol{x}$ given $\boldsymbol{y}$ is Gaussian

$$f_{x|y}(\boldsymbol{x}, \boldsymbol{y}) \sim N(\mathbf{E}(\boldsymbol{x}|\boldsymbol{y}), \mathbf{Cov}(\boldsymbol{x}|\boldsymbol{y})) = N\left(\boldsymbol{\mu}_x + \Sigma_{xy}\Sigma_y^{-1}(\boldsymbol{y} - \boldsymbol{\mu}_y), \Sigma_x - \Sigma_{xy}\Sigma_y^{-1}\Sigma_{yx}\right). \tag{3.187}$$

The conditional distribution is also known as the posterior distribution. The conditional mean, $\mathbf{E}(\boldsymbol{x}|\boldsymbol{y}) = \boldsymbol{\mu}_x + \Sigma_{xy}\Sigma_y^{-1}(\boldsymbol{y} - \boldsymbol{\mu}_y)$, is a linear function of $\boldsymbol{y}$, and updates $\boldsymbol{\mu}_x$ based on measurements of $\boldsymbol{y}$. The conditional covariance, $\mathbf{Cov}(\boldsymbol{x}|\boldsymbol{y}) = \Sigma_x - \Sigma_{xy}\Sigma_y^{-1}\Sigma_{yx}$, decreases from Σ_x but is independent of $\boldsymbol{y}$. This allows the prediction of estimation error to change before measurements are received. Additionally, the conditional and marginal covariances of $\boldsymbol{x}$ are equal only when $\Sigma_{xy} = 0$, which occurs when $\boldsymbol{x}$ and $\boldsymbol{y}$ are uncorrelated [7].

3.7.13 Mean Squared Error

The mean squared error (MSE) quantifies the expected magnitude of the error between random vector $\boldsymbol{x}$ and its estimate $\hat{\boldsymbol{x}}$. It is defined as

$$\text{MSE} = \mathbf{E}\left(\|\boldsymbol{x} - \hat{\boldsymbol{x}}\|^2|\boldsymbol{y}\right) \tag{3.188}$$

The MSE expression can be expanded as follows

$$\text{MSE} = \mathbf{E}\left[(\boldsymbol{x} - \hat{\boldsymbol{x}})^T(\boldsymbol{x} - \hat{\boldsymbol{x}})|\boldsymbol{y}\right] = \mathbf{E}\left(\|\boldsymbol{x}\|^2|\boldsymbol{y}\right) - 2\hat{\boldsymbol{x}}^T\mathbf{E}(\boldsymbol{x}|\boldsymbol{y}) + \|\hat{\boldsymbol{x}}\|^2.$$

To minimize MSE, we seek an $\hat{\boldsymbol{x}}$ that minimizes this expression. This estimator is known as the minimum mean squared error (MMSE) estimate. To find the MMSE estimator, we minimize the MSE expression with respect to $\boldsymbol{x}$

$$\text{MMSE} = \min_{\hat{x}}\left(\mathbf{E}\left(\|\boldsymbol{x}\|^2|\boldsymbol{y}\right) - 2\hat{\boldsymbol{x}}^T\mathbf{E}(\boldsymbol{x}|\boldsymbol{y}) + \|\hat{\boldsymbol{x}}\|^2\right).$$

and take the gradient of the MSE with respect to $\hat{\boldsymbol{x}}$ and setting it equal to zero gives

$$\nabla_{\hat{x}} = -2\mathbf{E}(\boldsymbol{x}|\boldsymbol{y}) + 2\hat{\boldsymbol{x}} = 0 \Rightarrow \hat{\boldsymbol{x}} = \mathbf{E}(\boldsymbol{x}|\boldsymbol{y}).$$

Thus, the MMSE estimator is the conditional expectation of $\boldsymbol{x}$ given $\boldsymbol{y}$. If we substitute $\hat{\boldsymbol{x}} = \mathbf{E}(\boldsymbol{x}|\boldsymbol{y})$ back into the MSE, then the MMSE is

$$\text{MMSE} = \mathbf{E}\left[(\boldsymbol{x} - \mathbf{E}(\boldsymbol{x}|\boldsymbol{y}))^T(\boldsymbol{x} - \mathbf{E}(\boldsymbol{x}|\boldsymbol{y}))|\boldsymbol{y}\right] = \mathbf{tr}(\mathbf{Cov}(\boldsymbol{x}|\boldsymbol{y})). \tag{3.189}$$

The MMSE is therefore achieved by using the conditional mean as the estimate, and the associated MSE is the trace of the conditional covariance matrix.

3.7.14 Linear Model with Noise

Assume we have a linear measurement model where $A \in \mathbb{R}^{m \times n}$, $\boldsymbol{x} \in \mathbb{R}^n$, $\boldsymbol{y} \in \mathbb{R}^m$ that includes additive Gaussian noise $\boldsymbol{w}$

$$\boldsymbol{y} = A\boldsymbol{x} + \boldsymbol{w}, \boldsymbol{x} \sim N(\mu_x, \Sigma_x), \boldsymbol{w} \sim N(\boldsymbol{0}, \Sigma_w) \text{are independent.}$$

Recall for an affine transformation, the transformed random vector is distributed as $A\boldsymbol{x} + \boldsymbol{b} \sim N(A\mu_x + \boldsymbol{b}, A\Sigma_x A^T)$ and so the joint distribution $f_{x|y}(\boldsymbol{x}, \boldsymbol{y})$ is

$$\begin{bmatrix} \boldsymbol{x} \\ \boldsymbol{y} \end{bmatrix} \sim N\left(\begin{bmatrix} \mu_x \\ A\mu_x \end{bmatrix}, \begin{bmatrix} \Sigma_x & \Sigma_x A^T \\ A\Sigma_x & A\Sigma_x A^T + \Sigma_w \end{bmatrix} \right). \tag{3.190}$$

Using the formula for the conditional Gaussian distribution (3.187) and Sherman–Morrison formula, the conditional covariance for the linear model is

$$\mathbf{Cov}(\boldsymbol{x}|\boldsymbol{y}) = \Sigma_x - \Sigma_x A^T (A\Sigma_x A^T + \Sigma_w)^{-1} A\Sigma_x = (\Sigma_x^{-1} + A^T \Sigma_w^{-1} A)^{-1}. \tag{3.191}$$

In our original covariance formula, we know that the covariance decreases when we observe a measurement. Also, the covariance serves as an indicator of estimation error. A smaller covariance implies a smaller expected error, indicating greater confidence in the estimate [7].

3.7.15 MAP and MMSE Estimates

The MMSE estimator selects the mean of the posterior distribution, while the maximum a posteriori (MAP) estimator identifies the value of $\mathbf{x}$ that maximizes the posterior probability. For Gaussian distributions, these two estimates coincide. However, for nonsymmetric distributions, such as the bimodal distribution shown in Figure 3.96, the MMSE and MAP estimates can differ significantly.

3.7.16 The Kalman Filter

Let us consider the linear dynamical system of the form

$$\begin{aligned} \boldsymbol{x}_{t+1} &= A\boldsymbol{x}_t + \boldsymbol{w}_t \\ \boldsymbol{y}_t &= C\boldsymbol{x}_t + \boldsymbol{v}_t \end{aligned}$$

where $\boldsymbol{x}_t \in \mathbb{R}^n$ is the state, $\boldsymbol{y}_t \in \mathbb{R}^p$ is the observed output, $\boldsymbol{w}_t \in \mathbb{R}^n$ is the process noise or state noise, and $\boldsymbol{v}_t \in \mathbb{R}^n$ is called the measurement noise. We will make the following modeling assumptions

- $\boldsymbol{w}_t \sim N(\boldsymbol{0}, \Sigma_w)$ is white Gaussian noise
- $\boldsymbol{v}_t \sim N(\boldsymbol{0}, \Sigma_v)$ is white Gaussian noise
- $\boldsymbol{v}_t$ and $\boldsymbol{w}_s$ are independent for all s and t, where time s is before time t
- the initial state is random $\boldsymbol{x}_0 \sim N(\mu_{x_0}, \Sigma_{x_0})$ and independent of $\boldsymbol{v}_t$.

From these assumptions the sensor noise $\boldsymbol{w}_t$ is independent of the state vector $\boldsymbol{x}_s$ for all times t and s. This implies the measurement noise introduced at any time does not depend on the past or current states. Additionally, the process noise $\boldsymbol{v}_t$ is assumed to be independent of $\boldsymbol{x}_0, \boldsymbol{x}_1, \ldots, \boldsymbol{x}_t$ and $\boldsymbol{y}_0, \boldsymbol{y}_1, \ldots, \boldsymbol{y}_t$. This system satisfies Markov property, such that the process $\boldsymbol{x}$ is Markov such that $\boldsymbol{x}_t|\boldsymbol{x}_0, \boldsymbol{x}_1, \ldots, \boldsymbol{x}_{t-1} = \boldsymbol{x}_t|\boldsymbol{x}_{t-1}$. Stated another way, if the state at time $t-1$ is known, any additional

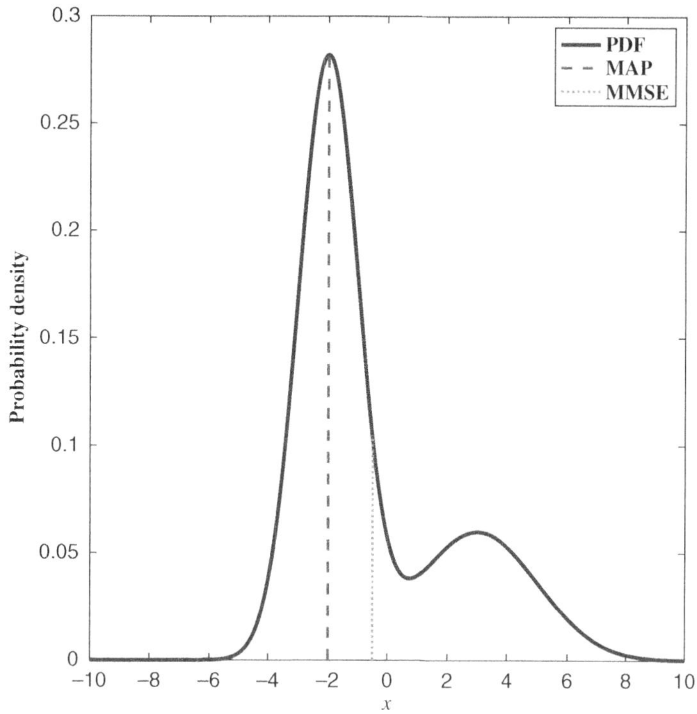

Figure 3.96 MAP and MMSE estimates on a bimodal distribution.

knowledge of earlier states (such as $\boldsymbol{x}_{t-2}, \boldsymbol{x}_{t-3}, \ldots, \boldsymbol{x}_0$) does not provide any further information about the state at time t [7].

3.7.16.1 State Estimation Notation

For state estimation, $\hat{\boldsymbol{x}}_{t|s}$ is the minimum mean square error (MMSE) estimate of $\boldsymbol{x}_t$ based on observations $\boldsymbol{y}_0, \ldots, \boldsymbol{y}_s$, denoted as $\hat{\boldsymbol{x}}_{t|s} = \mathbf{E}(\boldsymbol{x}_t|\boldsymbol{y}_0, \ldots, \boldsymbol{y}_s)$. The associated error covariance, $\Sigma_{t|s}$, is the conditional covariance of $\boldsymbol{x}_t$ given measurements of $\boldsymbol{y}_0, \boldsymbol{y}_1, \ldots, \boldsymbol{y}_s$. This is denoted as $\Sigma_{t|s} = \mathbf{Cov}(\boldsymbol{x}_t|\boldsymbol{y}_0, \ldots, \boldsymbol{y}_s)$ [7].

3.7.16.2 The Measurement Update

Let us derive the state estimate $\hat{\boldsymbol{x}}_{t|t}$ and $\Sigma_{t|t}$ in terms of predicted values $\hat{\boldsymbol{x}}_{t|t-1}$ and $\Sigma_{t|t-1}$. Using the measurement equation, the MMSE estimation problem becomes

$$\boldsymbol{y}_{t|t-1} = C\boldsymbol{x}_{t|t-1} + \boldsymbol{v}_t \Rightarrow \begin{bmatrix} \boldsymbol{x}_{t|t-1} \\ \boldsymbol{y}_{t|t-1} \end{bmatrix} = \begin{bmatrix} I & \mathbf{0} \\ C & I \end{bmatrix} \begin{bmatrix} \boldsymbol{x}_{t|t-1} \\ \boldsymbol{v}_t \end{bmatrix}$$

where $\boldsymbol{x}_{t|t-1} \sim N(\hat{\boldsymbol{x}}_{t|t-1}, \Sigma_{t|t-1})$ and $\boldsymbol{v}_t \sim N(\mathbf{0}, \Sigma_v)$. We can write this system as a jointly distributed Gaussian

$$\begin{bmatrix} \boldsymbol{x}_{t|t-1} \\ \boldsymbol{y}_{t|t-1} \end{bmatrix} \sim N\left(\begin{bmatrix} \hat{\boldsymbol{x}}_{t|t-1} \\ C\hat{\boldsymbol{x}}_{t|t-1} \end{bmatrix}, \begin{bmatrix} \Sigma_{t|t-1} & \Sigma_{t|t-1}C^T \\ C\Sigma_{t|t-1} & C\Sigma_{t|t-1}C^T \end{bmatrix} + \Sigma_v \right).$$

The MMSE estimate of $\boldsymbol{x}_{t|t-1}$ and its error covariance are the mean and the covariance of $\boldsymbol{x}_{t|t-1}|\boldsymbol{y}_{t|t-1} = \boldsymbol{x}_{t|t}|\boldsymbol{y}_{t|t} \sim N(\hat{\boldsymbol{x}}_{t|t},\ \Sigma_{t|t})$. Applying, the MMSE estimate (conditional form of the Gaussian) for the linear model gives us

$$\begin{aligned}\hat{\boldsymbol{x}}_{t|t} &= \mathbf{E}(\hat{\boldsymbol{x}}_{t|t-1}|\boldsymbol{y}_{t|t-1}) = \hat{\boldsymbol{x}}_{t|t-1} + \Sigma_{t|t-1}C^T(C\Sigma_{t|t-1}C^T + \Sigma_{\boldsymbol{v}})^{-1}(\boldsymbol{y}_t - C\hat{\boldsymbol{x}}_{t|t-1}) \\ \Sigma_{t|t} &= \mathbf{Cov}(\hat{\boldsymbol{x}}_{t|t-1}|\boldsymbol{y}_{t|t-1}) = \Sigma_{t|t-1} - \Sigma_{t|t-1}C^T(C\Sigma_{t|t-1}C^T + \Sigma_{\boldsymbol{v}})^{-1}C\Sigma_{t|t-1}.\end{aligned} \tag{3.192}$$

The Kalman gain, K_t, is defined as

$$K_t = \Sigma_{t|t-1}C^T(C\Sigma_{t|t-1}C^T + \Sigma_{\boldsymbol{v}})^{-1}. \tag{3.193}$$

3.7.16.3 Time Update

Using $\boldsymbol{x}_{t+1} = A\boldsymbol{x}_t + \boldsymbol{w}_t$ we can propagate the mean and covariance

$$\begin{aligned}\hat{\boldsymbol{x}}_{t+1|t} &= A\hat{\boldsymbol{x}}_{t|t} \\ \Sigma_{t+1|t} &= A\Sigma_{t|t}A^T + \Sigma_{\boldsymbol{w}}.\end{aligned} \tag{3.194}$$

3.7.16.4 Summary of the Kalman Filter Recursion

The steps are [7]

1. Initialization: starting with prior mean and covariance for the initial state

$$\hat{\boldsymbol{x}}_{0|-1} = \boldsymbol{\mu}_0,\ \Sigma_{0|-1} = \Sigma_0.$$

2. Measurement Update: then apply the measurement update

$$\begin{aligned}K_t &= \Sigma_{t|t-1}C^T(C\Sigma_{t|t-1}C^T + \Sigma_{\boldsymbol{v}})^{-1} \\ \hat{\boldsymbol{x}}_{t|t} &= \hat{\boldsymbol{x}}_{t|t-1} + K_t(\boldsymbol{y}_t - C\hat{\boldsymbol{x}}_{t|t-1}) \\ \Sigma_{t|t} &= \Sigma_{t|t-1} - K_tC\Sigma_{t|t-1}\end{aligned}$$

3. Time Update: then propagate the mean and covariance

$$\begin{aligned}\hat{\boldsymbol{x}}_{t+1|t} &= A\hat{\boldsymbol{x}}_{t|t} \\ \Sigma_{t+1|t} &= A\Sigma_{t|t}A^T + \Sigma_{\boldsymbol{w}}.\end{aligned}$$

4. Repeat for $t = 0, 1, 2\ldots$

3.7.16.5 Riccati Recursion

After applying the measurement and time updates to $\Sigma_{t|t-1}$, the Riccati recursion with initial conditions $\boldsymbol{\mu}_0$, Σ_0 gives us the state and covariance updates

$$\begin{aligned}\hat{\boldsymbol{x}}_{t+1|t} &= A\hat{\boldsymbol{x}}_{t|t-1} + A\Sigma_{t|t-1}C^T(C\Sigma_{t|t-1}C^T + \Sigma_{\boldsymbol{v}})^{-1}(\boldsymbol{y}_t - C\hat{\boldsymbol{x}}_{t|t-1}) \\ \Sigma_{t+1|t} &= A\Sigma_{t|t-1}A^T - A\Sigma_{t|t-1}C^T(C\Sigma_{t|t-1}C^T + \Sigma_{\boldsymbol{v}})^{-1}C\Sigma_{t|t-1}A^T + \Sigma_{\boldsymbol{w}}.\end{aligned} \tag{3.195}$$

The covariance update is independent of the measurement $\boldsymbol{y}$, allowing the estimation error covariance to be determined prior to observing any measurements. The Riccati recursion $\Sigma_{t|t-1}$ represents the state prediction error covariance at time t and progresses forward in time.

3.7.16.6 Steady-state Kalman Filter

When (C, A) is observable and $(A, \Sigma_{\boldsymbol{w}})$ is controllable the Riccati recursion converges to a steady-state value $\hat{\Sigma}$. The steady-state error covariance, $\hat{\Sigma}$, satisfies the Algebraic Riccati Equation (ARE)

$$\widehat{\Sigma} = A\widehat{\Sigma}A^T - A\widehat{\Sigma}C^T\left(C\widehat{\Sigma}C^T + \Sigma_v\right)^{-1}C\widehat{\Sigma}A^T + \Sigma_{\boldsymbol{w}}. \tag{3.196}$$

In steady-state, the Kalman filter becomes time-invariant, and the state update simplifies to

$$\widehat{\boldsymbol{x}}_{t+1|t} = A\widehat{\boldsymbol{x}}_{t|t-1} + L(\boldsymbol{y}_t - \widehat{\boldsymbol{y}}_{t|t-1}) \tag{3.197}$$

where $L = A\widehat{\Sigma}C^T\left(C\widehat{\Sigma}C^T + \Sigma_v\right)^{-1}$ is called the observer gain [7]. If we define the state estimation error, $\widetilde{\boldsymbol{x}}_{t|t-1} = \boldsymbol{x}_t - \widehat{\boldsymbol{x}}_{t|t-1}$, then we have

$$\widetilde{\boldsymbol{x}}_{t+1|t} = A\widetilde{\boldsymbol{x}}_{t|t-1} + \boldsymbol{w}_t - L(C\widetilde{\boldsymbol{x}}_{t|t-1} + \boldsymbol{v}_t) = (A - LC)\widetilde{\boldsymbol{x}}_{t|t-1} + \boldsymbol{w}_t - L\boldsymbol{v}_t. \tag{3.198}$$

The error dynamics are governed by $A - LC$ (where (C, A) is observable and $(A, \Sigma_{\boldsymbol{w}})$ is controllable). This error is driven by the process $\boldsymbol{w}_t - L\boldsymbol{v}_t$, which is an independent, identically distributed (IID) process with zero mean and covariance $\Sigma_{\boldsymbol{w}} + L\Sigma_v L^T$[7].

Example: Spring-mass-damper system [7].
Consider the mass-spring-damper system shown in Figure 3.97 with $k_i = 2,\ b_i = 0.1$, and $m_i = 1$. The six states represent the positions and velocities of the masses. Noisy measurements are available for masses 1 and 2, and our goal is to estimate the position of mass 3. The dynamics of the system are

$$\dot{\boldsymbol{x}}(t) = A\boldsymbol{x}(t) + B\boldsymbol{w}(t), \boldsymbol{y}(t) = C\boldsymbol{x}(t) + \boldsymbol{v}(t)$$

where

$$A = \begin{bmatrix} 0 & 0 & 0 & 1 & 0 & 0 \\ 0 & 0 & 0 & 0 & 1 & 0 \\ 0 & 0 & 0 & 0 & 0 & 1 \\ -4 & 2 & 0 & -0.2 & 0.1 & 0 \\ 2 & -4 & 2 & 0.1 & -0.2 & 0.1 \\ 0 & 2 & -2 & 0 & 0.1 & -0.1 \end{bmatrix}, B = \begin{bmatrix} 0 & 0 & 0 \\ 0 & 0 & 0 \\ 0 & 0 & 0 \\ 1 & 0 & 0 \\ 0 & 1 & 0 \\ 0 & 0 & 1 \end{bmatrix}, C = \begin{bmatrix} 1 & 0 & 0 & 0 & 0 & 0 \\ 0 & 1 & 0 & 0 & 0 & 0 \end{bmatrix}.$$

The system is discretized using a timestep of $T = 0.1$, resulting in

$$\boldsymbol{x}_{t+1} = A_d\boldsymbol{x}_t + B_d\boldsymbol{w}_t, \ \boldsymbol{y}_t = C_d\boldsymbol{x}_t + \boldsymbol{v}_t$$

with Gaussian noise $\boldsymbol{w}_t \sim N(0,\sigma_{\boldsymbol{w}}^2 I), \boldsymbol{v}_t \sim N(0,\sigma_v^2)$ where $\sigma_{\boldsymbol{w}} = 0.2$ and $\sigma_v = 0.1$.

Solution:
Figure 3.98 simulates a particular trajectory of the system and compares confidence intervals for the state estimates. The no-measurement confidence interval is computed solely using the time update step of the Kalman filter, given by $\boldsymbol{x}_{t+1} = A_d\boldsymbol{x}_t$ and $\Sigma_{t+1} = A_d\Sigma_t A_d^T + W$. The 90% confidence interval for mass 3's position and velocity does not improve over time without measurements. In contrast, the Kalman filter provides a more refined confidence interval by incorporating

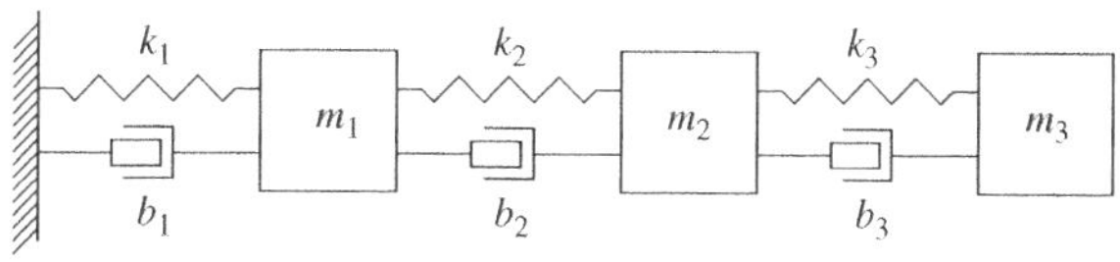

Figure 3.97 Spring-mass-damper system.

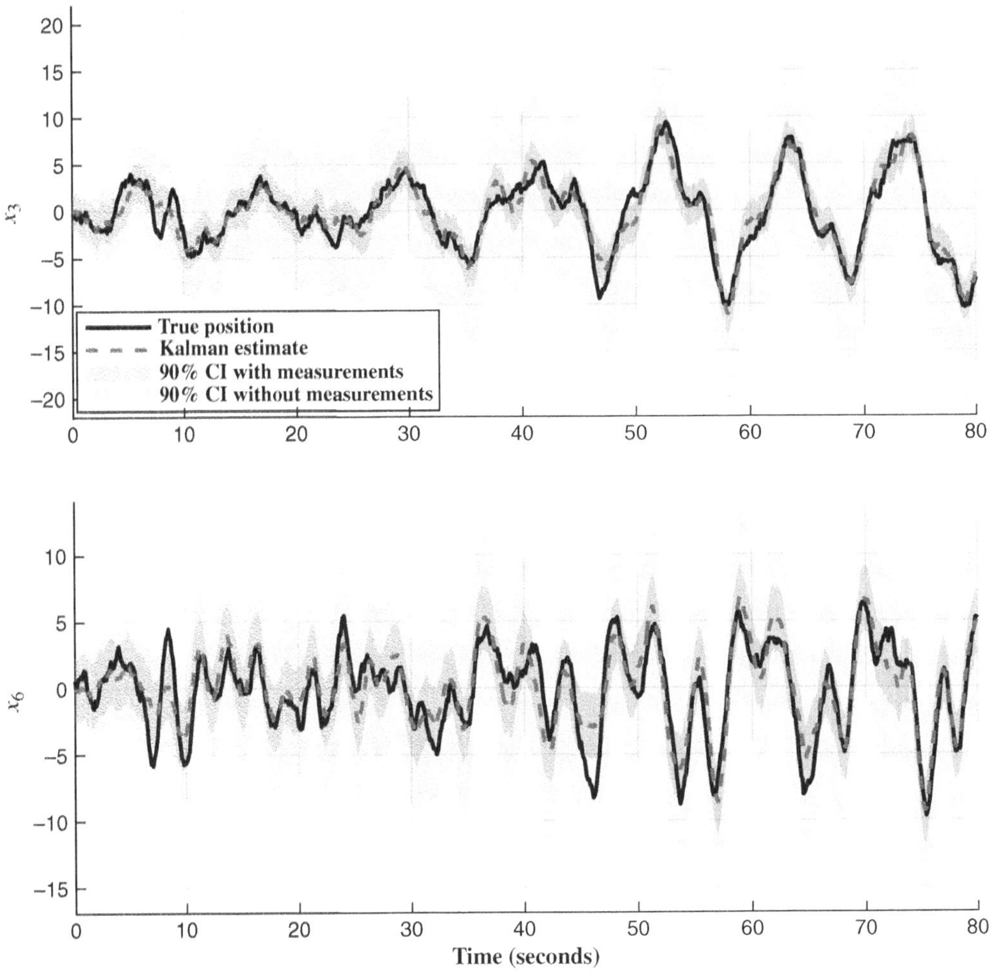

Figure 3.98 Simulation of spring-mass-damper system by using $k_i = 2$, $b_i = 0.1$, and $m_i = 1$. A discretization with a timestep of 0.1 was used. Process noise has a standard deviation $\sigma_w = 0.2$ and the measurement noise has $\sigma_v = 0.1$.

measurements. The Kalman filter typically achieves a tighter confidence interval due to its use of both prediction and update steps.

References

1 Gene F. Franklin, J. David Powell, and Abbas Emami-Naeini, *Feedback Control of Dynamic Systems*, 7th ed., Pearson, 2014.

2 Ken Dutton, Steve Thompson, and Bill Barraclough, *The Art of Control Engineering*, Addison-Wesley, 1997.

3 A. Aldo Faisal and Cheng Soon Ong, *Mathematics for Machine Learning*, Cambridge University Press, 2019.

4 Bong Wie, *Space Vehicle Dynamics and Control*, American Institute of Aeronautics and Astronautics, 1998.

5 Sigurd Skogestad, *Simple analytic rules for model reduction and PID controller tuning*, J. Process Control, 13(4), 291–309, 2003.

6 Sanjay Lall, *Lecture notes for E207A*, Department of Aeronautics and Astronautics, Stanford University, Stanford, CA, 2004.

7 Stephen Boyd, *Lecture notes for EE363*, Department of Electrical Engineering, Stanford University, Stanford, CA, 2008.

8 Claire Tomlin, *Lecture notes for E205*, Department of Aeronautics and Astronautics, Stanford University, Stanford, CA, 2003.

4

Sources of Aberrations

An optical beam is a directed flow of light distinguished by a specific spatial profile and directionality. An optical beam may encounter various sources of aberration that can affect both its spatial profile and direction. This chapter examines the various sources of aberrations frequently observed in beam control systems.

4.1 Vibration and Jitter

Optical vibration and jitter refer to the periodic and random oscillations and perturbations in the position of an optical beam, particularly in the angular position of the beam that affects the beam propagation performances. There are many factors that can cause optical vibrations and jitter, including vibration in optical components, fluctuations in the optical path, mechanical vibrations, thermal changes in the environment, and atmospheric turbulence. Managing optical vibration and jitter is crucial for ensuring the precision and reliability of optical beam control systems.

4.1.1 Platform Jitter

Platform jitter is the common source of optical jitter encompassing many aberration sources present in the beam control system platform. The platform jitter is a result of random vibration of many different frequency components as well as narrowband frequency vibrations due to the rotating components and structural vibrations of the platform. Platform jitter is often represented by the power spectral density (PSD) of vibration signal power against spectrum of frequencies. For a window of $-N \leq n \leq N$ with the vibration signal x_n that is sampled at discrete times $t_n = t_0 + (n\,\Delta t)$ for a total measurement period $T = (2N + 1)\Delta t$, the power spectral density becomes

$$S_{xx}(f) = \lim_{N\to\infty} \frac{(\Delta t)^2}{T} \left| \sum_{n=-N}^{N} x_n e^{-j2\pi f_n \Delta t} \right|^2 \tag{4.1}$$

Figure 4.1 shows the example of acceleration PSD plot of a ground vehicle driving on a highway. Parseval's theorem states that the variance of the platform jitter can be estimated by integrating the power spectral density with respect to frequency such that

$$\sigma_x^2 = \int_0^{\infty} S_{xx}(f)\,df \tag{4.2}$$

Optical Beam Control: Imaging Satellites and Laser Systems, First Edition.
Brij N. Agrawal, Jae Jun Kim and Sachin Agrawal.

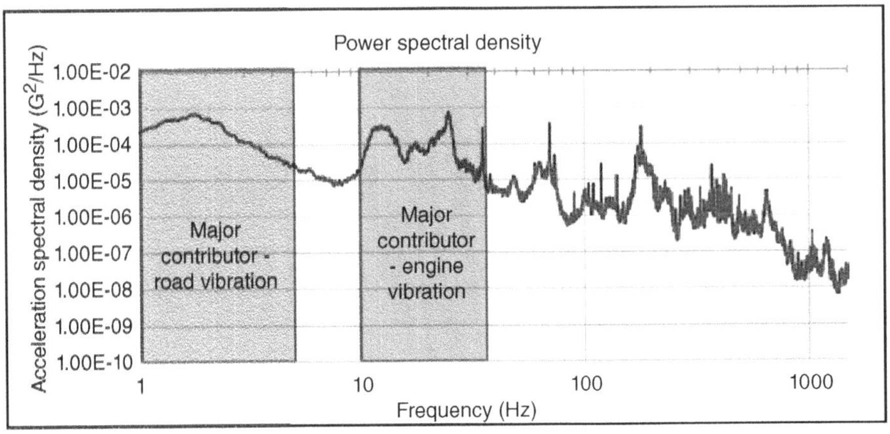

Figure 4.1 Example power spectral density of a highway vehicle.

To reduce the effect of platform vibration, optical beam control systems often employ a vibration isolation table. Assuming a linear transfer function of the vibration isolation table $G(s)$, PSD of the isolated jitter on the table $S_{yy}(f)$ is related to the platform jitter $S_{xx}(f)$ as

$$S_{yy}(f) = |G(f)|^2 S_{xx}(f) \tag{4.3}$$

where the complex variable s is related to the frequency f by $s = j2\pi f$. The residual jitter after the vibration isolation can be computed from Eq (4.2) such that $\sigma_y^2 = \int_0^\infty S_{yy}(f)\, df$. Equation 4.3 may also be used to determine the PSD of the actively controlled jitter, where $G(s)$ denotes the sensitivity transfer function of the control system in this case.

4.1.2 Atmospheric Jitter

Atmospheric jitter refers to the rapid fluctuations of the position of the beam caused by turbulence in the atmosphere, mostly due to large-scale turbulent cells that deviate the beam from the direction of propagation. Atmospheric jitter can lead to image blurring, loss of beam signal quality, and challenges in tracking celestial objects. The effect of atmospheric jitter in optical beam propagation is further discussed in Section 4.3.

4.2 Optical Aberrations

Optical aberrations resulting from the optical system design and optical components can be classified into two categories: monochromatic aberrations and chromatic aberrations. Monochromatic aberrations arise from the geometric design of lenses or mirrors and can occur during both reflection and refraction of light. In contrast, chromatic aberrations are a consequence of light dispersion when multiple wavelengths are present in the optical beam.

4.2.1 Defocus and Spherical Aberration

Defocus aberration refers to an axial deviation from the best focus location as shown in Figure 4.2. Defocus is a common aberration that can be fixed by changing the position of the lens or detector, or sometimes by changing the lens's curvature.

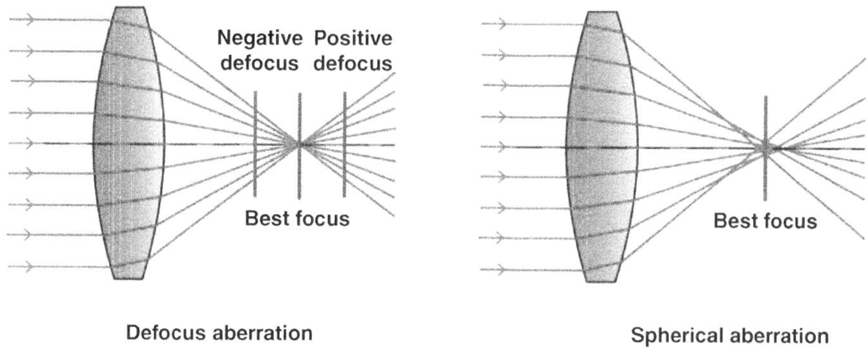

Figure 4.2 Defocus aberration and spherical aberration.

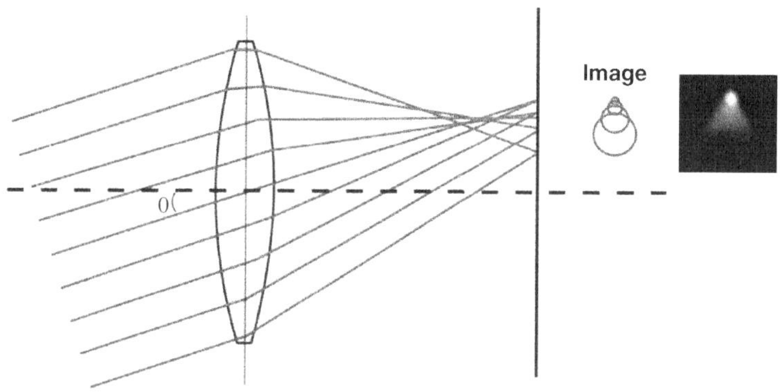

Figure 4.3 Coma aberration.

Spherical aberration is an optical effect that occurs when light rays entering a lens at different points are not focused to the same point on the optical axis. Spherical aberration is caused by using a lens or a mirror with a spherical surface. Spherical aberration can also occur when light passes through an interface between substances with different refractive indexes. Spherical aberration can be eliminated in optical systems using a parabolic primary mirror or a combination of mirrors and lenses with appropriate shapes.

4.2.2 Coma Aberration

Coma shown in Figure 4.3 refers to an aberration caused by an off-axis optical beam in which the point source object will appear to have a tail (coma) like a comet. Coma aberration is sensitive to the aperture size and could be reduced by adjusting the position of the aperture stop or using a combination of lenses symmetric about a central aperture stop.

4.2.3 Astigmatism Aberration

Astigmatism is an optical aberration that occurs when a lens or mirror is unable to focus light rays in different directions in the same plane. Depending on the angle of the off-axis rays entering the lens, a point image is oriented in either of two different directions, tangentially or sagittally,

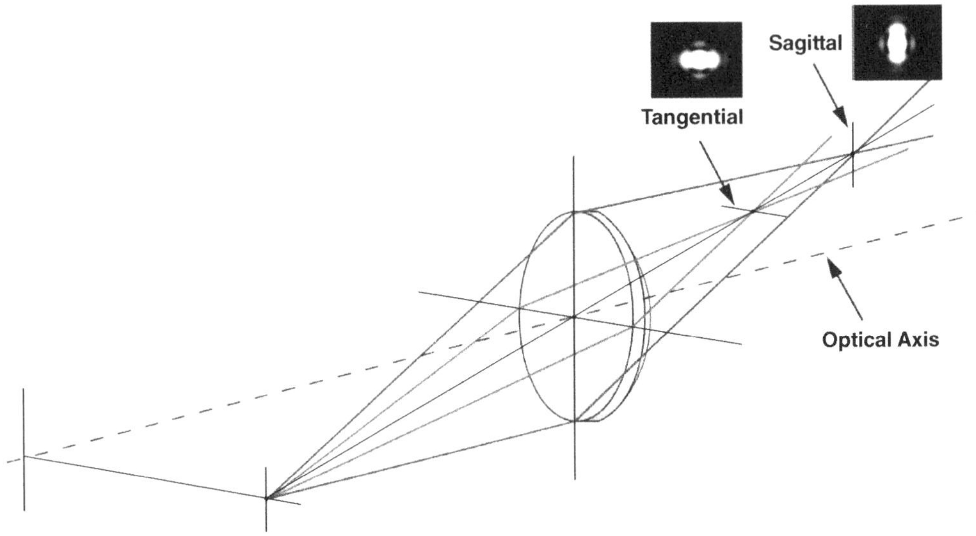

Figure 4.4 Astigmatism aberration.

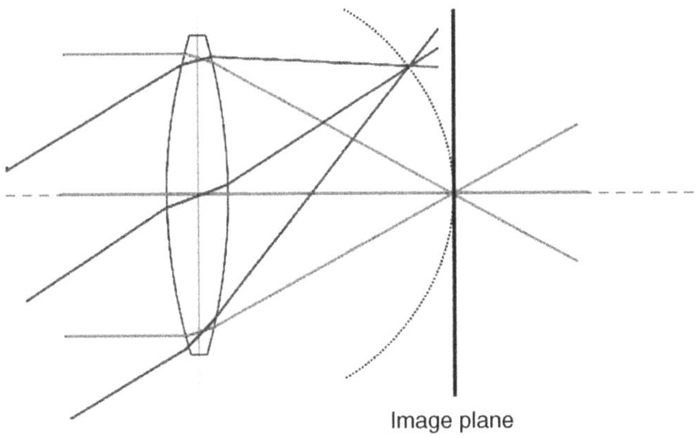

Figure 4.5 Field curvature.

as shown in Figure 4.4. The effect of astigmatism aberrations can be minimized by using multiple mirrors and lenses with appropriate shapes, aperture sizes, and indices of refraction.

4.2.4 Field Curvature

Field curvature shown in Figure 4.5 (also known as Petzval field curvature) describes the optical aberration in which an optical beam with different field angles cannot be brought properly into focus on a flat image plane. This causes parts of the image to be in focus while others are out of focus, depending on where they fall in the field. Field curvature usually affects the sharpness of an image across its plane and is most noticeable when the depth of field is shallow or when the lens is used at wide apertures.

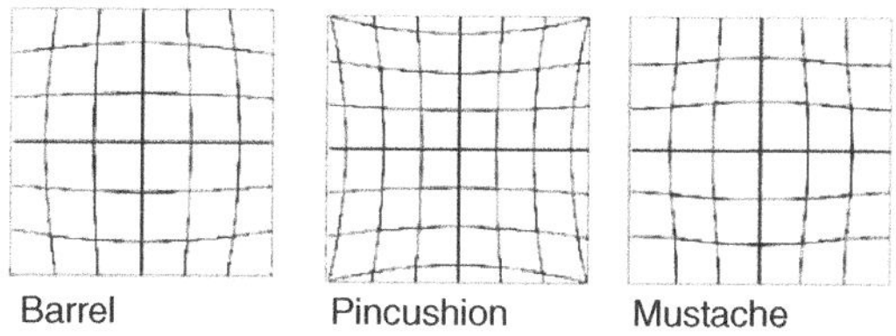

Figure 4.6 Different types of image distortion.

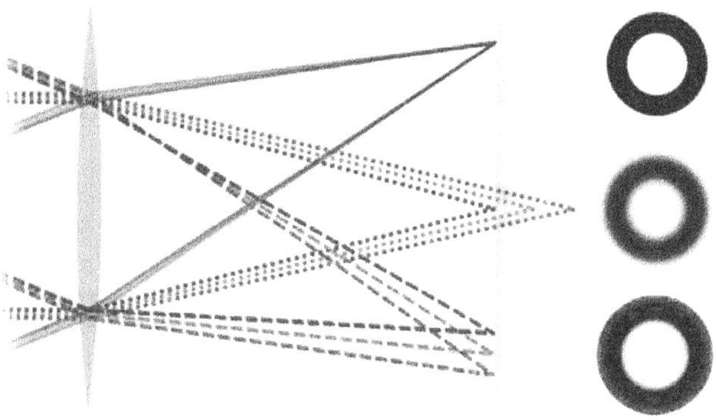

Figure 4.7 Chromatic aberration.

4.2.5 Image Distortion

Image distortion refers to a variation in magnification based on the field angle that changes the shape of an image. Image distortion alters the shape of objects, causing them to appear stretched, compressed, or warped. The main cause of image distortion is related to the optical design and the type of optics used. Image distortion is distinct from other optical aberrations that affect image sharpness without altering the shape or structure of the object in the image. Figure 4.6 shows different types of image distortion.

4.2.6 Chromatic Aberration

Chromatic aberration, shown in Figure 4.7 and known as color fringing, occurs when an optical system cannot focus all wavelengths of light onto a single point. This causes color distortions or mismatches, especially in high-contrast situations. Chromatic aberration is mainly caused by dispersion, when the refractive index varies depending on the wavelength of light. Chromatic aberration could be reduced by using low-dispersion glass elements and achromatic lenses to bring different wavelengths of light to the same focal point.

4.2.7 Optical Surface Errors

Optical surface error refers to the deviation of a surface of lenses and/or mirrors from its ideal shape. The quality of optical surfaces is critical because even slight changes can cause significant optical beam aberrations. Optical surface errors are typically measured using the reflected wavefront error tests shown in Figure 4.8. The common measure of optical surface errors is summarized in Table 4.1.

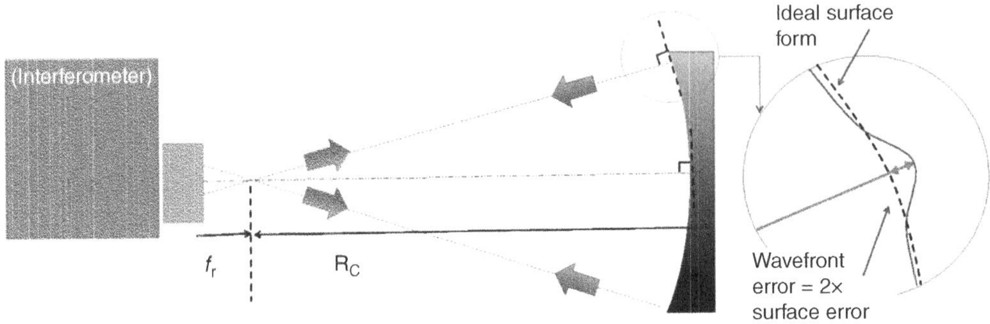

Figure 4.8 Reflected wavefront error test *Source*: [2].

Table 4.1 Measure of optical surface errors.

Measure of surface errors	Description
Peak-to-valley	Maximum distance between the highest and lowest points on the surface, typically measured in waves as a fraction of the test light's wavelength
Root mean square	Statistical measurement of the surface's deviations that provides a more comprehensive description of the surface's quality, typically measured in waves
Slope error	The angular deviation of the surface's normal from its ideal shape
Roughness	High spatial frequency deviations from its ideal shape that cause scattering of light, typically measured in Å
Scratch-dig	Scratches are any markings or tearing of the optical surface. Digs are pits, small rough spots, bubbles, or inclusions in the optical surface. Scratch and dig are specified as SS–DD, where SS represents the maximum allowable scratch width and DD represents the maximum allowable dig diameter, given in microns.

4.3 Air Turbulence

Air turbulence, also referred to as atmospheric turbulence, is the turbulent motion of the atmosphere caused by moisture and temperature gradients. This motion causes fluctuations in the index of refraction, leading to a variety of effects on an optical beam, including temporal irradiance fluctuations (scintillation) and phase fluctuations.

4.3.1 Kolmogorov Theory

Kolmogorov theory is a statistical framework that describes the behavior of turbulent fluid flows and the transfer of energy within them. It was developed in the 1940s by Russian mathematician Andrey Kolmogorov. According to Kolmogorov theory, the atmosphere can be considered as a viscous fluid with two distinct states of motion: laminar and turbulent. In laminar flow, the movement of fluid is orderly, and the velocity is consistent across different layers. However, in turbulent flow, the fluid's velocity and other properties become irregular and disordered. The interaction and blending of different regions of the fluid, where fast-moving areas mix with slower-moving areas, creates complex, unpredictable flow patterns called turbulent eddies.

The energy cascade theory, illustrated in Figure 4.9, explains how energy in a turbulent flow moves from large eddies to smaller eddies. Energy is initially introduced at the largest scales of

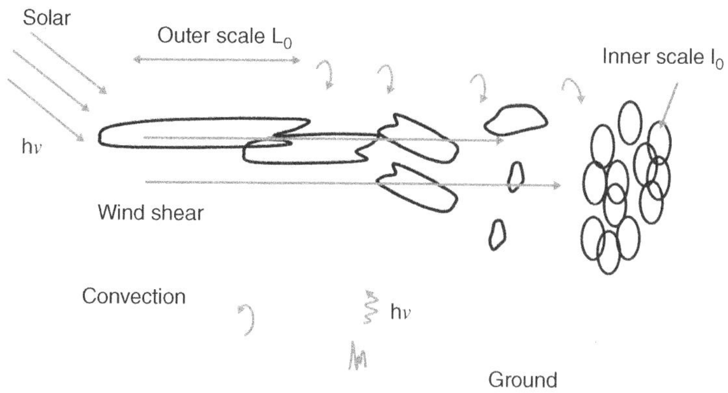

Figure 4.9 Energy cascade concept with outer scale and inner scale turbulent eddies.

motion called "outer scale (L_0)." These large-scale turbulent eddies are the first structures to form and contain the most energy. They are created by external factors, such as wind and temperature gradients, and have a significant amount of coherent motion. As the large-scale eddies move and interact, their energy is transferred to smaller eddies. The cascade of energy from large to small eddies happens in a region called the inertial range. In this range, the eddies are not directly affected by viscosity, and the energy simply moves down through the scales without being dissipated. Eventually, as the eddies become smaller, they reach a point where the size of the eddies is so small that the effects of viscosity become significant. This is called the "inner scale" (l_0), where the smallest turbulent eddies exist before the viscosity of the fluid causes energy dissipation in the smallest eddies. The outer scale L_0 typically ranges from tens to hundreds of meters, while the inner scale l_0 is as small as a few millimeters.

4.3.2 Air Turbulence Parameters

4.3.2.1 Refractive Index Structure Parameter

The refractive index structure parameter, commonly referred to as C_n^2, is a measure of the fluctuations in the refractive index of the atmosphere caused by turbulent eddies at various scales. C_n^2 is a fundamental parameter widely used in modeling and compensating for turbulence-induced optical wavefront distortions, including beam wander, scintillation, phase aberrations, and image degradation. Refractivity of air due to temperature fluctuations is written as

$$N \equiv (n-1) \times 10^6 = 77.6\left(1 + \frac{7.52 \times 10^{-3}}{\lambda^2}\right)\left(\frac{P}{T}\right) \tag{4.4}$$

where P is pressure in millibars, T is temperature in K, λ is wavelength in microns, and n is index of refraction. The above equation can be written for the index of refraction n as

$$n = 1 + 77.6\left(1 + \frac{7.52 \times 10^{-3}}{\lambda^2}\right)\left(\frac{P}{T}\right) \times 10^{-6} \tag{4.5}$$

Since λ is a small number for optical beams, the refractivity of air is mostly dependent on temperature fluctuations, which can be approximated as

$$\delta N \cong -77.6 \times \left(\frac{P}{T^2}\right)\delta T \tag{4.6}$$

Since air turbulence is a random process that needs to be described by statistical quantities, a statistical description of the turbulence is given by the structure function of the index of refraction, which represents the mean square refractive index difference between two points in space. If turbulence is considered locally homogeneous and isotropic within the inertial subrange, then refractive-index structure function is given by

$$D_n(R) = \left\langle \left[n\left(\vec{R_1}\right) - n\left(\vec{R_2}\right)\right]^2 \right\rangle = C_n^2 R^{\frac{2}{3}} \tag{4.7}$$

where $n\left(\vec{R_1}\right)$ and $n\left(\vec{R_2}\right)$ denote the refractive index at two points separated by distance R and C_n^2 is the refractive index structure parameter. Similarly, the phase structure function, which represents the mean square phase difference between two points, is defined as

$$D_\phi(R) = \left\langle \left[\phi\left(\vec{R_1}\right) - \phi\left(\vec{R_2}\right)\right]^2 \right\rangle = \left[2.91\left(\frac{2\pi}{\lambda}\right)^2 \int_{\text{Path}} C_n^2(z)dz\right] R^{\frac{5}{3}} \tag{4.8}$$

In the more general case, a wave structure function, which includes amplitude effects in addition to the phase, is needed to provide a generalized way to describe the variations of optical waves due to atmospheric turbulence.

The refractive index structure function is related to the power spectral density of refractive index by the following expression.

$$\Phi_n(\kappa) = \frac{1}{4\pi^2\kappa^2}\int_0^\infty \frac{\sin(\kappa R)}{\kappa R} \cdot \frac{\partial}{\partial R}\left(R^2 \frac{\partial D_n(R)}{\partial r}\right) dR \tag{4.9}$$

where κ denotes the spatial wavenumber of the fluctuations. Under the assumption of statistically homogeneous and isotropic turbulence, the power spectrum becomes

$$\Phi_n(\kappa) = 0.033 C_n^2 \kappa^{-\frac{11}{3}} \tag{4.10}$$

which is referred to as the Kolmogorov spectrum. The Kolmogorov spectrum is valid only over the inertial subrange

$$\frac{1}{L_0} < \kappa < \frac{1}{l_0} \tag{4.11}$$

As a measure of atmospheric turbulence strength, typical values of C_n^2 range from 10^{-15} in condition of weak turbulence to 10^{-13} when turbulence is strong. If measured over a 24-hour period, the C_n^2 shows a diurnal cycle as shown in Figure 4.10 with a well-defined peak during mid-day hours and

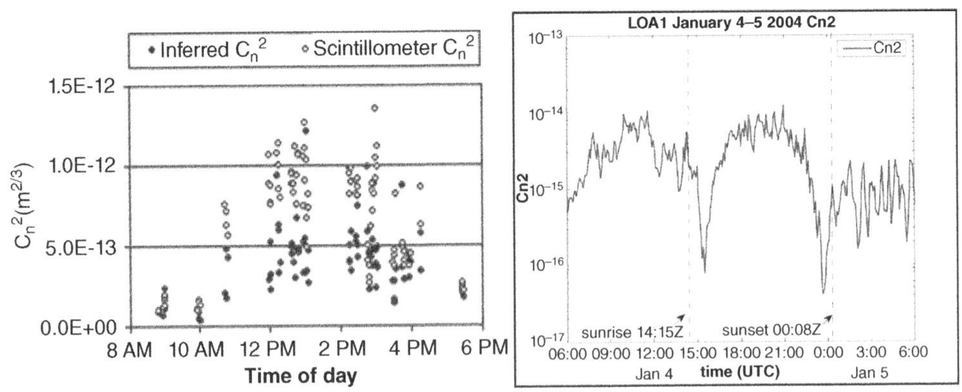

Figure 4.10 Diurnal cycle of C_n^2.

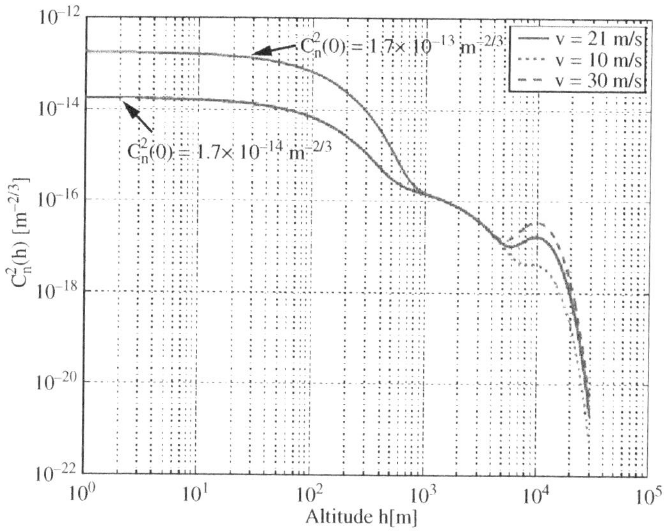

Figure 4.11 H–V model for C_n^2.

minima near sunrise and sunset. Over a short time interval and constant height above the ground, it may be reasonable to assume that C_n^2 is constant.

For vertical and slant propagation paths, the refractive-index structure parameter varies as a function of the height above ground. Several C_n^2 profile models are used to accurately describe the variation in optical turbulence strength with altitude. One of the most widely used models is the Hufnagel–Valley (H–V) model, shown in Figure 4.11, with its corresponding equation given by

$$C_n^2(h) = 0.00594\left(\frac{v_{pw}}{27}\right)^2 \cdot \left(10^{-5}h\right)^{10}\exp\left(\frac{-h}{1000}\right) + 2.7 \cdot 10^{-16}\exp\left(\frac{-h}{1500}\right) + C_n^2(0)\exp\left(\frac{-h}{100}\right) \tag{4.12}$$

where v_{pw} is the rms windspeed in m/s.

4.3.2.2 Fried's Atmospheric Coherence Length

Fried's atmospheric coherence length, also known as Fried's parameter (r_0), is a widely used metric for quantifying the strength of atmospheric turbulence at a given location. The Fried parameter characterizes the size of the area over which the optical beam remains relatively undistorted by atmospheric turbulence.

Figure 4.12 illustrates observing an object at a zenith angle ζ measured from the vertical path z where the guide star is located. The Fried parameter is defined as

$$r_0 = \left[0.423k^2 \sec\zeta \int_{\text{Path}} C_n^2(z)\,dz\right]^{-\frac{3}{5}} \tag{4.13}$$

where $k = 2\pi\lambda$ is the wavenumber. At the visible and infrared wavelengths, the Fried's parameter r_0 typically varies from 20 cm for weak turbulence to 1 cm for strong turbulence.

In a vacuum, optical system performance is limited by the diffraction set by the effective size of the aperture. In turbulence, optical system performance is further limited by atmospheric turbulence. Table 4.2 summarizes the effect of turbulence compared to the diffraction-limited optical performance in angular resolution, imaging spot size, and coherent beam spot size.

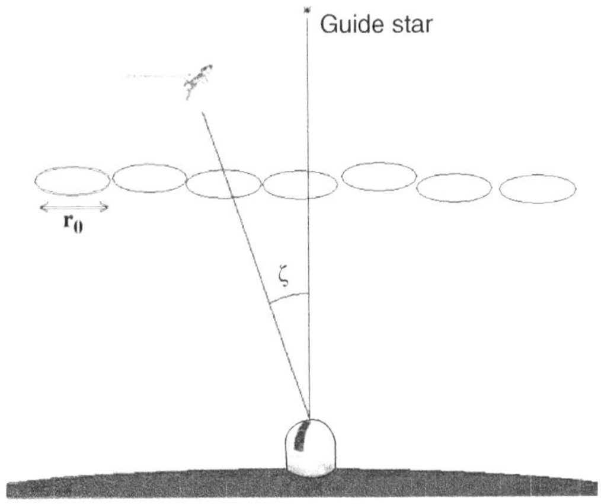

Figure 4.12 Illustration of fried parameter.

Table 4.2 Optical system performance limited by diffraction versus atmospheric turbulence.

	Diffraction	Turbulence
Angular resolution	$\frac{\lambda}{D}$	$\frac{\lambda}{r_0}$
Imaging spot size (f: focal length)	$\lambda\frac{f}{D}$	$\lambda\frac{f}{r_0}$
Coherent beam spot size at range R	$\frac{\lambda}{D}R$	$\frac{\lambda}{r_0}R$

4.3.2.3 Greenwood Frequency

The Greenwood frequency is a measure of the rate at which Kolmogorov atmospheric turbulence changes with time. The Greenwood frequency determines how quickly the adaptive optics system must respond to compensate for turbulence. As a rule of thumb, the closed-loop bandwidth of the adaptive optics system should be at least 10 times the Greenwood frequency. Greenwood frequency is defined by

$$f_G = 2.31\lambda^{-\frac{6}{5}}\left[\sec\zeta\int_{\text{Path}} C_n^2(z)\upsilon_w(z)^{\frac{5}{3}}dz\right]^{\frac{3}{5}} \tag{4.14}$$

where υ_w is the wind velocity as a function of altitude.

4.3.2.4 Isoplanatic Angle

The isoplanatic angle defines the size of the region around a given point in the sky where the optical beam aberration due to atmospheric turbulence is uniform. Within this angular area, the light from different parts of the object such as a star or a planet experience roughly the same optical beam aberrations due to turbulence. Outside of this angle, the distortion may vary significantly, as the turbulence effects can differ depending on the line of sight and the distance the light travels through different layers of the atmosphere. Isoplanatic angle θ_0 is related to C_n^2 by the following equation.

$$\theta_0 = 0.058\lambda^{\frac{6}{5}}(\sec\zeta)^{-\frac{8}{5}}\left[\int_{\text{Path}} C_n^2(z)z^{\frac{5}{3}}dz\right]^{-\frac{3}{5}} \tag{4.15}$$

4.3.2.5 Rytov Parameter

The Rytov parameter, also known as log amplitude variance, is a measure of the variance of the phase of the wave as it propagates through atmospheric turbulence. The Rytov parameter is often used to characterize the severity of the turbulence's impact on the light wave's phase and intensity. Rytov parameter is given by the following equation.

$$\sigma_\chi^2 = 4.806\lambda^{-\frac{7}{6}}(\sec\zeta)^{11/6}\int_{\text{Path}} C_n^2(z)z^{\frac{5}{6}}dz \tag{4.16}$$

4.3.3 Non-Kolmogorov Atmospheric Turbulence

Kolmogorov turbulence theory is an idealized theory that assumes local isotropic turbulence. When the turbulence is anisotropic, where the turbulence properties depend on direction or inhomogeneous where turbulence varies across space, atmospheric turbulence often does not follow the Kolmogorov spectrum exactly. Some experiments also show that there are deviations from Kolmogorov's 5/3 power law, especially when the turbulence is affected by large-scale weather patterns, wind shear, or varying temperature and pressure gradients.

4.3.4 Effect of Air Turbulence in Optical Beam Propagation

Atmospheric turbulence caused by the refractive index variations of the air at various scales has a significant impact on optical beam propagation. The effect of atmospheric turbulence on optical beam propagation is as follows.

4.3.4.1 Beam Jitter

Beam jitter (also known as beam wander) is mostly due to the large-scale turbulent eddies cells that deviate from the beam around the direction of propagation. It has a time constant on the order of

$$\tau = \frac{\text{Beam size}}{\text{Wind speed}} \tag{4.17}$$

A shorter time constant means the atmospheric tilt changes rapidly. For a coherent beam with a diameter D, the variance of the beam tilt angle (in rad^2) due to the atmospheric jitter is related to Fried's parameter as follows

$$\sigma_{\theta G}^2 = 0.170\left(\frac{D}{r_0}\right)^{\frac{5}{3}}\left(\frac{\lambda}{D}\right)^2 \tag{4.18}$$

$$\sigma_{\theta Z}^2 = 0.182\left(\frac{D}{r_0}\right)^{\frac{5}{3}}\left(\frac{\lambda}{D}\right)^2 \tag{4.19}$$

where $\sigma_{\theta G}$ denotes the standard deviation of G tilt, which is defined as the average tilt over the aperture that can be measured by calculating the centroid of the beam intensity distribution, and $\sigma_{\theta Z}$ denotes the standard deviation of Z tilt that includes more precise wavefront tilt information with the presence of higher order wavefront aberrations. The difference between G tilt and Z tilt is illustrated in Figure 4.13 for the case of a 1D wave.

4.3.4.2 Beam Spreading

Atmospheric turbulence causes an optical beam to spread beyond its initial divergence and diffraction limits. Short-term beam spreading (i.e. over a short time window) is primarily due to small turbulent eddies that distort the wavefront of the propagating optical beam. Long-term beam spreading is the result of both short-term beam spreading and beam jitter, as shown in Figure 4.14.

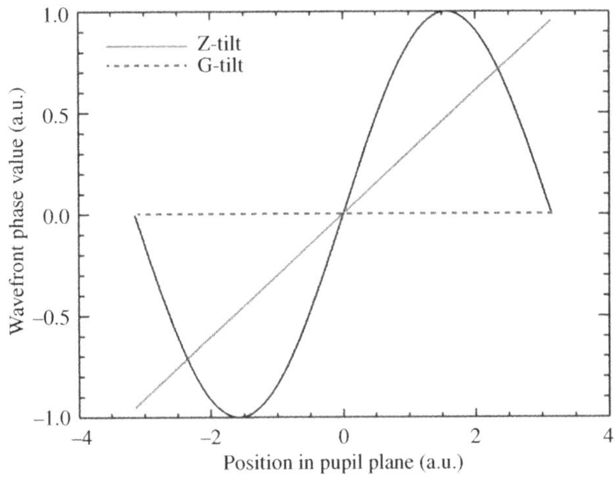

Figure 4.13 G tilt versus Z tilt for 1D Wave.

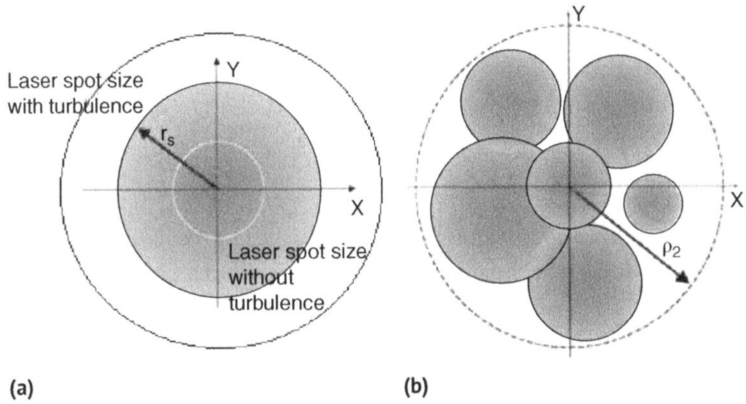

Figure 4.14 Beam spread due to atmospheric turbulence (a) short-term beam spread, (b) long-term beam spread.

4.3.4.3 Scintillation

Atmospheric turbulence introduces random fluctuations in the refractive index of the air. These fluctuations cause phase variations across the wavefront, which can reduce the spatial coherence of the beam. As the beam undergoes diffraction and scattering due to turbulence, phase variations can lead to varying interference effects across the wavefront, which manifest as scintillation, as depicted in Figure 4.15. Interference can cause bright spots, due to constructive interference, as well as dark spots, due to destructive interference, in the far-field pattern.

Scintillation includes both the temporal variation in received irradiance, such as star twinkling, and spatial variation within a receiver aperture, such as speckle. The degree of scintillation is described by the scintillation index, which is defined as the variance of irradiance fluctuations scaled by the square of the mean irradiance.

$$\sigma_I^2 = \frac{\langle I^2 \rangle}{\langle I \rangle^2} - 1 \tag{4.20}$$

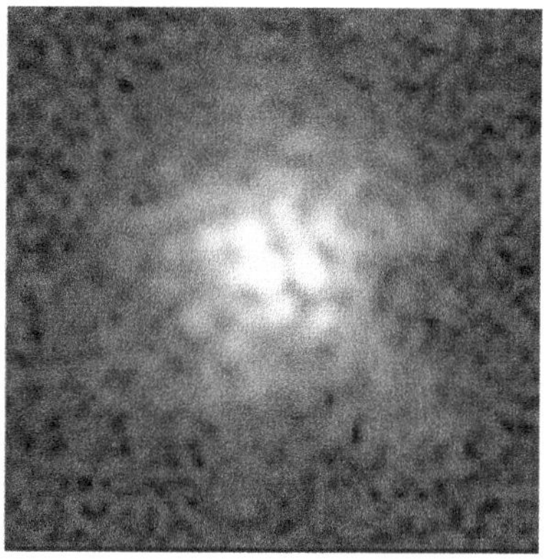

Figure 4.15 Scintillated optical beam.

Assuming a coherent optical beam, the diffraction angle due to the inner scale of the turbulence is

$$\theta = \frac{\lambda}{l_0} \tag{4.21}$$

For a propagation distance R, scintillation starts to appear when the beam spread due to diffraction is larger than the inner scale of the turbulence such that

$$R\theta = \frac{R\lambda}{l_0} > l_0 \; or \; \sqrt{R\lambda} > l_0 \tag{4.22}$$

This condition differentiates between tilt and scintillation of the optical beam. In the case of a coherent optical beam, variance of intensity at a propagating distance L for a plane wave and a spherical wave is written as

$$\sigma_\chi^2 = 0.56k^{7/6}\int_0^L C_n^2\left(z\right)\left(L-z\right)^{\frac{5}{6}} dz \text{ for } a \text{ plane wave} \tag{4.23}$$

$$\sigma_\chi^2 = 0.56k^{7/6}\int_0^L C_n^2\left(z\right)\left(z/L\right)^{5/6}\left(L-z\right)^{5/6} dz \text{ for a spherical wave} \tag{4.24}$$

where σ_χ^2 is the variance of intensity. For a constant C_n^2, e.g. horizontal path, the variance of intensity becomes

$$\sigma_\chi^2 = 0.307C_n^2k^{7/6}L^{11/6} \text{ for a plane wave} \tag{4.25}$$

$$\sigma_\chi^2 = 0.124C_n^2k^{7/6}L^{11/6} \text{ for a spherical wave} \tag{4.26}$$

4.4 Measure of Aberrations

4.4.1 Point Spread Function

The point spread function (PSF) describes how a point source of light is represented in an image produced by an optical system. The PSF can be also interpreted as spatial "impulse response" of the optical system at focus. In an ideal optical system with a circular aperture, the PSF is the Airy disk characterized by a central bright spot surrounded by concentric rings of decreasing intensity.

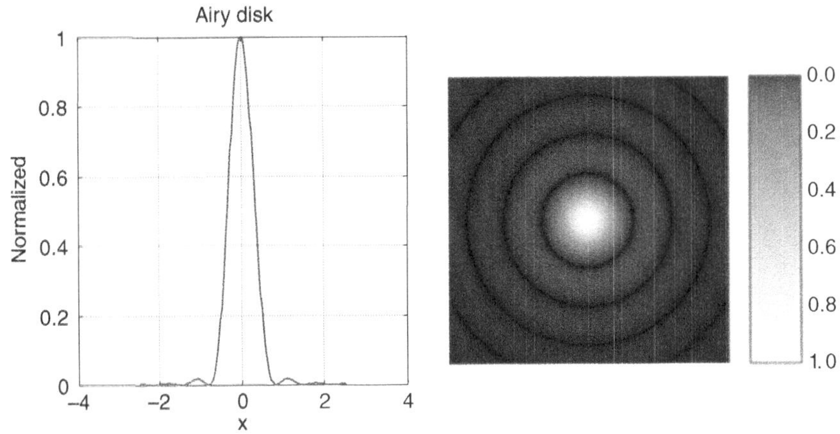

Figure 4.16 Airy disk point spread function. *Source*: Sakurambo/Wikimedia Commons/Public Domain.

For coherent, monochromatic light, the Airy disk shown in Figure 4.16 is described by the following equation:

$$I(\theta) = I_0 \left(\frac{2J_1(x)}{x}\right)^2 \tag{4.27}$$

where θ is the diffraction angle, $x = \frac{2\pi}{\lambda} r \sin(\theta)$ and $J_1(x)$ is the first-order Bessel function.

The pupil function $P(x, y)$ describes the complex transmission or reflection characteristics of the optical system's aperture, typically expressed as a function of spatial coordinates in the aperture plane.

$$P(x, y) = A(x, y) \exp(i\phi(x, y)) \tag{4.28}$$

where $A(x, y)$ is the amplitude and $\phi(x, y)$ is the phase of the pupil function. The pupil function and the PSF are related as

$$PSF(x', y') = |\mathcal{F}\{P(x, y)\}|^2 \tag{4.29}$$

4.4.2 Optical Transfer Function and Modular Transfer Function

The optical transfer function (OTF) describes how well an optical system can transfer different spatial frequencies from the object to the image. It essentially measures the system's ability to reproduce the detail and contrast of the original scene in the resulting image. The OTF is defined as the Fourier transform of the PSF such that

$$OTF\,(v) = \mathcal{F}(PSF) \tag{4.30}$$

The modulation transfer function (MTF) is given by the absolute value of the optical transfer function such that

$$MTF(f_x, f_y) = |OTF\,(f_x, f_y)| \tag{4.31}$$

MTF curves can be developed for different parts of a system, e.g. optics, detectors (i.e. pixel geometry), etc. The net MTF for system is the cascade (pointwise multiplication) of all subsystem MTFs as shown in Figure 4.17.

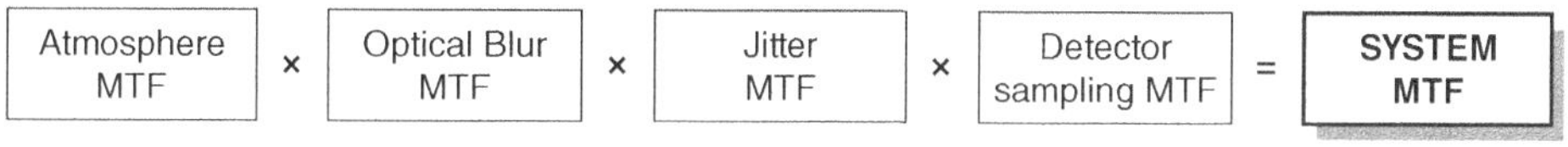

Figure 4.17 MTF multiplication.

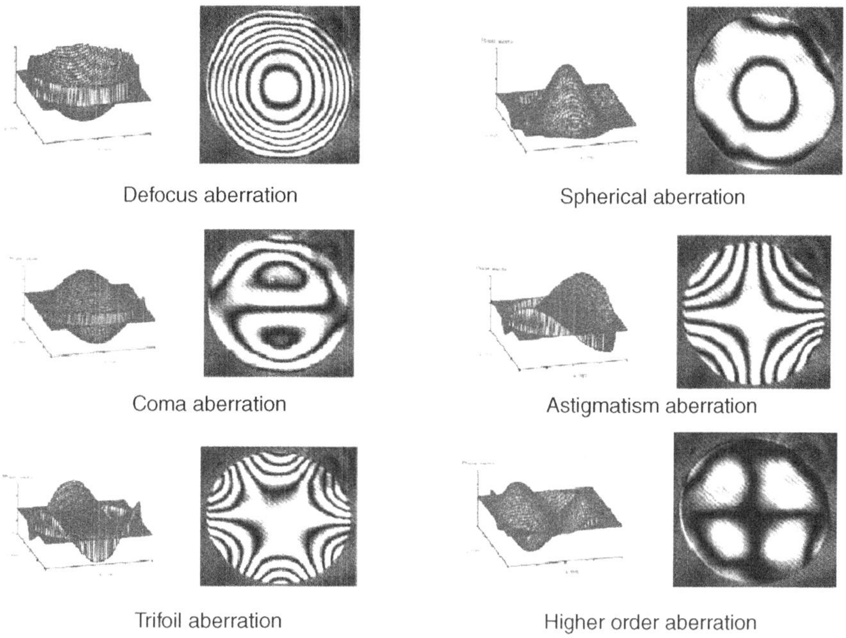

Figure 4.18 Example wavefront aberration. *Source*: Brij N. Agrawal.

4.4.3 Encircled and Ensquared Energy

Traditional measurements of MTF are well correlated with optical performance, however, in some cases, the characterization of encircled (ensquared energy) could also serve as a measure of optical performance, particularly for telescopes and lenses that focus light into a circular or square area that corresponds to an image sensor. Encircled energy is defined as

$$E(r) = \frac{\text{Energy within radius } r}{\text{Total energy}} \tag{4.32}$$

In optical systems and astronomy, encircled energy describes the fraction of light or energy that is contained within a specific region, typically within a certain radius, in relation to the total energy or light output of the system.

4.4.4 Wavefront Error

Wavefront error is commonly used to represent optical aberrations as well as atmospheric turbulence. In geometrical optics, the wavefront is defined as a continuous surface that is orthogonal to all rays in the light beam, representing the propagation of light as it moves through space. Figure 4.18 shows various types of wavefront aberrations measured using an interferometer.

The zonal representation and modal representation are two different ways to express and analyze wavefront errors. In the zonal representation, the wavefront is described in terms of zones or regions across the aperture or optical field. The wavefront is divided into several concentric rings or rectangular zones, and each zone is treated as having a uniform phase shift or distortion. The distortion in each zone is typically characterized by the average phase deviation from the ideal wavefront for that zone.

In modal representation, the wavefront is expressed as a sum of orthogonal modes or basis functions that describe the shape of the wavefront. These modes are usually chosen to represent common wavefront distortions, such as Zernike polynomials shown in Figure 4.19, which are commonly used to describe wavefront aberrations in optics.

$$Z_n^m(\rho,\ \varphi) = R_n^m(\rho)\cos(m\varphi) \tag{4.33}$$

$$R_n^m(\rho) = \sum_{k=0}^{\frac{n-m}{2}} \frac{(-1)^k (n-k)!}{k!\left(\frac{n+m}{2}-k\right)!\left(\frac{n-m}{2}-k\right)!}\rho^{n-2k} \tag{4.34}$$

where $R_n^m(1) = 1$

R. J. Noll [4] developed a description of the average spatial content of turbulence-induced wavefront distortion in terms of Zernike polynomials and showed that the piston-removed wavefront distortion for Kolmogorov turbulence averaged over an aperture of diameter D expressed as a mean square value in units of radians^2 of optical phase is

$$\langle \phi^2 \rangle = 1.0299\left(\frac{D}{r_0}\right)^{5/3} \tag{4.35}$$

Various Zernike mode contributions in the atmospheric turbulence are summarized in Table 4.3.

4.4.5 Strehl Ratio

A simple definition of Strehl ratio is the ratio of the central intensity of the achieved PSF (image of a non-resolved source) to a PSF having no aberrations. Using the extended Marechal

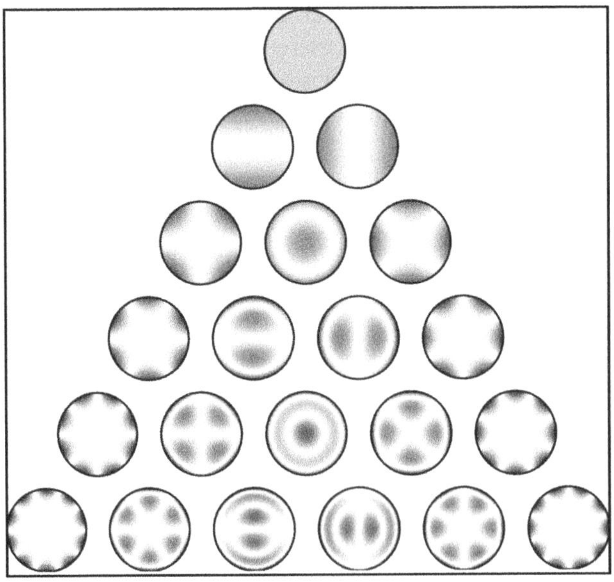

Figure 4.19 Example Zernike polynomials.

Table 4.3 Contribution of different Zernike modes for atmospheric turbulence.

Aberration	Contribution	Residual distortion
All terms	$1.0299(D/r_0)^{5/3}$	$1.0299(D/r_0)^{5/3}$
Z_2 X-tilt	$0.4479(D/r_0)^{5/3}$	$0.582(D/r_0)^{5/3}$
Z_3 Y-tilt	$0.4479(D/r_0)^{5/3}$	$0.134(D/r_0)^{5/3}$
Z_4 De-focus	$0.0232(D/r_0)^{5/3}$	$0.111(D/r_0)^{5/3}$
Z_5 X-astigmatism	$0.0232(D/r_0)^{5/3}$	$0.0880(D/r_0)^{5/3}$
Z_6 Y-astigmatism	$0.0232(D/r_0)^{5/3}$	$0.0649(D/r_0)^{5/3}$
Z_7 3rd order coma	$0.0061(D/r_0)^{5/3}$	$0.0587(D/r_0)^{5/3}$
Z_8 3rd order coma	$0.0061(D/r_0)^{5/3}$	$0.0525(D/r_0)^{5/3}$
Z_9 Coma	$0.0062(D/r_0)^{5/3}$	$0.0463(D/r_0)^{5/3}$
Z_{10} Coma	$0.0062(D/r_0)^{5/3}$	$0.0401(D/r_0)^{5/3}$
Z_{11}	$0.0024(D/r_0)^{5/3}$	$0.0377(D/r_0)^{5/3}$
Z_{12}	$0.0025(D/r_0)^{5/3}$	$0.0352(D/r_0)^{5/3}$

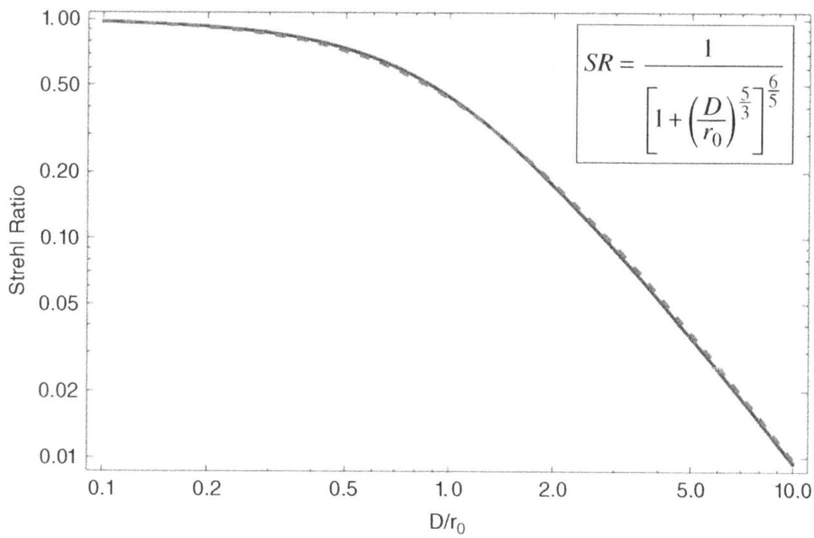

Figure 4.20 Strehl ratio versus D/r_0.

approximation, for an rms phase aberration σ_ϕ of 2 radians or less, the Strehl ratio is represented by the formula:

$$SR = e^{-\sigma_\phi^2} \tag{4.36}$$

For uncompensated turbulence, the Strehl ratio and resolving power (the volume under the OTF curve) are shown in Figure 4.20 along with the formula in Eq. (4.37).

Strehl ratio can also be derived in terms of Fried's coherence length r_0

$$SR = \frac{16}{\pi}\int_0^1 u\left[\cos^{-1}u - \sqrt{1-u^2}\right]\exp\left[-3.44\left(\frac{D}{r_0}\right)^{\frac{5}{3}} u^{\frac{5}{3}}\right]du \tag{4.37}$$

As with the MTF, the net Strehl ratio of an optical system can be approximated by the product of the individual Strehl ratios corresponding to each independent source of optical aberration.

References

1 https://vru.vibrationresearch.com/lesson/what-is-the-psd/.
2 https://apertureos.com/off-axis/specifying-wavefront-vs-surface-error-aspheres/.
3 OKO guide to adaptive optics, https://www.okotech.com/images/pdfs/catwww4.pdf.
4 Noll, R. J. (1976). Zernike polynomials and atmospheric turbulence. J. Opt. Soc. Am., 66(3), 207–211. https://doi.org/10.1364/JOSA.66.000207.

5

Vibration Isolation and Jitter Control

5.1 Introduction

Jitter in imaging satellites and laser systems refers to the small, rapid, and often unintended variations or fluctuations in the pointing direction or alignment of the satellite or laser beam. In imaging satellites, jitter can degrade the quality of captured images by causing blurring or distortion, especially in high-resolution imaging where precise pointing is critical. In laser systems, jitter can affect the accuracy and effectiveness of laser targeting or communication by causing the laser beam to wander off-target. For high performance imaging and laser systems, jitter requirements are very tight and significant effort is spent in design to meet these performance requirements.

Figure 5.1 shows where the line-of-sight (LOS) pointing as a function of time is shown in the left graph. The right picture compares the images taken when significant micro-vibration perturbations were present, when some corrective measures had been applied afterward, and when the imager LOS was undisturbed during the picture-taking.

As seen in Figure 5.1, the pixel size in focal plane of the imaging satellite is $25 \times 25\ \mu m^2$. The allowable motion is 5 μm, one-fifth of the pixel size. As will be shown in Chapter 7, this results in allowable angular motion of focal plane motion divided by focal length. The impact of jitter on

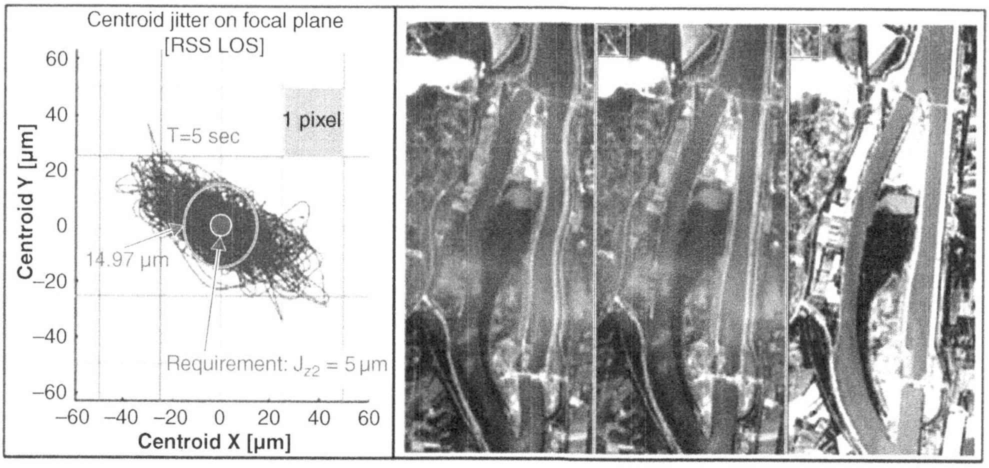

Figure 5.1 Example of LOS pointing errors and the resulting effects on image quality. *Source*: NASA/Public Domain.

Optical Beam Control: Imaging Satellites and Laser Systems, First Edition.
Brij N. Agrawal, Jae Jun Kim and Sachin Agrawal.

image resolution will be discussed in Chapter 7. The impact of jitter on laser system performance will be discussed in Chapter 8. The focus of this chapter is on techniques to control vibration and jitter.

Vibration sources on an imaging spacecraft include reaction or momentum wheels (discrete frequencies related to wheel speeds and harmonics), control moment gyros (CMG) (discrete frequencies related to CMG speed and harmonics), solar drive motors, thermal shocks (especially solar arrays transitioning from eclipse to non-eclipse and vice versa), thruster firing (create jitter), cryogenic coolers (discrete frequencies), and slew maneuvers (affect entire spacecraft structure and associated frequencies).

Different techniques have been developed for reducing jitter for optical payloads. One technique is to increase the damping of the spacecraft support structure to reduce vibration response by adding passive and/or active damping. The second technique is to introduce isolation between vibration source and the spacecraft. The third technique is to isolate the optical payloads from the spacecraft. The fourth technique is to reduce jitter in the optical beam by using fast steering mirrors (FSM) and jitter sensors. For spacecraft requiring very low jitter, a combination of these techniques is used. Figure 5.2 [3] shows these different vibration control techniques.

Vibration and jitter control techniques can be divided into two groups: Passive and active. Passive control relies on the inherent damping properties of materials or mechanical systems to reduce vibrations or jitters. It typically involves adding passive components such as dampers, isolators, or absorbers to the system. Passive control systems are simpler and often cheaper to implement compared to active control systems.

They work by dissipating the energy of the vibrations, thereby reducing their amplitude or impact on the system. However, passive systems are generally less adaptable to changing conditions and may not provide precise control as active systems. Active control involves the use of sensors and actuators to actively monitor and counteract vibrations or jitter in real-time. Sensors detect vibrations or jitters, and actuators generate forces or motions to counteract them. Active control systems

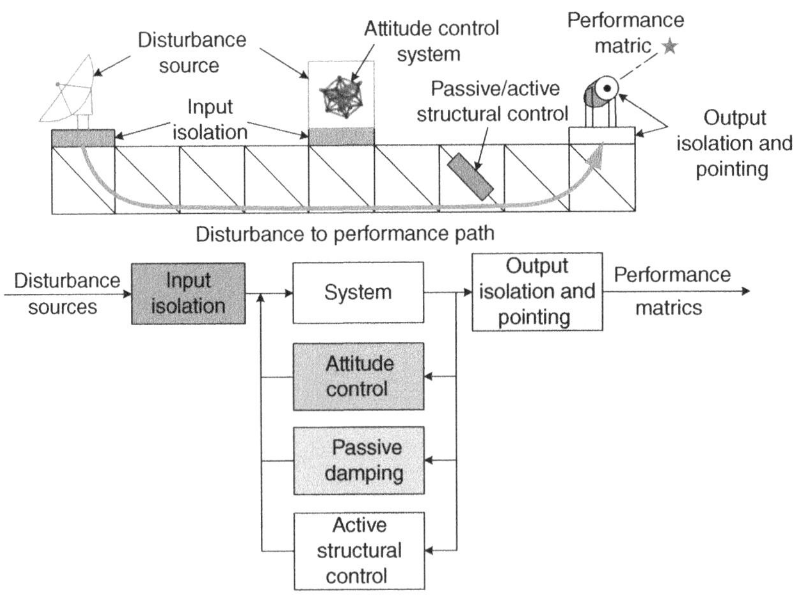

Figure 5.2 Vibration control techniques.

can adapt in real-time to changing conditions, making them more effective in some situations where passive control methods may not suffice. They can provide more precise control and are often capable of suppressing a wider range of vibration frequencies. However, active control systems are typically more complex and expensive to implement compared to passive systems.

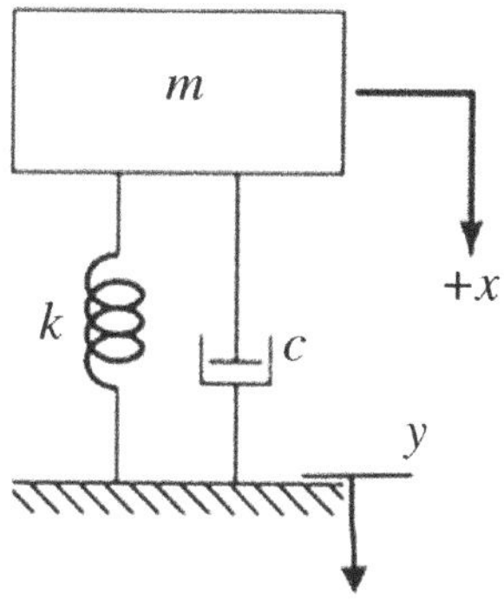

Figure 5.3 Single degree of freedom system under base excitation.

5.2 Passive Vibration and Jitter Control

Passive vibration and jitter control techniques can be explained by the following simple example. Let us consider a mechanical system as shown in Figure 5.3.

It consists of a mass, m, attached by means of spring, (K), and a viscous dashpot, (C), to a movable support. The mass is constrained to translational motion in the x direction. The position of the mass at any instant is fully described by a single parameter, x, which is the displacement of the mass from the position of static equilibrium. The support is moving by displacement y. The equation of motion of the mass is

$$m\ddot{x} + c(\dot{x} - \dot{y}) + k(x - y) = 0 \tag{5.1}$$

Let the base excitement be

$$y = Y_0 \sin\omega t \tag{5.2}$$

Let the response of the mass be

$$x = X_p \sin(\omega t - \phi) \tag{5.3}$$

where X_p is the response amplitude and ϕ is the phase difference from the base excitation. Solving Eq. 5.1, we get the vibration amplitude transmission from the platform to the mass m is

$$T_d = \frac{X_p}{Y_o} = \frac{\sqrt{(2\zeta r)^2 + 1}}{\sqrt{(1 - r^2)^2 + (2\zeta r)^2}} \tag{5.4}$$

where

$\omega_n = \sqrt{k/m}$, natural frequency, rad/s.

The ratio $\zeta = c/c_c$ is defined as the fraction of critical damping and is

$$\zeta = \frac{c}{2\sqrt{mk}} = \frac{c}{2m\omega_n} \tag{5.5}$$

$r = \frac{\omega}{\omega_n}$, frequency ratio.

The following figure shows vibration transmissibility as a function of frequency ratio and damping.

From Figure 5.4, it is clear that the transmissibility is less than unity, i.e., vibration isolation, when the frequency ratio is higher than $\sqrt{2}$. So, a vibration-sensitive platform should be supported on low-frequency support structure. Support structure should have high damping to reduce vibration transmission with a frequency ratio equal to 1. So, the frequency and damping of the support structure should be selected based on the disturbance frequency range.

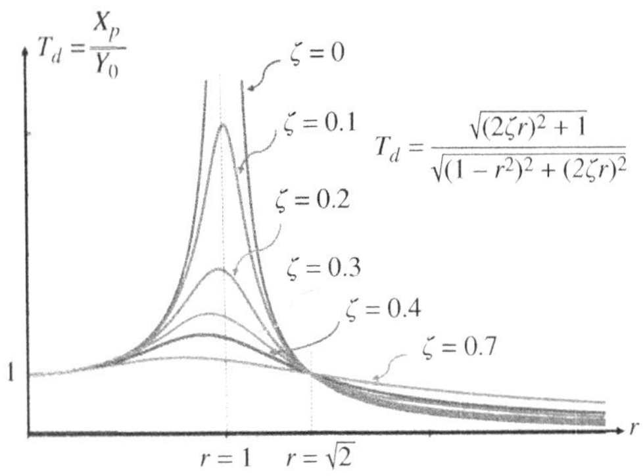

Figure 5.4 Transmissibility of vibration as a function of frequency ratio.

Table 5.1 Passive damping and associated design methods.

	Type of treatment					
	Strut/link dampers	**Constrained layers**	**Tuned-mass dampers**	**Joint/inter face dampers**	**Embedded dampers**	**Unconstrained or free layers**
Target modes	Global	Member bending and extension	Narrow frequency range, any mode shape	Local or global	Member bending and extension	Member extension and bending
Primary design method	Modal strain energy	Modal strain energy	Complex eigen values	Joint test or modal strain energy	Modal strain energy	Hand calc. modal strain energy
Special features	Removable, light weight	Flexible, wide bandwidth	Low cost, low volume	Low weight, low volume	Embedded, low outgassing	Low cost, easy design

5.2.1 Passive Vibration Control

All passive damping treatments share a common goal: absorb significant amounts of strain energy in modes of interest and dissipate this energy through some energy-dissipation mechanism. The effectiveness of all passive damping methods varies with frequency and temperature, though some more than others. For each of the basic passive damping mechanisms, there are several choices for implementation, which can be divided into two major categories: discrete and distributed. Table 5.1 summarizes the primary passive damping concepts along with their typical uses.

5.2.1.1 Viscoelastic Devices

Viscoelastic materials are widely used for damping in both commercial and aerospace applications. Viscoelastic materials are elastomeric materials whose long-chain molecules cause them to convert mechanical energy into heat when they are deformed. Complex arithmetic provides a convenient means for keeping track of the phase angle by which an imposed cyclic stress leads to the resulting cyclic strain. Figure 5.5 shows typical viscoelastic material configurations.

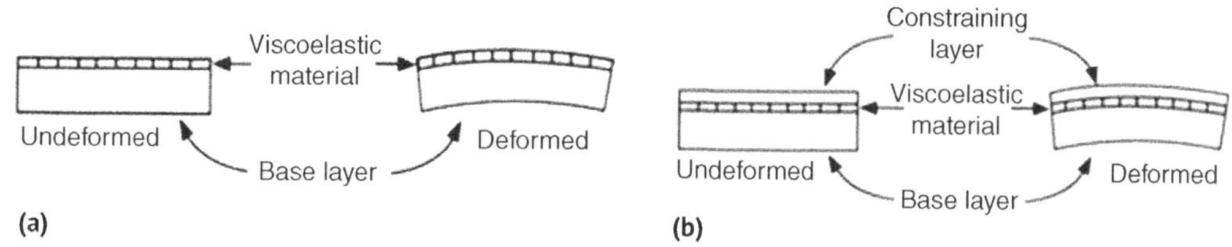

Figure 5.5 Typical viscoelastic material configuration (a) Free layer treatment and (b) constrained layer treatment. *Source*: [3]/IEEE.

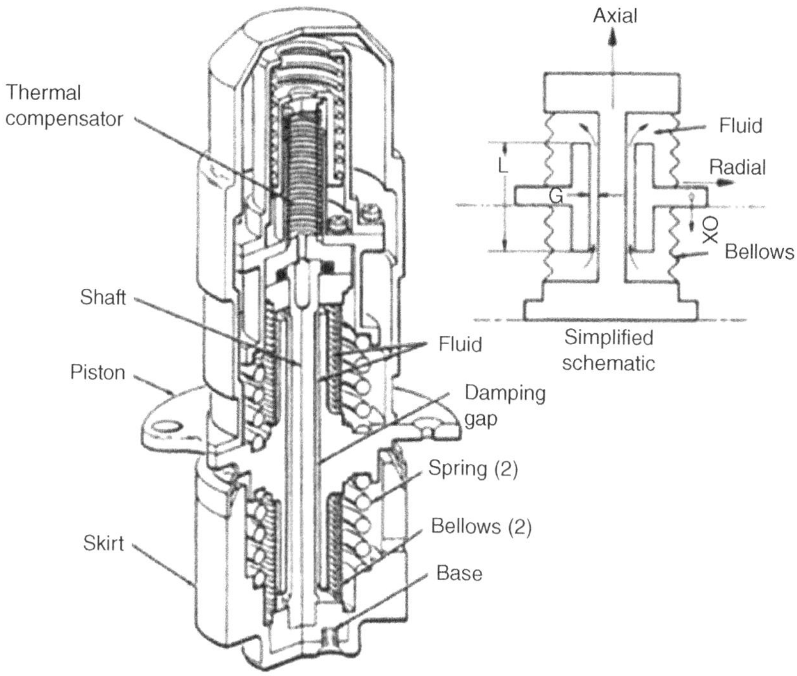

Figure 5.6 Basic viscous fluid damper. *Source*: [3]/IEEE.

5.2.1.2 Viscous Devices

A basic viscously damped isolator is depicted in Figure 5.6. Fluid of a specified viscosity is hermetically sealed within two sets of bellows and constrained to flow between them through an orifice when motion occurs. The fluid flow through the orifice provides the damping. A third set of bellows is included to accommodate thermal expansion and contraction of the fluid. Springs are included to provide the desired isolator stiffness. For safety, the device is encased in an outer skirt to protect against contamination from ruptured bellows. This unit isolator can be simply modeled as a spring in parallel with a spring-damper. By changing the damping (fluid viscosity) of the isolator, the in-band peaking of the isolation characteristic can be changed, but at the cost of isolation bandwidth. Variations in temperature have a similar effect because of their effect on fluid viscosity.

The D-strut is one of the most common applications of viscous systems. In a classical spring-mass system, i.e., one that can be modeled with a single spring in parallel with a dashpot, the amplitude of damping varies with a change in the damping coefficient; however, a more

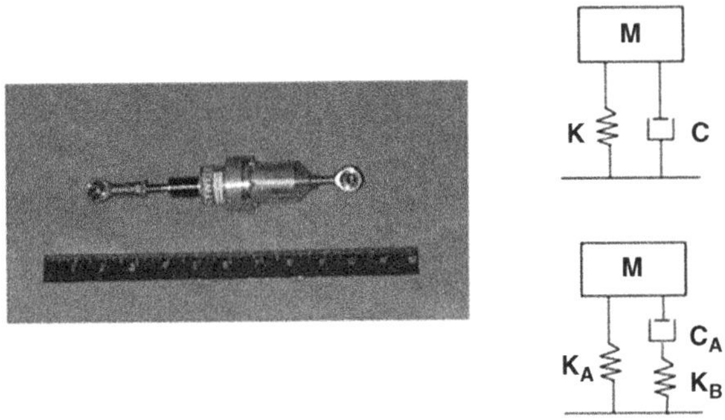

Figure 5.7 Passive D-strut. *Source*: [3]/IEEE.

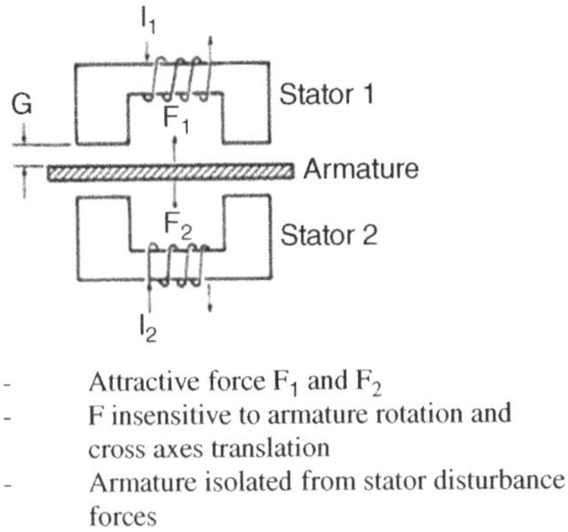

Figure 5.8 Magnetic isolation description. *Source*: [3]/IEEE.

general representation of a viscous damper or D-strut includes a second spring in series with the dashpot, as shown in Figure 5.7. The series spring (K_B) is usually very stiff compared to the parallel spring (K_A) and is often ignored; however, in many cases it can be used to improve performance. Its existence converts the system from a two-parameter to a three-parameter system, enhancing the opportunity for tuning the system or making the system adaptable. In the three-parameter system, changing the damping coefficient changes the frequency at which damping occurs, rather than the amplitude of the damping. At low frequency, the impedance is given by static stiffness K_A, and at high frequency the stiffness increases to $K_A + K_B$. In the transition frequency range, the impedance is complex, and the phase is leading. The magnitude of damping is related to the phase angle θ or the loss factor η. Note that the peak magnitude of η is related to the ratio of stiffness parameters K_B/K_A and is independent of the damping coefficient C_A. The frequency at which the peak phase angle (ϕ) or loss factor (η) occurs is ω_0, and is inversely proportional to C_A [1, 2, 3].

5.2.1.3 Magnetic Devices

Magnetic isolation is achieved by using two opposing electromagnets acting on an armature plate, as shown in Figure 5.8. The key to technology is in noncontacting interference. This avoids the coupling paths, which are inherent in mechanical isolators, enabling magnetic suspensions to readily achieve a very large dynamic isolation range (greater than 80 dB). A typical magnetic isolator consists of two opposing horseshoe electromagnets acting on an armature plate. The electromagnets are mounted to the base, the passive armature is attached to the payload, and they are separated by a large air gap. Advantages of magnetic isolation are:

- Noncontacting interface
- Faster isolation roll-off of degrees of freedom (typically 100 dB/decade)
- Ease of adjustment (mode changes, structural uncertainties)
- Automatic alignment and centering (unbalance compensation)
- Noncontaminating materials

5.2.1.4 Passive Piezoelectrics

Piezoelectric ceramic materials have the unique ability to produce a strain when subjected to an electrical charge, and, conversely, they produce a charge when strained mechanically (piezoelectric theory explained in subsequent section). This has made them popular as actuators and sensors in active vibration control systems (ACVS). This dual transformation ability also makes them useful as passive structural dampers. In passive energy dissipation applications, the electrodes of the piezoelectric are shunted with a passive electric circuit. The electrical circuit is designed to dissipate the electrical energy that has been converted from mechanical energy by the piezoelectric.

5.2.1.5 Tuned Mass Dampers

The use of tuned mass dampers (TMD), as shown in Figure 5.9, is another widely used passive vibration damping treatment. These devices are viscously damped second-order systems appended to a vibrating structure. Proper selection of the parameters of these appendages tunes the TMD to one of the natural frequencies of the flexible structure, resulting in the addition of damping to that resonance. TMDs can target any mode, including the first, and add a considerable amount of damping to it.

The mass of the TMD and its stiffness are chosen to make the natural frequency of the TMD match the resonant frequency of the beam to be damped. The damping effectiveness of a TMD is dependent on its damping ratio and the frequency it is tuned at. The lack of large enough damping ratio in the TMD results in breaking the resonant mode, to which the device is tuned for, to two underdamped modes – a phenomenon known as "mode splitting." A damping ratio between 20 and 45 provides an effective damping treatment using a TMD. The capability of tuning mass dampers to damp out a particular mode of vibration is due to the 90° phase lead that the TMD adds to the open-loop system at the natural frequency of that mode, along with a gradual change in its phase angle around that frequency.

Analytical explanation of TMD is as follows.

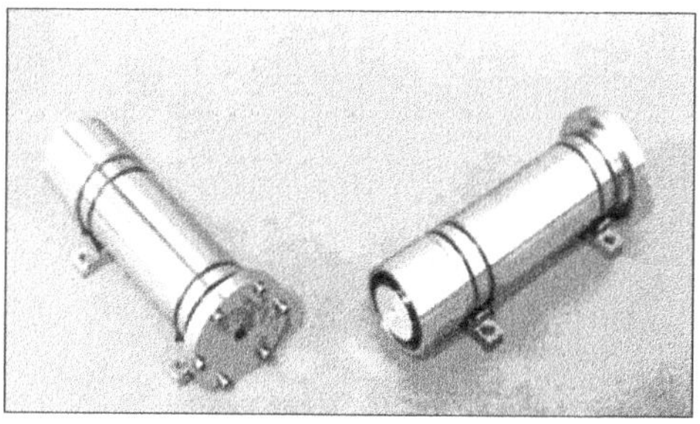

Figure 5.9 Tuned mass dampers. *Source:* [3].

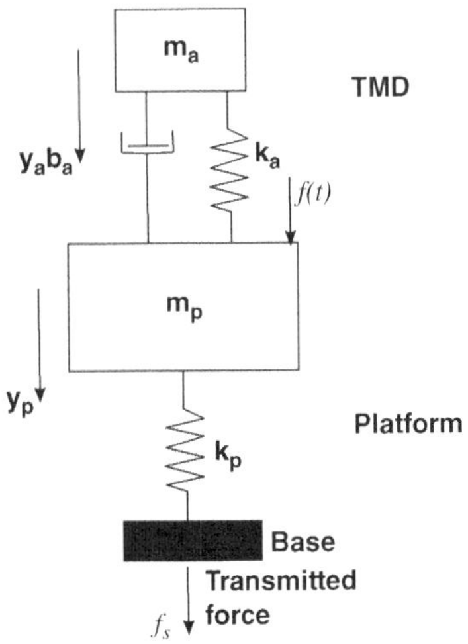

Figure 5.10 Tuned mass configuration.

Figure 5.10 shows a free-body diagram of a platform with mass m_p and stiffness K_p, subjected to disturbance $f(t)$, displacement y_p, and transmitted force to the base is f_s. On the top of the platform is a tuned mass damper with mass m_a, stiffness K_a, viscous damping coefficient b_a, and displacement y_a.

The equations of motion for the tuned mass damper are

$$m_p\ddot{y}_p + k_p y_p - b_a(\dot{y}_a - \dot{y}_p) - K_a(y_a - y_p) = f(t)$$
$$m_a\ddot{y}_a + K_a(y_a - y_p) = 0 \tag{5.6}$$

Let us assume the damping of the tuned mass damper b_a is zero. Let us assume that the disturbance to the system of $f(t)$ is sinusoidal at a frequency. Then both platforms and the damper will vibrate at the same frequency. Let us assume these vibrations are given by Y_a sinωt for the damper and Y_p sinωt for the platform. Substituting these displacements in Eq. 5.6 for damper, we see that the platform motion is zero if the disturbance frequency ω is equal to natural frequency of the damper ω_a given by

$$\omega_a = \sqrt{k_a/m_a}$$

It shows that if we tune damper frequency to disturbance frequency for undamped damper, the platform motion can be zero. However, in this case damper motion will be very high. For damped tuned mass damper, the response of the platform is given by

$$\frac{y_p}{\delta_{st}} = \sqrt{\frac{(2\zeta g)^2 + (g^2 - \alpha^2)^2}{(2\zeta g)^2(g^2 - 1 + \mu g^2)^2 + (\mu\alpha^2 g^2 - (g^2 - 1)(g^2 - \alpha^2))^2}} \tag{5.7}$$

where

$$\omega_p = \sqrt{k_p/m_p}$$

Frequency ratio, g, is given by

$$g = \frac{\omega}{\omega_a}$$

Critical damping ratio of the damper ζ is given by

$$\zeta = \frac{b_a}{2\sqrt{k_a m_a}}$$

Static displacement δ_{st} is given by

$$\delta_{st} = \frac{f(0)}{k_p}$$

The mass ratio μ is defined as

$$\mu = \frac{m_a}{m_p}$$

and the non-dimensional frequency α is given by

$$\alpha = \frac{\omega_a}{\omega_p}$$

For certain TMD configurations, the platform response is given in Figure 5.11.

The response of the platform with an undamped ($\zeta = 0$) TMD mounted to it is shown in Figure 5.11. Although the two degree of freedom system has no platform response at the forcing frequency equal to TMD natural frequency, $g = 1$, there are now two resonances on either side near $g = 0.8$ and $g = 1.3$. The spring-mass described here is called a tuned vibration absorber. It is not practical as its TMD response will be too large, and there are two other resonances, but it serves as a baseline for the TMD. Upon the inclusion of damping ($\zeta > 0$), the responses become smaller, and the spring-mass-damper is called a TMD. The rest of the curves in Figure 5.11 are for different TMD configurations. As the mass ratio, μ, is increased, more damping is imparted to the platform. However, it will increase for mass-sensitive platforms such as satellites. Therefore, for a particular application, TMD design is optimized.

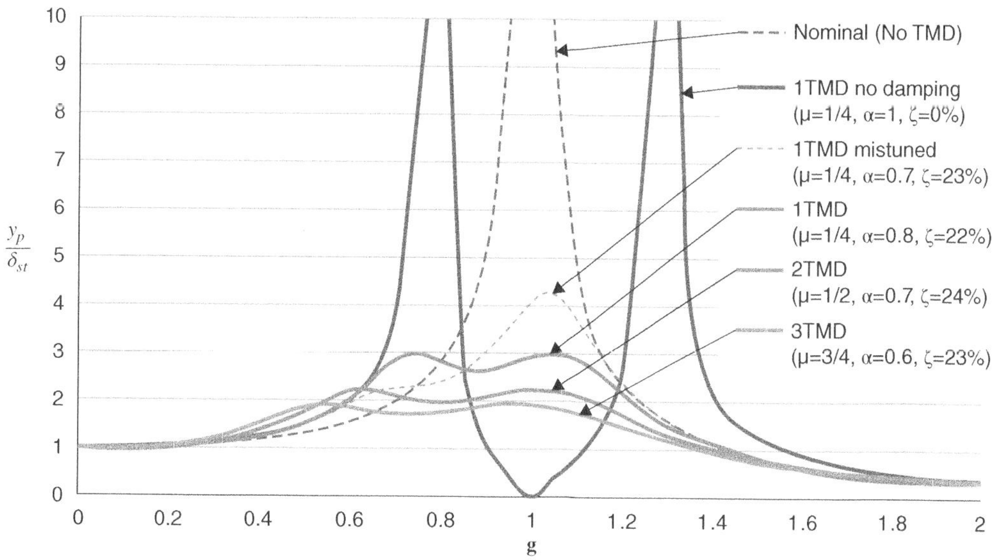

Figure 5.11 Platform response for various TMD configurations.

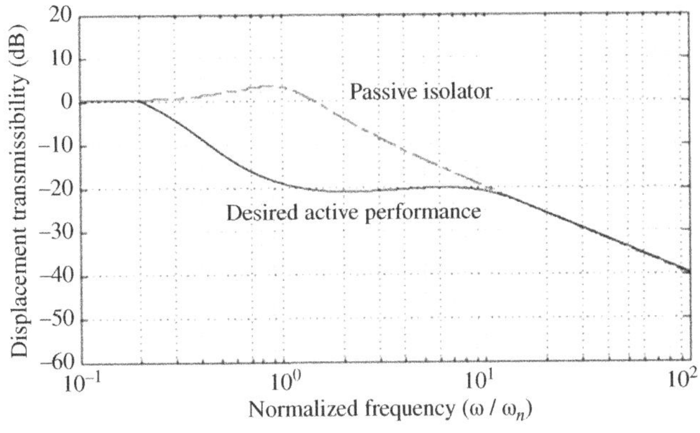

Figure 5.12 Generalized active/passive performance comparison.

5.3 Active Vibration and Jitter Control

In active vibration and jitter control, inertial reference unit (IRU) provides jitter-free reference laser, and a sensor measures the jitter in the telescope and control algorithms command FSM to correct it. The image beam or laser beam passes through the same path as the reference laser beam and so the jitter due to telescope is corrected. Before discussing active vibration and jitter control systems, we will discuss actuators, sensors, and control algorithms. The comparison of performance of passive and active vibration control is given in Figure 5.12. For higher frequencies, passive isolators do a good job. For low frequencies, active vibration control is needed.

5.3.1 Actuators

5.3.1.1 Voice Coils

Voice coil actuators are versatile direct drive, hysteresis-free, cog-free, non-commuted, limited motion servo motors with linear control characteristics. However, they are used for large motion. The basic setup, as shown in Figure 5.13, is a permanent magnet field assembly in conjunction with a coil winding that produces a force proportional to the current which is applied to the coil. The magnets are of the same polarity. When current is applied to the circumferentially wound coil, it interacts with the radial magnetic field of the permanent magnet assembly via the Lorentz force principle to create an axial force (mutually perpendicular to the vectors of the current flow and the magnetic field) between the coil and the magnetic assemblies. The polarity of the current-producing voltage applied to the two terminals of the coil dictates the direction of the force on the coil and therefore stroke movement. The air gap between the magnet and coil assemblies is usually 0.01–0.015 inches. Because there is no contact between the moving parts, voice coil-controlled devices are more reliable than hydraulic or pneumatic devices.

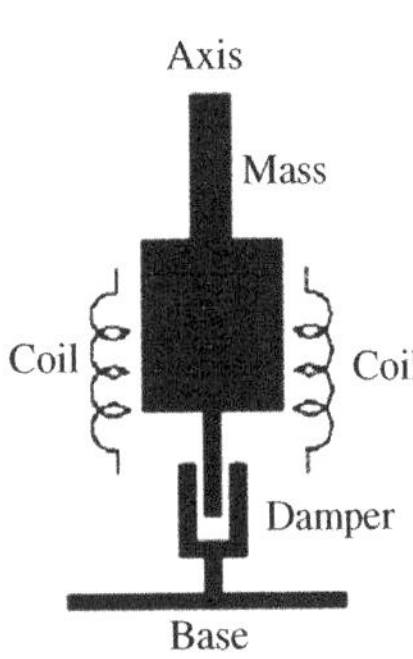

Figure 5.13 Voice coil schematic.

5.3.1.2 Piezoceramic

Piezoelectricity is a property of certain classes of crystalline materials. When mechanical pressure is applied to one of these materials, the crystalline structure produces a voltage proportional to the pressure.

Conversely, when an electric field is applied to one of these materials, the crystalline structure changes shape, producing dimensional changes in the material. These materials are used both for actuators and sensors. The piezoelectric ceramics are more versatile. The hard, dense ceramics can be tailored to specific applications. The hard, dense ceramics can be manufactured in almost any given shape or size. They are chemically inert and immune to moisture and other atmospheric conditions. In addition, the mechanical and electrical axes of these ceramics can be precisely oriented in relation to the shape of the ceramic. These axes are set during "poling," the process that induces piezoelectric properties in the ceramics. The poling field can be applied so that the ceramic exhibits piezoelectric responses in various directions or combinations of directions. Poling, or application of a large coercive field on the order of 50 V/mil, aligns crystal in the ceramic. Poling causes the ceramic to grow in the field direction and to shrink laterally, roughly according to Poisson's ratio, as shown in Figure 5.14.

The poling process completely changes the dimensions of a ceramic element. After the poling process is complete, a voltage lower than the poling voltage changes the dimensions of a ceramic element as long as the voltage is applied. The voltage with the same polarity as the poling voltage causes additional expansion along the poling axis and contraction along the lateral axes 1 and 2. Application of a voltage of opposite polarity to that of the poling voltage, reverse field, causes the ceramic to shrink along the three axes until a negative level is reached. The poling action then switches, and the ceramic grows again. Hence, the "butterfly" curve, as shown in Figure 5.14, exists between field and strain. A voltage with the same polarity as the poling voltage results from a compressive force applied parallel to the poling axis or from a tensile force applied perpendicular to the poling axis. A voltage with the opposite polarity results from a tensile force applied parallel

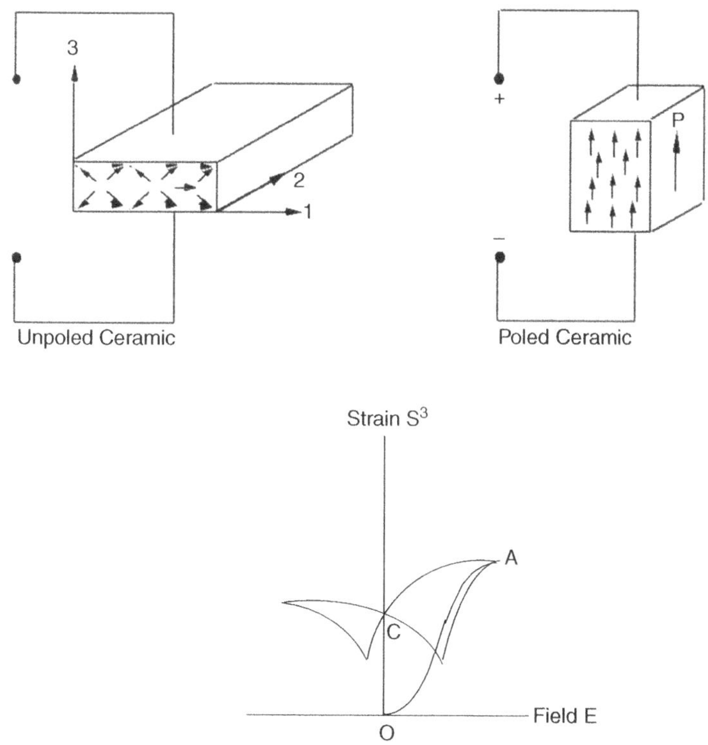

Figure 5.14 Piezoceramic poling.

to the poling axis or from a compressive force applied perpendicular to the poling axis. Therefore, piezoceramics can be used both as a sensor and an actuator.

For piezoceramic lateral strain coefficient d31, coulombs/N, m/V is 1.8×10^{-10}, Young's modulus, E, N/m^2 is 6.3×10^{10}, Ed31, coulombs/m^2, or N/m/V is 11.34, coefficient of thermal expansion, strain/k is 5, and absolute permittivity D, Farad/m or N/V^2 is 1.5×10^{-8}.

5.3.1.2.1 Lateral Mode Piezoelectric Sensors

The method as shown in Figure 5.15 is to bond or embed a piezo wafer to a surface and monitor the electric potential on electrodes. Piezo is poled perpendicular to the plated surfaces. A piezoelectric sensor produces a charge when strained laterally and is given by

$$Q = AEd_{31}(\varepsilon_1 + \varepsilon_2) \tag{5.8}$$

where

Q = produced charge

A = lateral area

E = Young's modulus

d_{31} = lateral charge coefficient

ε_1, ε_2 = strains along axes 1 and 2, respectively.

Capacitance is given by

$$C = \frac{DA}{t} \tag{5.9}$$

where

C = capacitance

D = dielectric constant (permittivity)

t = thickness

Voltage produced, V, is given by

$$\begin{aligned} V &= \frac{Q}{C} \\ &= t\left(\frac{Ed_{31}}{D}\right)\left(\varepsilon_1 + \varepsilon_2\right) \end{aligned} \tag{5.10}$$

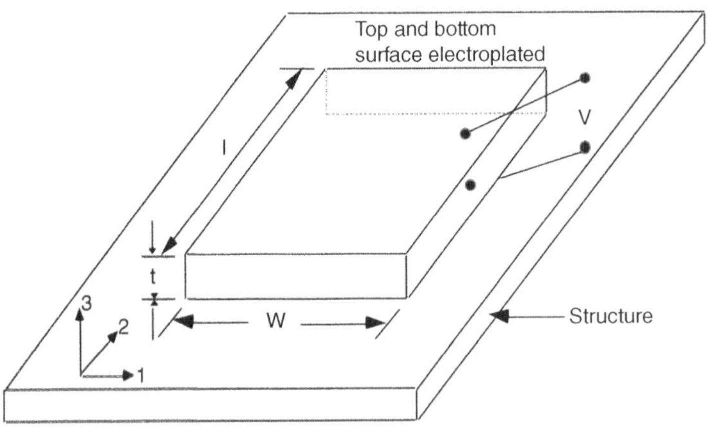

Figure 5.15 Lateral mode piezoelectric sensors.

5.3.1.2.2 Piezoelectric Stack Actuators

Configuration of piezoelectric stack actuator is shown in Figure 5.16.

A stack is defined as a single or multiple-layered piezoelectric element which is relatively long in z-axis, along pole axis. In Figure 5.16, there is an applied force F, and a spring is in parallel. The objective is to find the resultant displacement of the actuator when voltage is applied.

The unconstrained strain (when there is no external resistance stiffness) of the actuator in z (or 3) direction is given by

$$\varepsilon_{pe} = \frac{d_{33}V}{L_a} \tag{5.11}$$

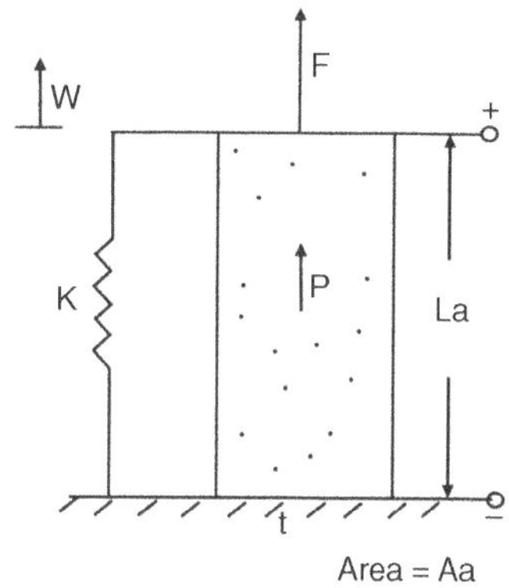

Figure 5.16 Piezoelectric stack actuators.

Thus, the unconstrained displacement of the actuator is

$$W_a = d_{33}V \tag{5.12}$$

When a stiffness resists the motion of the actuator to W, the actuator exerts external force, and it is given by

$$F_a = \frac{d_{33}VK}{1 + K/K_a} \tag{5.13}$$

Applying a force balance, we get

$$F = KW - \frac{E_a(W_a - W)}{L_a}A_a \tag{5.14}$$

where K is stiffness of external spring. Solving the above equation for displacement,

$$W = \frac{d_{33}V + F/K_a}{1 + K/K_a} \tag{5.15}$$

where K_a = actuator spring constant, given by

$$K_a = E_aA_a/L_a \tag{5.16}$$

If external force $F = 0$, then the application of voltage V will result in deflection

$$W = \frac{d_{33}V}{1 + K/K_a} \tag{5.17}$$

The actuator force is then given by WK,

$$F_a = \frac{d_{33}VK}{1 + K/K_a} \tag{5.18}$$

If the actuator displacement is constrained to zero, the blocked force of the actuator is given by,

$$F_a = \frac{d_{33}E_aA_aV}{L_a} \tag{5.19}$$

These actuators are valued for their fast response times, high precision, compact size, and lack of magnetic interference. They are commonly used in applications where precise and rapid control of motion or force is required.

5.3.1.3 Fast Steering Mirrors

The FSM is a mirror whose orientation is controlled by a set of actuators. The actuators used are normally voice coil actuators for large-angle motion and piezoelectric actuators for small and fine angle motion. The throw range is normally ± 0.1 mrad (0.0057°) to $\pm 1.5°$ depending on the actuators used. The typical bandwidth is 800 Hz–3 KHz.

As an example, Newport 300–01 voice coil actuated 1-inch FSM provides two-axis motion with closed-loop control to 580 Hz. The angular range is $\pm 1.5°$ and angular resolution is ≤ 1 μrad rms. FSM has two operational modes. One is an open loop, where FSM simply converts the input signal voltage to the current for the voice coil motor controlling the mirror angle. Second is closed-loop mode, where the mirror angle measured by the built-in angle position sensor is feedback to the controller to form an internal feedback control loop.

Piezoelectric actuated FSM are faster than voice coil FSM and their resolution is in nanoradian range and steering range of up to 100 mrad. As an example, PI Instruments 334 miniature piezo tip/tilt mirrors, S-334 1SL at Naval Postgraduate School (NPS) are 10 mm in diameter. The closed-loop tilt angle is 25 mrad and resolution is 1 μrad.

5.3.1.3.1 FSM Dynamics

It is desirable to have a plant model of FSM that includes digital-analog conversion and sampling effect of the command and sensor output. Conducting experiments is a straightforward and accurate way to obtain this information. The experimental set is shown in Figure 5.17.

The block diagram of FSM closed-loop system is shown in Figure 5.18.

Using the z-transform, the discrete transfer function for the FSM inner-loop is $P(e^{j\omega T}) = P(z)|_{z=j\omega T}$, where T is the sample rate. This transfer function can be obtained by computing the ratio of the FSM plant output, $Y(z)$, to the digital control input, $U(z)$

$$\frac{Y(z)}{U(z)} = P\left(z\right). \tag{5.20}$$

As an example, for OKO FSM closed-loop mode, the dynamics are shown in Figure 5.19.

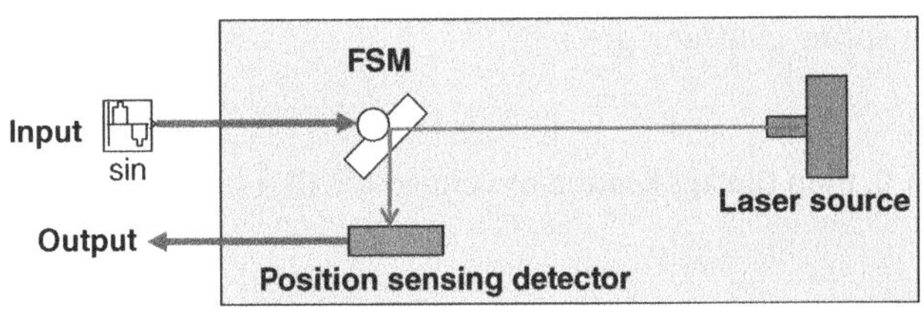

Figure 5.17 Open-loop response.

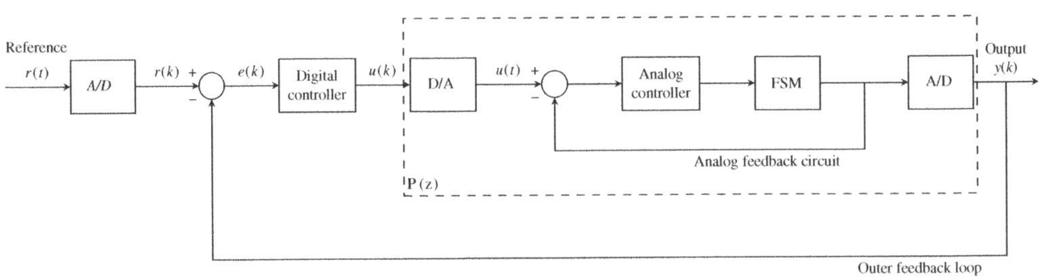

Figure 5.18 FSM closed-loop diagram.

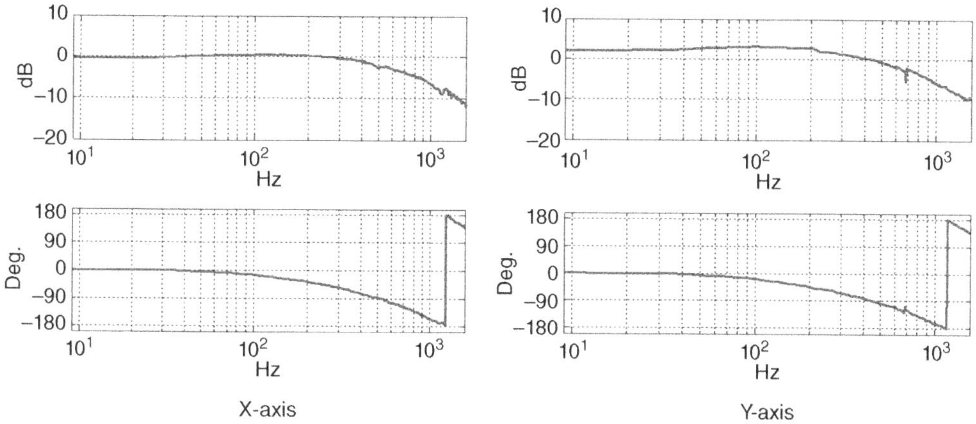

Figure 5.19 OKO FSM dynamics.

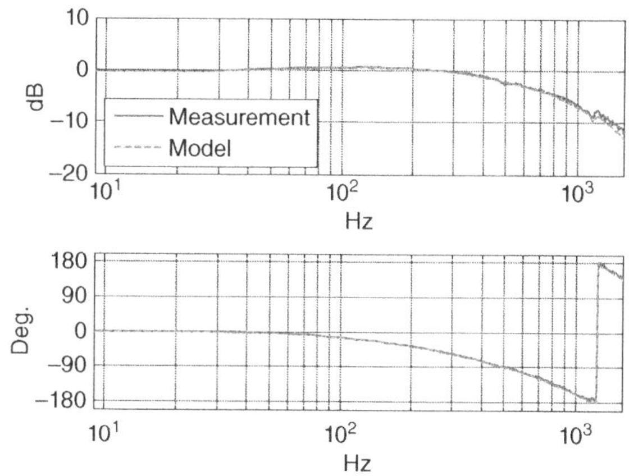

Figure 5.20 Comparison between measurement and plant model for transfer function.

From experimental results, plant models can be obtained. An eighth order discrete-time plant model obtained for x-axis by curve fitting with sample rate $T = 10026$ Hz is

$$P(z) = \frac{0.093112(z + 0.2616)(z - 0.9524)(z^2 - 1.633z + 0.8723)(z^2 - 1.45z + 0.9576)}{(z - 0.9411)(z - 0.778)(z^2 - 0.9137z + 0.3594)(z^2 - 1.637z + 0.8746)(z^2 - 1.442z + 0.9522)} \tag{5.21}$$

Figure 5.20 shows a close comparison between measurement and plant model for transfer function.

5.3.1.4 Jitter Sensor

For jitter sensors, optical position sensors, known as Position Sensing Modules or PSMs are commonly used. NPS Jitter Control Testbed used On-Trak Photonics Inc. PSMs. They have a detection area of 10 mm × 10 mm. Each duolateral, dual axis PSM (model PSM2-10) requires an amplifier, the OT-301, to output the x and y position (in volts) of the centroid of the laser beam spot. The combination of amplifier and detector is called a position sensing detector (PSD). Various beam splitters and folding mirrors are used to direct the beam as desired. The detectors have a minimum sensitivity of 50–100 μW, which drives the selection of the beam splitters and the determination of laser power. Their frequency response for the gain settings normally used is 15 kHz.

The optimum beam diameter for accurate detection on the PSM is between 1 and 3 mm, and the maximum allowed intensity should be less than 300 W/cm^2. The output range of the PSM2-10/OT-301 detector is $\pm$10 V and the OT-301 amplifier has a noise level of 1 mV. The minimum resolution of the PSM2-10 with the OT 301 amplifier is, therefore, 0.5 μm.

5.3.2 Control Algorithms

In order for active jitter control, control algorithms are used to command control the FSM. There are two types of control algorithms. The first approach is a feedback only approach based on measuring the jitter. This approach uses control laws, such as position integration derivative and linear quadratic gaussian (LQG), to determine the command for the FSM. The second approach uses adaptive filters to estimate the disturbance signal and feed it forward to correct the residual jitter error introduced with using the first approach.

The first approach, feedback control, is the classical approach discussed in Chapter 3. In the second approach, Adaptive filters have been used extensively over the past several decades in the field of active noise and vibration control [4, 5]. The basic principle of an adaptive filter is that the controller gains can be varied throughout the control process to adapt to changing parameters and can therefore cancel disturbances more effectively than passive methods. Transverse FIR filter structure used for the adaptive filter control method is first described, followed by the weight update algorithms and experimental results.

5.3.2.1 Transverse Filter

An Lth order transverse FIR filter has the structure shown in Figure 5.21 [4–6]. Each of the stages, or taps, delays the input signal by one sample, and this filter is sometimes called a tapped-delay line. The filter output in terms of the current sample, n, is expressed as follows:

$$y(n) = \sum_{i=0}^{L} w_i(n)x(n-i) = \mathbf{w}^T(n)\mathbf{x}(n) \tag{5.22}$$

where $\mathbf{w}^T(n) = [w_0(n) \quad w_1(n) \quad \cdots \quad w_L(n)]$ is the filter weight vector of length $L+1$, $\mathbf{x}(n) = [x(n) \quad x(n-1) \quad \cdots \quad x(n-L)]$ is the a vector of delayed inputs, and $y(n)$ is the filter output.

The output $y(n)$ is compared with the desired output for disturbance $d(n)$. The error $e(n)$ is given by

$$e(n) = d(n) - y(n) = d(n) - \mathbf{w}^T(n)\mathbf{x}(n) \tag{5.23}$$

5.3.2.1.1 Weight Update Algorithms

Least mean square (LMS) is a common algorithm widely used to update the filter weights. In the LMS algorithm, the cost function, $J(n)$, is the expected value of the mean square error (MSE)

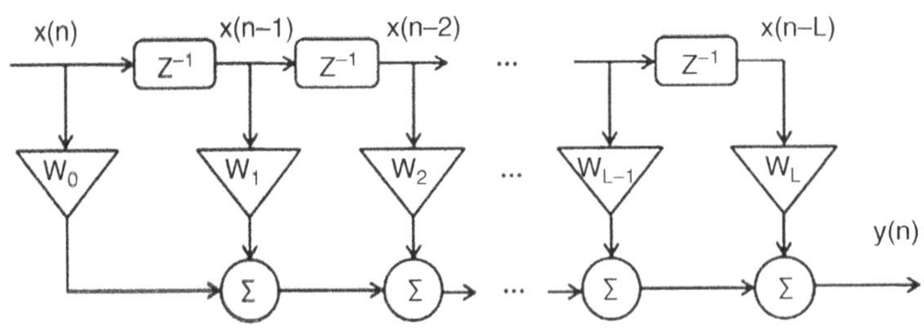

Figure 5.21 Transverse FIR filter structure.

denoted by $E[e(n)^2]$. When the statistics of the disturbance and the reference signal are available, the weights that minimize $E[e(n)^2]$ can be computed. In practice, however, such a priori information is often unavailable. It is assumed that $x(n)$ and $d(n)$ are correlated. The LMS algorithm replaces the expectation by the sample that just arrived, and the MSE is approximated by the instantaneous squared error, $J(n) = e(n)^2$. The LMS algorithms uses the steepest-gradient descent method to update the weights in the direction of lowest error. The difference equation for updating weights can be expressed as

$$\mathbf{w}(n+1) = \mathbf{w}(n) - \frac{\mu}{2}\frac{\partial J(n)}{\partial \mathbf{w}(n)} = \mathbf{w}(n) + \mu e(n)\mathbf{x}(n) \tag{5.24}$$

where μ is the convergence coefficient that controls the speed of convergence to steady-state weight values.

There is another method for weight update called recursive least squares (RLS). In this method, the cost function is given by

$$J(n) = \sum_{i=1}^{n} \lambda^{n-i} e(i)^2 \tag{5.25}$$

where the forgetting factor, $0 < \lambda \le 1$, allows more recent data to be weighted more heavily and data long past to be forgotten. A value of $\lambda = 1$ implies all previous error history is included in the cost function, while smaller values exclude more past errors.

In a system for disturbance rejection, a there is a primary plant where disturbances generate vibration/jitter. There is a secondary plant like the FSM that corrects the jitter. As we discussed earlier, secondary plants also have dynamics. Filtered-x is a technique to include the effect of the secondary path to avoid potential instability. As shown in Figure 5.22, the reference $x(n)$ signal is passed through a model of the secondary plant, $\hat{S}(z)$ before it is fed to the adaptive algorithm. It can be applied to any weight update algorithm, including LMS and RLS. Filtered-X LMS adaptive filter is often referred to as FXLMS. In a Filtered-X adaptive filter, the reference signal $x(n)$ for updating filter coefficients uses $x'(n)$ given by $\mathbf{x}'(n) = \hat{S}(z)\mathbf{x}(n)$, whereas the input of the FIR filter is still $x(n)$ as the output of the filter goes through the actual secondary path.

5.3.3 Active Optics Beam Jitter Control

To develop improved techniques for optical beam jitter control, a laser jitter control testbed was developed at the NPS, as shown in Figures 5.23 and 5.24. The components are mounted on a Newport optical bench, which can be floated to isolate the components from external vibrations.

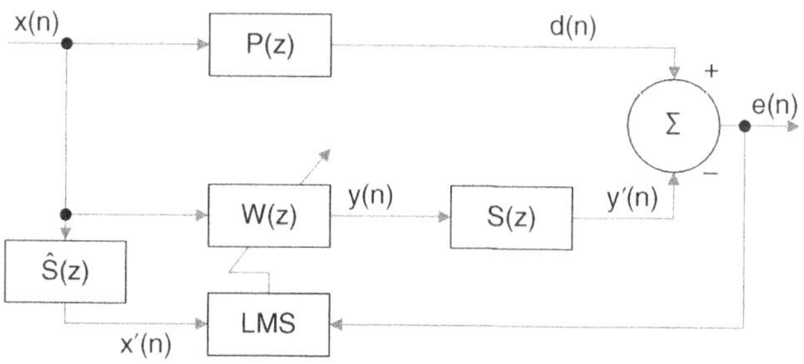

Figure 5.22 Filtered-X LMS system representation.

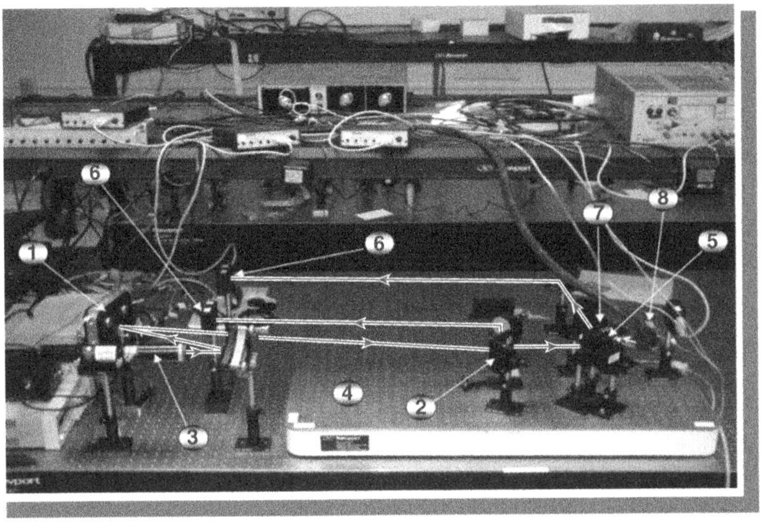

Figure 5.23 Laser jitter control testbed. *Source*: [5]/American Astronautical Society/Public domain.

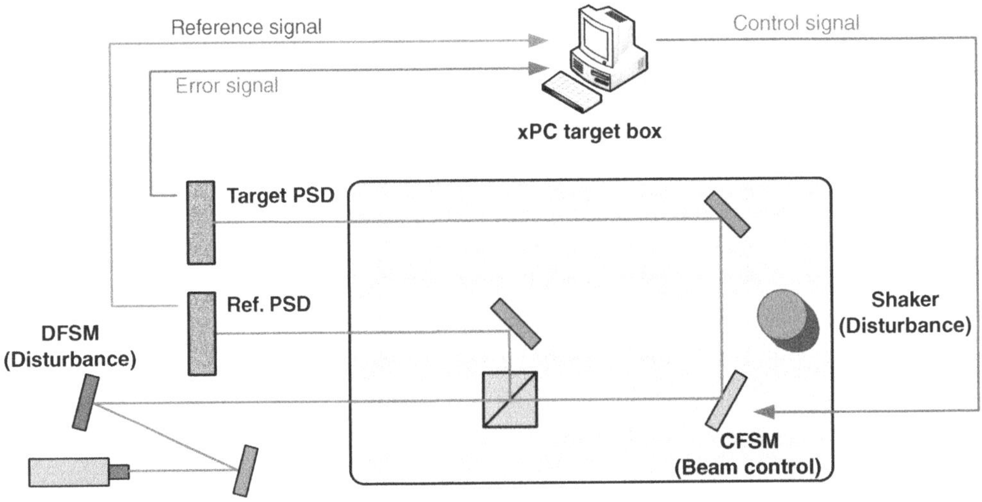

Figure 5.24 Schematic of the laser jitter control testbed [3]. *Source*: [11]/American Astronautical Society/Public domain.

The laser beam originates from a source and passes through a disturbance injection fast steering mirror (DFSM). The DFSM corrupts the beam using random or periodic disturbances simulating disturbances that might originate with the transmitting station or tip and tilt errors which the beam may suffer as it passes through the atmosphere. A vibration isolation platform is used to mount the relay system and to isolate the relay system from the optical bench. A control FSM, designated the CFSM, is used to correct the disturbed beam. The corrected beam is then reflected off the platform to the target PSD. The disturbed laser beam is measured by reference PSD. In Figure 5.23, (1) refers to disturbance fast steering mirror, (2) refers to beam splitter, (3) refers to laser, (4) refers to optical bench, (5) refers to control steering mirror, (6) refers to position sensing detectors for reference and target, (7) refers to accelerometer, and (8) refers to a shaker.

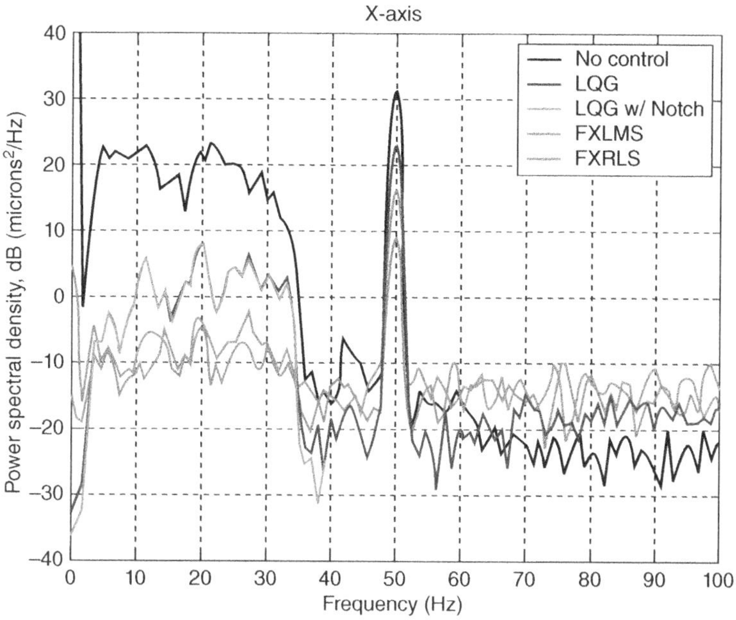

Figure 5.25 Power spectral density for jitter control methods. *Source*: [3]/IEEE.

5.3.3.1 Experimental Results

Experiments were performed to verify the control methods including the FXLMS. The results of FXLMS and Filtered-X Recursive Least Square (FXRLS) methods are compared with LQG control techniques with/without classical notch filters for sinusoidal disturbance rejection. Figure 5.25 shows the power spectral density of the signal across the frequency spectrum from 0 to 100 Hz. The mean squared error plot shown in Figure 5.26 shows that the FXRLS method manages to further reduce the mean square error of the signal compared to the FXLMS and LQG designs. For the sinusoidal disturbance at 50 Hz, the LQG with a notch filter provides the best performance. However, to apply this technique, the disturbance frequency must be accurately known, and an error or change in disturbance frequency will degrade the performance significantly. Overall FXRLS provides superior performance.

5.3.4 Slew Maneuvers

Imaging satellites slew from taking pictures from one target to another [7]. At the end of the slew, due to motion from the flexible appendages, it takes some time, called the settling time, before the satellites achieve desired stable pointing. The objective of the research was to evaluate different torque profiles and control techniques for fast slews to minimize settling time. Initially, the single axis flexible simulator, as shown in the Figure 5.27, was developed at NPS [7] to do research in this area. Later, a three-axis flexible simulator was developed with CMGs.

The simulator shown in Figure 5.27 consists of a rigid central body floating on a granite table by air pads. It has an air bearing so that it can rotate in one axis without friction. The central body has a reaction wheel as an actuator, an angular sensor and a rate gyro. The center body is connected to L-shaped flexible structure supported on air pads. The torque profiles considered in the study are bang-bang, versine, and input shaping.

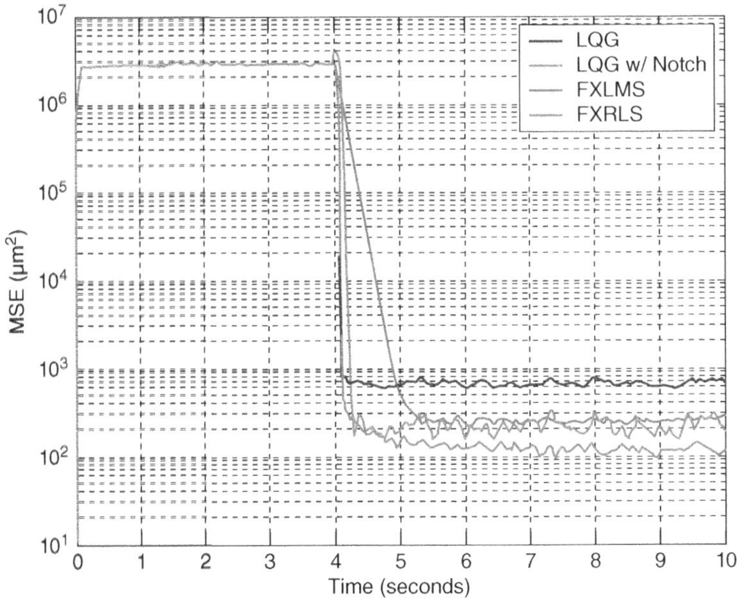

Figure 5.26 Power spectral density for jitter control methods.

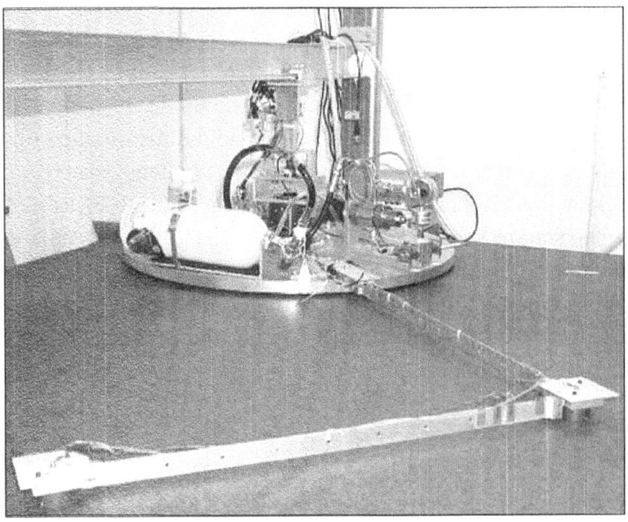

Figure 5.27 Single axis flexible spacecraft simulator. *Source*: [11]/American Astronautical Society/Public domain.

5.3.4.1 Bang–Bang Profile

$$\begin{cases} 1, & 0 \leq t \leq t_1 \\ -1, & t_1 < t \leq t_2 \\ 0, & t_2 < t \end{cases} \tag{5.26}$$

In the bang–bang torque profile, as shown in Figure 5.28, positive maximum torque is applied for half of the period and negative maximum torque is applied for the remaining half of the period.

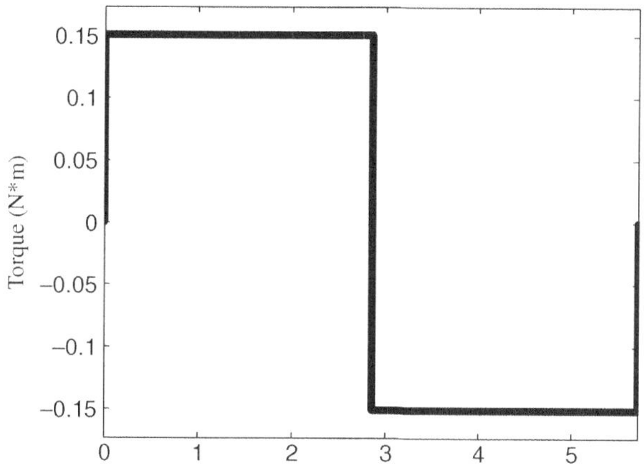

Figure 5.28 Bang–bang torque profile.

This torque profile provides a minimum slew time; however, it results in a large settling time for a flexible spacecraft.

5.3.4.2 Versine Profile

$$\begin{cases} \frac{1}{2}\left[1 - \cos\left(\frac{\pi t}{t_A}\right)\right], & 0 \le t \le t_A \\ 1, & t_A \le t \le t_B \\ \frac{1}{2}\left[1 - \cos\left(\frac{\pi(t_1 - t)}{t_A}\right)\right], & t_B \le t \le t_1 \\ -\frac{1}{2}\left[1 - \cos\left(\frac{\pi(t - t_1)}{t_A}\right)\right], & t_1 < t \le t_C \\ -1, & t_C \le t \le t_D \\ -\frac{1}{2}\left[1 - \cos\left(\frac{\pi(t_2 - t)}{t_A}\right)\right], & t_D \le t \le t_2 \\ 0, & t_2 < t \end{cases} \tag{5.27}$$

In the versine torque profile, as shown in Figure 5.29, the sudden changes in torque are smoothed using a cosine function. This results in a significant decrease in settling time for the flexible spacecraft. However, it does increase slew time. Overall, slew time is decreased. This torque profile is commonly used for flexible imaging spacecraft.

5.3.4.3 Input Shaping

The goal of input shaping is to provide a command which results in zero residual vibration. As shown in Figure 5.30, a vibration can be eliminated by applying impulses of appropriate amplitude and phase such that they exactly cancel the mode. For example, the command issued at time t_1, in this case an impulse, starts a vibration which decays as a function of the modal damping. The second impulse is phased such that it is applied at the vibrating mode's half-period point. The net vibration following the second impulse is zero.

The torque command is created by convolving a sequence of impulses, called the input shaper, with any desired command. The amplitudes and time locations of the impulses are determined by the system's natural frequencies and damping ratios.

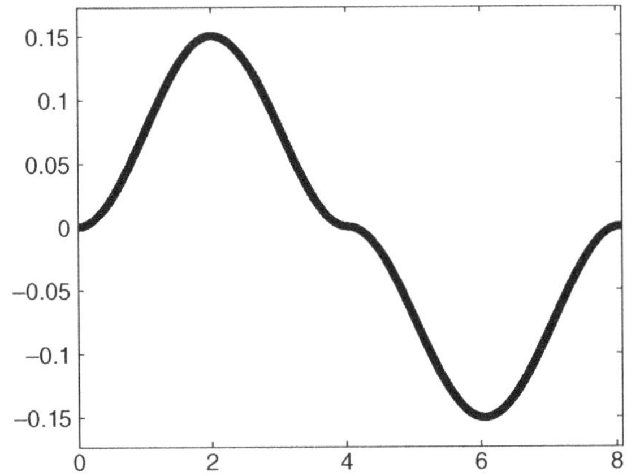

Figure 5.29 Versine torque profile.

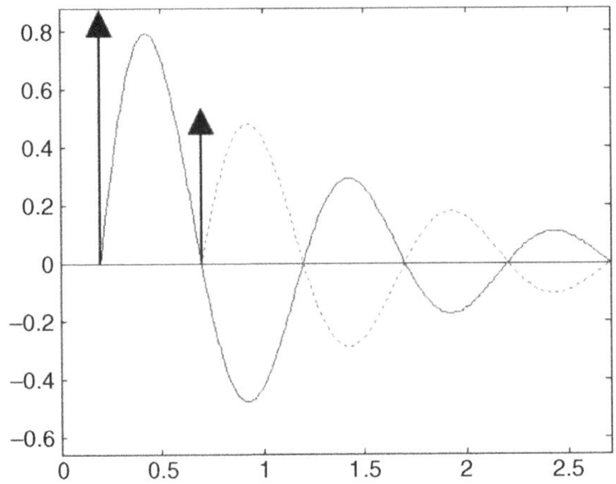

Figure 5.30 Input shaping.

There are three types of zero vibration (ZV) sequence for the impulses. The shortest impulse train is a two impulse ZV sequence. It is robust, however, only for small variation (5%) in modal frequency. In order to increase the shaper's robustness, we differentiate the vibration equation with respect to the natural frequency, and it is called the zero vibration derivative (ZVD) shaper. It consists of three impulses. It has robustness up to 20% in natural frequency. If the vibration equations are differentiated again, we get zero vibration derivative derivative shaper (ZVDDS). In this case we need four impulses. It allows uncertainty up to 40%. Experiments were performed using versine torque profile and input shaping for one mode and two modes using ZVD.

5.3.4.3.1 Input Shaping Simulation

Figure 5.31 shows that analytical simulations show that for input shaping ZVD, maximum vibrations after 15 seconds settling time are reduced by 95.1% for single-mode input shaping and 96.1% for two-mode input shaping. For Versine torque profile without Input Shapting, vibrations are significantly higher, resulting in longer settling time.

5.3.4.3.2 Input Shaping Experimental Results

Figure 5.32 shows experimental results for input shaping ZVD. Maximum vibrations after 20 seconds are reduced by 82.1% for one-mode input shaping and 81.9% for two-mode input shaping.

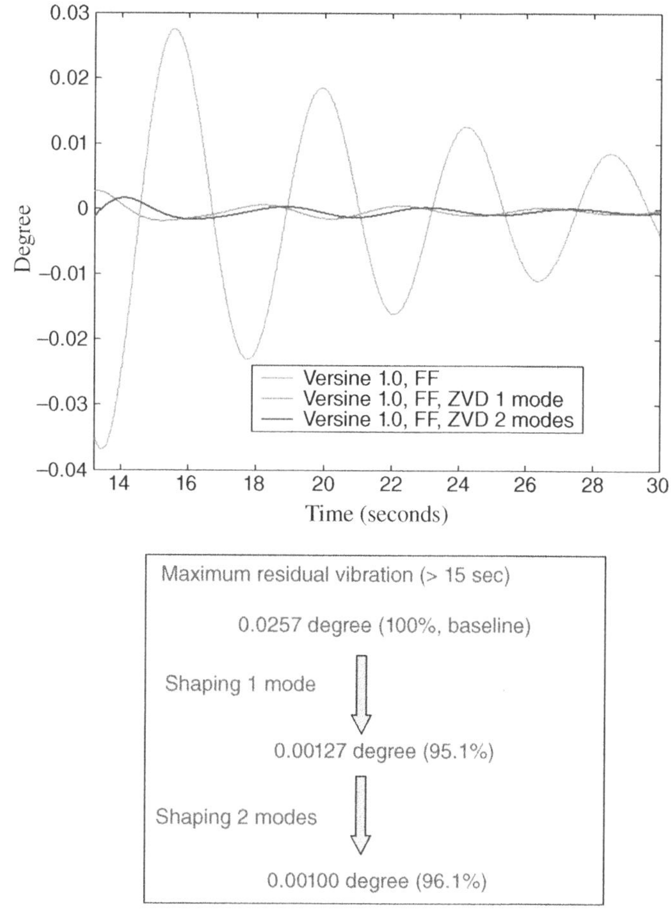

Figure 5.31 Input shaping simulation results for one and two modes.

5.3.5 Reference Laser Beam

In order to measure jitter, we need to have an inertially jitter free laser beam that goes the same path as our object beam, like an image beam or high energy laser beam. Since the platform will have jitter, we need an inertially stable platform with zero jitter. There are two approaches to achieve a jitter free reference beam: inertially stable platform and strap down, as shown in Figure 5.33.

In the inertial stable platform, the laser and gyro are mounted on a platform that is connected to the base platform of the telescope by voice coil actuators. The voice coils are actuated by the signal from the gyro to make the gyro output zero. Since the gyro measures inertial acceleration, its zero output means the platform is inertially stable and the laser beam is jitter free. In the strap down approach, the gyro and laser beam are mounted on the base platform. The gyro measures the base motion and commands the base steering mirror (BSM), to move in an opposite direction to eliminate the motion of the base platform from the laser beam. The inertial stable platform gives better performance as the dynamics of BSM does not participate in the performance. A typical performance for base motion rejection is given on Figure 5.34.

An inertial stable platform gives better high-frequency response and less maximum error. The strapdown is simple and cheaper.

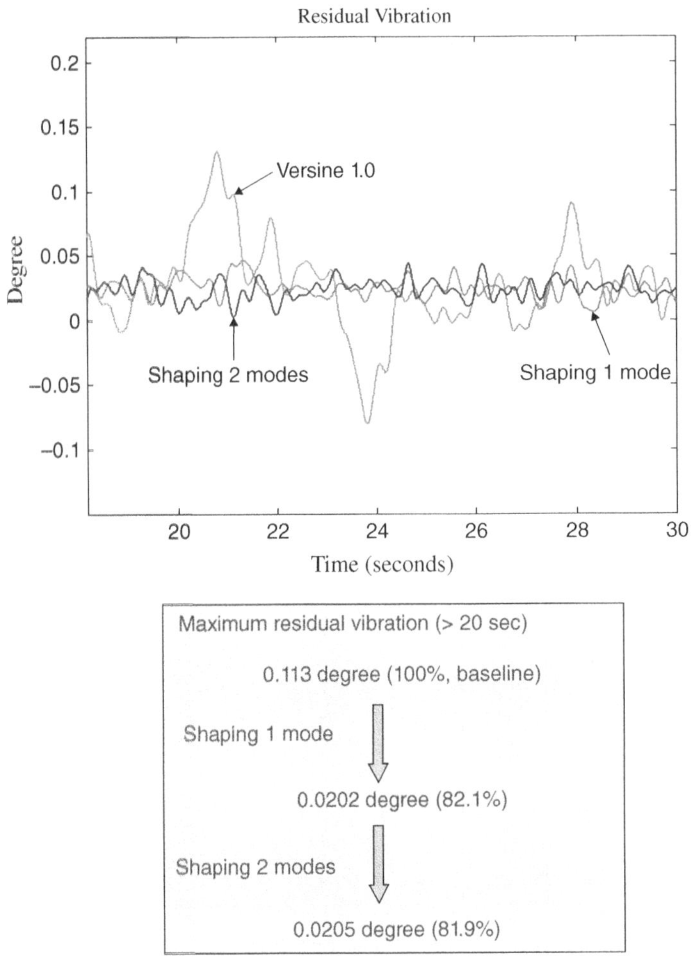

Figure 5.32 Input shaping experimental results for one and two modes.

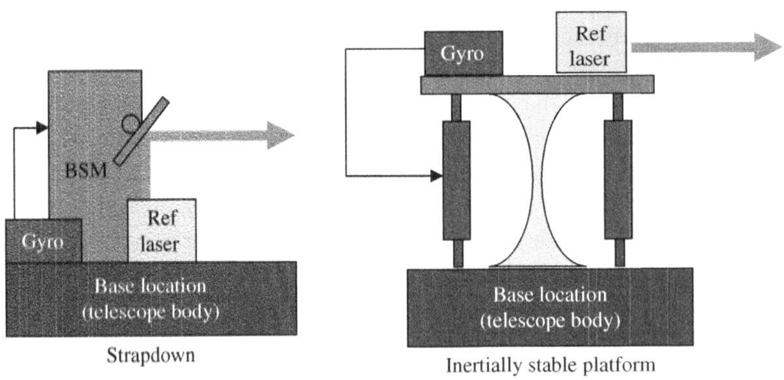

Figure 5.33 Inertially stable platform and strapdown platform.

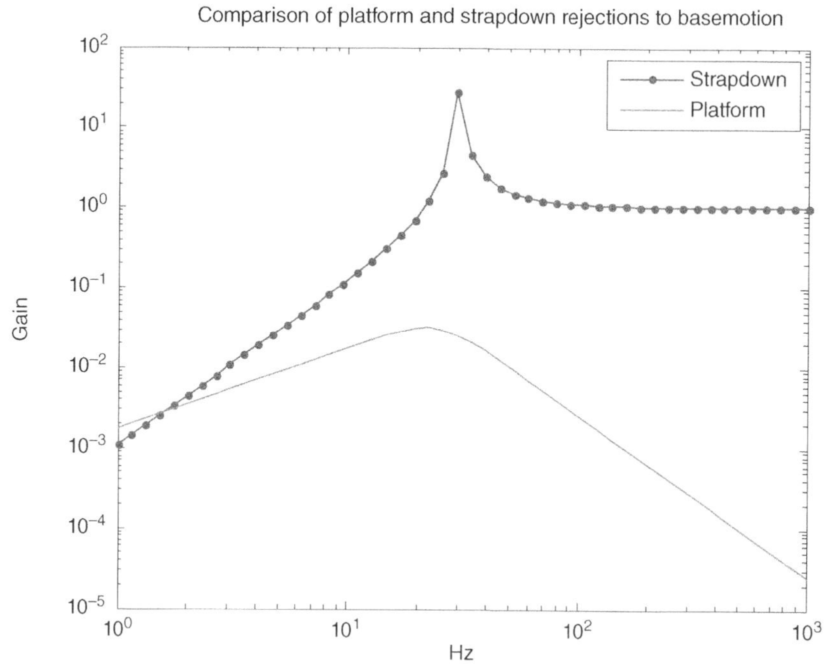

Figure 5.34 Comparison of the error rejection transfer function for the strapdown and inertial stable platform.

An implementation of active jitter control measurement and control is given in Figure 5.35 for both imaging satellites and laser systems.

As shown in Figure 5.35, the reference laser beam is on an inertially stable platform mounted on the telescope. A jitter-free reference laser beam originates from there. The reference laser beam travels through the telescope in a similar path to incoming image beam or outgoing laser beam, resulting in similar LOS (line of sight) jitter. The jitter is measured by a LOS jitter sensor, which feeds into a jitter rejection controller. The command from jitter rejection controller goes to jitter rejection FSM to reject the jitter of incoming image or outgoing laser beam.

5.4 Active Vibration Isolation

Vibration isolation platforms are used to isolate the optical payload from spacecraft vibration. Control algorithms have been developed for two platforms at the NPS, ultra quiet platform (UQP) and the precision pointing hexapod (PPH) [8].

5.4.1 Ultra Quiet Platform

The UQP, as shown in Figure 5.36, is used for testing control algorithms for vibration isolation of an imaging payload. It is configured similar to a six degrees of freedom "cubic" Stewart platform where the struts are arranged as if they were on the edges of a cube, providing for three orthogonal pairs of actuators. In such arrangements, control in six degrees of freedom is possible using linear actuators, and the coupling between actuators is minimized. The UQP is mounted on a spacecraft

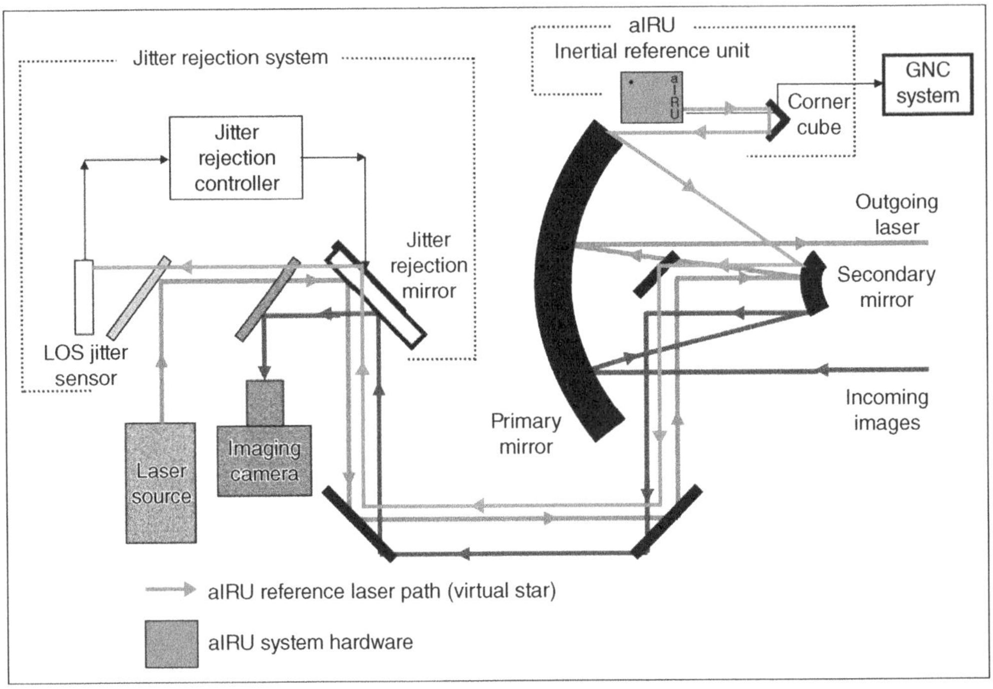

Figure 5.35 Jitter measurement and correction in optical beam.

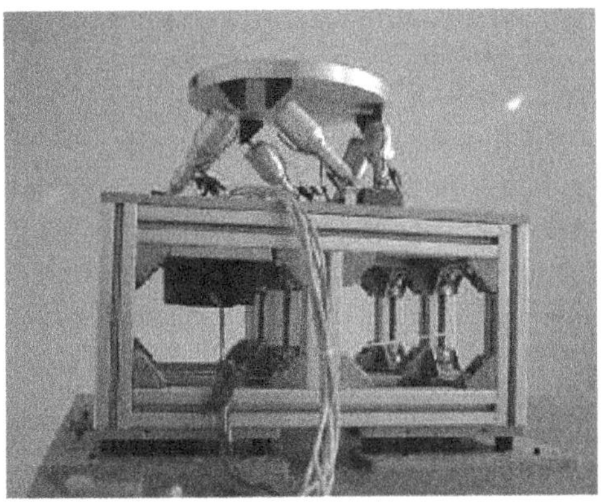

Figure 5.36 Ultra quiet platform (UQP). *Source*: [3]/IEEE.

mockup, to which an Aura Bass shaker is mounted, serving as the disturbance source. The entire experiment sits on sixteen rubber feet attached to a 3800 lb. Newport RS4000 isolation table, which uses four Newport I-2000 series laminar flow isolator pneumatic pedestals to help further isolate the experiment from floor vibration. Each strut consists of a piezoceramic stack actuator (PZT) and a geophone sensor. The PZT converts control signal voltages to physical movement of the strut. The maximum displacement of the actuator is 50 μm, which is sufficient for vibration applications, but not for platform pointing/steering. The Geospace GS-11D geophone sensors consist of wire coils supported by soft springs under the influence of a magnetic field, which provides a signal

Figure 5.37 Precision pointing hexapod. *Source*: [3]/IEEE.

proportional to velocity. The source of disturbance for the disturbance rejection experiment is an Aura Bass shaker (Model AST-1B-4, 25 W, 4 ohm) [8].

The experiment requires power amplification for the actuators and signal conditioning for the geophone sensors. These are provided via a PCB Piezotronics 790A06 six-channel power amplifier (2000 V, 100 mA), and a CSA engineering AVCS signal-conditioning unit, respectively. The control function is performed by the combination of a dSPACE DSP system and a host PC. Coding for the control algorithms is performed in the MATLAB/Simulink environment using C-coded "S-functions" to perform the more specialized tasks.

5.4.2 Precision Pointing Hexapod

The PPH experimental platform, as shown in Figure 5.37, is used for testing control algorithms for both vibration isolation of an imaging payload and fine steering. The PPH is based on an arrangement of six self-supporting electromagnetic voice coil actuators with in-line accelerometers that could enable control of higher vibration. Lower frequency steering and vibration isolation is provided using laser/photodiode based two-axis position detecting system and eddy current position sensors. The system is capable of delivering a significant range of motion in all six degrees of rigid body motion. It can deliver over 5.7 mm of axial/position travel, 20 mm of lateral motion, 2.5° of tilt motion, and 10° of twist [8].

References

1 https://www.nasa.gov/mission_pages/hubble/story/index.html.
2 https://www.harris.com/solution/james-webb-space-telescope-jwst.
3 Agrawal, Brij, "Jitter Control for Imaging Spacecraft." In *2009 4th International Conference on Recent Advances in Space Technologies (RAST)*, 615–620.
4 Kuo, Sen M., and Dennis R. Morgan, Active noise control systems, New York: John Wiley & Sons, Inc., 1996.
5 Watkins, R. Joseph and Brij N. Agrawal. Use of least mean squares filter in control of optical beam jitter, AIAA J. Guid. Control Dyn. 30(4), 1116–1122, 2007.
6 Hayes, Monson H., Statistical Digital Signal Processing and Modeling, Wiley, 1996.

7 Chen, Hong-Jen and Brij Agrawal, Method of Slewing the Spacecraft to Minimize Settling Time, Paper No. AIAA-2002-4656, AIAA Guidance, Navigation, and Control, Monterey, CA, August 2002.
8 Agrawal, B. N., and H. J. Chen, "Active Vibration Isolation on Spacecraft Using Smart Struts." IAF-00-1.4.05, In *51st International Astronautical Congress*, October 2–6, 2000, Rio de Janeiro, Brazil.

6

Adaptive Optics

6.1 Introduction

Adaptive optics (AO) is a rapidly growing multidisciplinary field encompassing physics, chemistry, electronics, and computer science. AO systems are used to correct the wavefront of a beam of light. Historically, these systems have their roots in international astronomy and US defense communities. Astronomers realized that if they could compensate for the aberrations caused by atmospheric turbulence, they would be able to generate high-resolution astronomical images; with sharper images comes an additional gain in contrast, which is also advantageous for astronomers since it means that they can detect fainter objects that would otherwise go unnoticed. While astronomers were trying to overcome the blurring effects of atmospheric turbulence, defense contractors were interested in ensuring that photons from their high-power lasers would be correctly pointed so as to destroy strategic targets. More recently, due to advancements in the sophistication and simplicity of AO components, researchers have utilized these systems to make breakthroughs in areas such as microscopy, laser communication, vision correction, and retinal imaging.

AO, as the name implies, adjusts the optical elements, such as mirror shape, in real time to counteract distortions in light wavefront caused by such atmospheric turbulence and in imaging satellites due to imperfection in the shape of the primary mirror. AO has three main components: wavefront sensor, control system, and deformable mirror (DM). Wavefront sensor measures distortion in the light wavefront. Control system calculates the necessary adjustment, such as voltage to deformable actuators, to correct distortions. Deformable mirror, with actuators, can change shape rapidly to counteract distortion, aligning the wavefront more closely with what would have been received without any disturbance. AO normally has a fourth element, a reference laser beam that follows the same path as the object light beam to measure the wavefront error that is used by control system to change the shape of the DM. Figure 6.1 shows the operation of AO for a ground telescope.

As shown in Figure 6.1, a telescope is used to image a planet. Without AO, the target light will be distorted by the atmosphere and image of the planet captured by the camera will be distorted. In AO, the reference beam is created by a sodium guide star that is created by a laser beam from the ground that excites atoms in sodium layer at an altitude of approximately 95 km. The reference beam goes through the same atmosphere, and it is also distorted. The wavefront sensor measures the wavefront error in the reference beam, which is analyzed, and the control system sends the command to actuators on the back of the DM to change shape to compensate for wavefront error and this continues until the wavefront sensor measures an error close to zero. Now the target light will be distorted by the atmosphere and then corrected by the DM and the camera will get an undistorted image.

Optical Beam Control: Imaging Satellites and Laser Systems, First Edition.
Brij N. Agrawal, Jae Jun Kim and Sachin Agrawal.

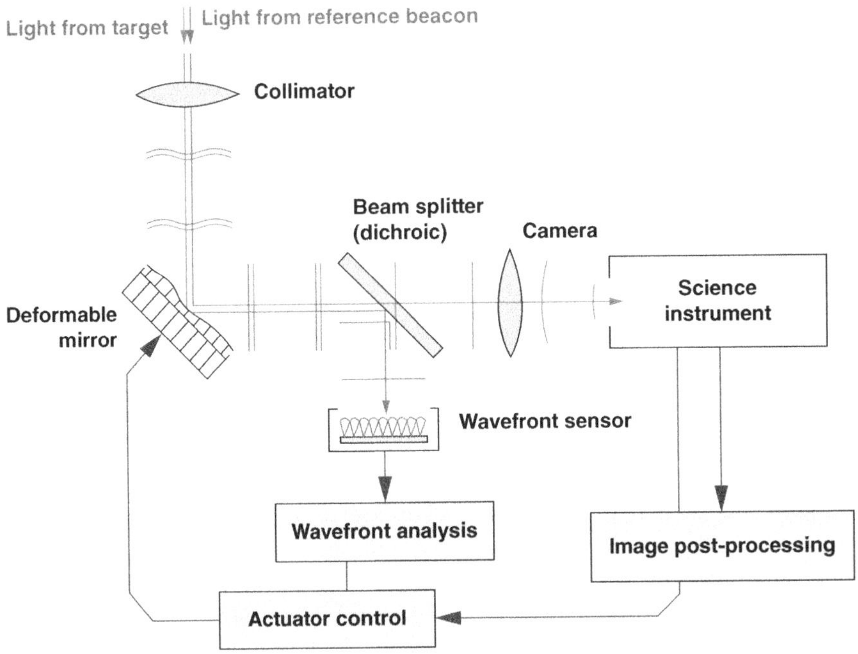

Figure 6.1 AO for ground telescopes.

AO has several applications.

1. Astronomy
   AO is widely used in large ground-based telescopes to correct for atmospheric turbulence, allowing astronomers to obtain much clearer and sharper images of celestial objects.
2. Ophthalmology
   AO is used to capture high-resolution images of the retina, providing insights into structures at the cellular level. This is valuable for diagnosing and monitoring diseases like glaucoma, diabetic retinopathy, and age-related macular degeneration.
   In laser eye surgeries, AO can improve treatment precision by compensating for individual eye aberrations, making corrective procedures like LASIK more effective.
3. Microscopy
   AO is applied in advanced microscopes to correct distortions that occur when light passes through dense, inhomogeneous biological tissues, enhancing image clarity and resolution.
4. Free-space Optical Communication
   AO is used to improve the performance of free-space laser communication systems, particularly for satellites and drones. By compensating for atmospheric interference, AO enhances the reliability and bandwidth of these communication links.
5. Military and Defense
   AO is used in imaging and surveillance systems to enhance image quality, particularly when looking through the atmosphere over long distances. In systems like high-energy lasers, AO can help maintain beam quality and direction by compensating for atmospheric distortion, making the beams more effective at long ranges.
6. Environmental Monitoring and Remote Sensing
   AO enhances the clarity and resolution of satellite imaging systems, which is useful for environmental monitoring, climate research, and remote sensing of Earth's resources.

7. Laser-based Manufacturing
   In some manufacturing processes that use high-power lasers, AO can be used to ensure the laser maintains a precise focus, improving the quality and accuracy of cutting, etching, and other laser-based fabrication techniques.

6.2 Wave Front Sensors

The role of the wavefront sensor in an AO system is to measure the wavefront deviations from a reference wavefront. There are three basic configurations of wavefront sensors commonly used: Shack–Hartmann wavefront sensor, Curvature sensor, and Pyramid sensor.

6.2.1 Shack–Hartman Wavefront Sensor

A Shack–Hartmann wavefront sensor (SHWS) uses a lenslet array to divide an incoming beam into a bunch of smaller beams, each of which is imaged onto a CCD camera, which is placed at the focal plane of the lenslet array. If a uniform plane wave is incident on a SHWS, as shown in the upper figure of Figure 6.2, a focused spot is formed along the optical axis of each lenslet, yielding a regularly spaced grid of spots in the focal plane. However, if a distorted wavefront, as shown in lower image of Figure 6.2, is used, the focal spots will be displaced from the optical axis of each lenslet. The amount of shift of each spot's centroid is proportional to the local slope (i.e. tilt) of the wavefront at the location of that lenslet.

The wavefront slope at lenslet i, j along axis x, $S_{ij}(x)$, and along axis $S_{ij}(y)$ are given as shown in Figure 6.3 as follows:

$$
\begin{aligned}
S_{ij}(x) &= \frac{1}{f}\Delta x_{ij} \\
S_{ij}(y) &= \frac{1}{f}\Delta y_{ij}
\end{aligned}
\tag{6.1}
$$

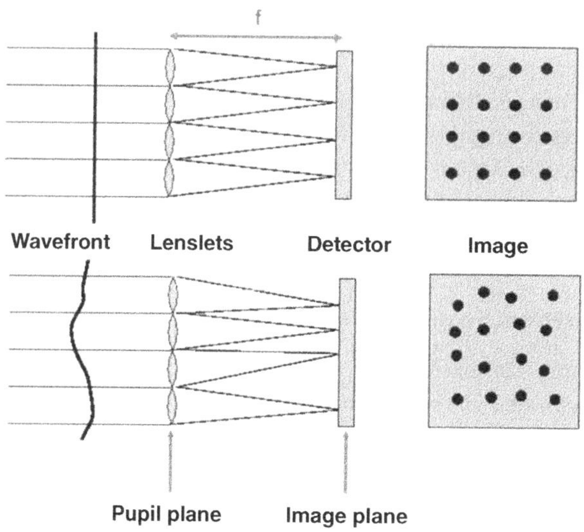

Figure 6.2 Wavefront sensor.

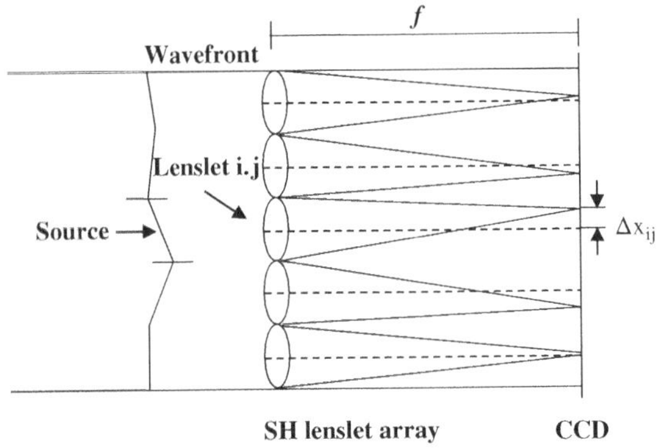

Figure 6.3 Operation of lenslet array.

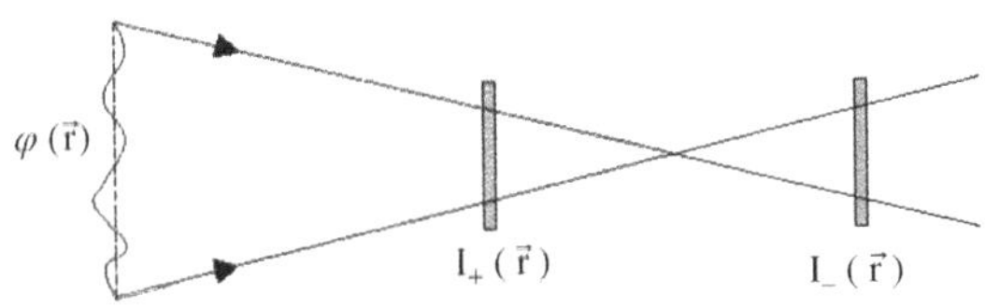

Figure 6.4 Curvature wavefront sensor.

Shack–Hartman sensor is one of the most commonly used wavefront sensors. Reconstruction of wavefront from slope will be covered in more detail in Section 6.3 of this chapter.

The four parameters that greatly affect the performance of a given SHWS are the number of lenslets (lenslet diameter, which typically ranges from ~100 to 600 μm), dynamic range, measurement sensitivity, and the focal length of the lenslet array (typical values range from a few millimeters to about 30 mm). The number of lenslets restricts the maximum number of Zernike coefficients that a reconstruction algorithm can reliably calculate; studies have found that the maximum number of coefficients that can be used to represent the original wavefront is approximately the same as the number of lenslets. When it comes to measurement sensitivity θ_{min} and dynamic range θ_{max}, these are competing specifications (see Figure 6.3 to the right). The former determines the minimum phase that can be detected, while the latter determines the maximum phase that can be measured.

Sensitivity is moderate and performs well for small to moderate aberrations, but is less sensitive to very fine wavefront changes. Dynamic range is high and can handle large aberrations due to the large number of lenslets covering the detector. Speed is very fast, suitable for real-time applications. It is a simple optical setup and is widely used due to ease of implementation and robustness. It is commonly used in astronomy, laser systems, and eye surgery, where quick and reliable measurements are critical.

6.2.2 Curvature Wavefront Sensor

The curvature wavefront sensor (CWFS) measures wavefront errors by capturing intensity distributions in planes located just before and after the focal plane, as shown in Figure 6.4. These intensity patterns provide information on the curvature of the wavefront, which can be related to the second derivative (or Laplacian) of the wavefront error.

CWFSs are especially useful in situations where high sensitivity is required, such as in low-light or high-turbulence environments (e.g. astronomy and laser communications). They offer several advantages. CWFSs are highly sensitive to both small and large wavefront distortions. They can be easier to implement because they do not require a grid of lenses (as in Shack–Hartmann sensors) or complex pyramidal optics (as in pyramid wavefront sensors, PWFS). CWFSs can capture a wide range of distortions, making them suitable for AO where continuous adjustments are required. Curvature sensors are often preferred in AO for large telescopes, where they help correct atmospheric distortion in real time, improving image resolution and detail in ground-based astronomical observations.

Here is a general outline for deriving the wavefront error from CWFS data:

6.2.2.1 Intensity Difference and Curvature

Let I_+ and I_- represent the intensity distributions on two detectors placed symmetrically on either side of the focal plane (one in front, one behind). The difference in intensity between these two planes is directly related to the Laplacian (or curvature) of the wavefront error $W(x, y)$:

$$\Delta I(x, y) = I_+(x, y) - I_-(x, y) \tag{6.2}$$

This intensity difference $\Delta I(x, y)$ is proportional to the Laplacian of the wavefront phase $W(x, y)$:

$$\nabla^2 W(x, y) = k \cdot \frac{\Delta I(x, y)}{I_0} \tag{6.3}$$

where: $\nabla^2 W(x, y)$ is the Laplacian of the wavefront error $W(x, y)$, k is a constant that depends on the setup (wavelength of light and distance between planes), and I_0 is the average intensity in the focal plane.

6.2.2.2 Poisson Equation for Wavefront Reconstruction

The wavefront $W(x, y)$ can be reconstructed by solving the Poisson equation obtained from the above relation:

$$\nabla^2 W(x, y) = k \cdot \frac{\Delta I(x, y)}{I_0} \tag{6.4}$$

Solving this equation gives the wavefront error $W(x, y)$ across the aperture. Numerical methods, such as Fourier transform methods or finite difference approaches, are commonly used to solve this equation.

6.2.3 Pyramid Wavefront Sensor

A Pyramid wavefront sensor (PWFS) consists of a pyramid-shaped prism placed at the focal plane of an optical system, which splits the incoming light beam into four separate beams, creating four distinct pupil images on a detector. By analyzing the intensity distribution across these pupil images, the sensor can measure the wavefront gradients and reconstruct the overall wavefront aberrations present in the optical system.

PWFS is highly sensitive to wavefront changes, especially for high-order (finer) aberrations. Dynamic range is lower than SHWS; it can struggle with extremely large aberrations due to the finite dynamic range of the detector. Speed is moderate to fast. It is capable of real-time correction but may require more processing than SHWS due to complex intensity analysis. High precise alignment and calibration are necessary and the optical setup is more complex than SHWS or CWFS. Applications are favored in AO systems where very fine wavefront correction is required, such as high-resolution imaging in telescopes, microscopy, and retinal imaging.

A generalized approach to determine the wavefront for PWFS is as follows.

6.2.3.1 Intensity Difference Calculation

Each of the four quadrants formed by the pyramid divides the incoming light into different paths, leading to four separate intensity patterns (I_1, I_2, I_3, and I_4) on the detector. The wavefront slope in each direction (x and y) is inferred from the differences in these intensities.

Define:

$$\begin{aligned} I_x \frac{(I_1 + I_4) - (I_2 + I_3)}{I_1 + I_2 + I_3 + I_4} \\ I_y \frac{(I_1 + I_2) - (I_3 + I_4)}{I_1 + I_2 + I_3 + I_4}. \end{aligned} \tag{6.5}$$

Here I_x and I_y are normalized intensity differences representing the wavefront slope in the x and y directions, respectively.

Solving this Poisson equation yields the wavefront error $W(x, y)$ across the aperture. Numerical techniques like Fourier transforms, finite difference methods, or iterative methods (such as the conjugate gradient method) are commonly used for this purpose.

6.2.3.2 Reconstructing the Wavefront Error

The wavefront error $W(x, y)$ (or phase) can be derived from the slopes I_x and I_y. This reconstruction involves solving a Poisson equation, which relates the Laplacian of the wavefront $W(x, y)$ to the divergence of the slope vector field.

$$\nabla^2 W(x, y) = \frac{\partial I_x}{\partial x} + \frac{\partial I_y}{\partial y}. \tag{6.6}$$

Solving this Poisson equation yields the wavefront error $W(x, y)$ across the aperture. Numerical techniques like Fourier transforms, finite difference methods, or iterative methods (such as the conjugate gradient method) are commonly used for this purpose.

6.2.4 Choosing a Sensor

For high-speed applications like laser systems and surgeries, the Shack–Hartmann sensor is usually the best choice due to its speed and simplicity. For high-sensitivity needs in detecting small aberrations, especially in large telescope systems, the curvature wavefront sensor is effective. For high-resolution, fine wavefront correction where sensitivity to high-order aberrations is critical, the pyramid wavefront sensor is advantageous, although it may be limited by dynamic range and require precise calibration.

6.3 Wave Reconstruction

In this section, we will discuss the construction of the wavefront from the wavefront gradient data provided by a SHWS or PWFS. The zonal method and modal method are two common approaches used for wavefront reconstruction.

In the zonal method, the wavefront is reconstructed locally, zone by zone, using direct numerical integration of the gradient data provided by the Shack–Hartmann Sensor.

It works with discrete grid points corresponding to the lenslet array. The wavefront shape is determined independently in small regions. In the modal method, the wavefront is reconstructed globally by fitting the data to a series of basic functions, such as Zernike polynomials. The entire wavefront is described by a set of coefficients for the chosen basis functions.

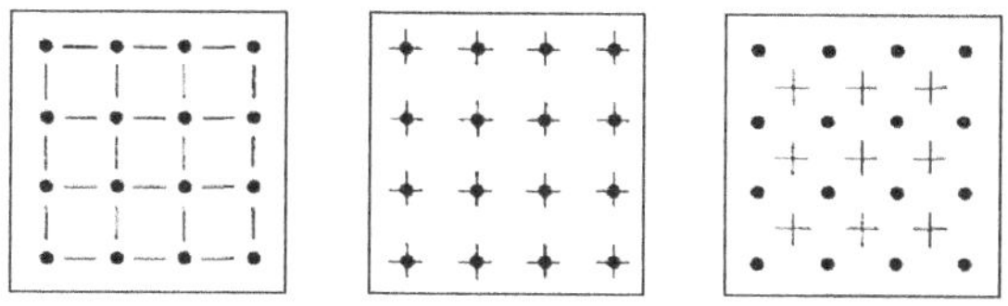

(a) Hudgin (b) Southwell (c) Fried

Figure 6.5 Slope measurement grid patterns. *Source*: [4].

6.3.1 Zonal Method

The formulation of the problem depends on the grid pattern used, which, in turn, depends on where the actual slope measurements are taken [4]. There are three grid patterns commonly used, Hudgin, Southwell, and Fried, as shown in Figure 6.5.

In Figure 6.5, the horizontal dashes indicate positions of x-slope sampling. The vertical dashes are the y-slope sampling positions. The dots are the estimated phase points. Hudgin grid could be applicable for shearing interferometer where the beam is first divided in order to apply, separately, the x and y shears. Southwell and Fried grid patterns are applicable to Shack–Hartman sensor, where each lenslet measures both x and y slopes at the same point.

6.3.1.1 Hudgin Grid Pattern

In this configuration, there are N^2 phase points and $2N(N-1)$ slope measurement points. The relationship between slope S and phase φ is given by the following equations.

$$\begin{aligned} S_i^x &= (\phi_{i+1} - \phi_i)/h && i = 1, \cdots, N-1, \\ S_{i,j}^y &= \left(\phi_{i,j+1} - \phi_{i,j}\right)/h && i = 1, \cdots, N,\ j = 1, \cdots, N-1 \end{aligned} \tag{6.7}$$

where $h = D/N$ and D is the width of the aperture. Each phase point represents an equal subregion of area h^2 in the aperture.

6.3.1.2 Southwell Grid Pattern

For Southwell grid pattern, the relationship between Slope S and phase φ is given by the following equations.

$$\begin{aligned} \frac{S_{i+1,j}^x + S_{i,j}^x}{2} &= \frac{\left(\phi_{i+1,j} - \phi_{i,j}\right)}{h} && i = 1, \cdots, N-1,\ j = 1, \cdots, N \\ \frac{S_{i,j+1}^x + S_{i,j}^x}{2} &= \frac{\left(\phi_{i,j+1} - \phi_{i,j}\right)}{h} && i = 1, \cdots, N,\ j = 1, \cdots, N-1. \end{aligned} \tag{6.8}$$

6.3.1.3 Fried Grid Pattern

For Fried grid pattern, the relationship between slope S and phase φ is given by the following equations.

$$\begin{aligned} S_{ij}^x &= \frac{\left[(\phi_{i+1,j} + \phi_{i+1,j+1})/2 - (\phi_{ij} + \phi_{i,j+1})/2\right]}{h}, \\ S_{ij}^y &= \frac{\left[(\phi_{i,j+1} + \phi_{i+1,j+1})/2 - (\phi_{ij} + \phi_{i+1,j})/2\right]}{h}. \end{aligned} \tag{6.9}$$

6.3.2 Least Squares Solution

The relationship between slope and phase can be written in matrix form as follows.

$$\boldsymbol{s} = B\boldsymbol{\phi}. \tag{6.10}$$

Since piston mode cannot be determined by Shack–Hartmann sensor, there will be an infinite number of phase solutions with the given relationship. This means that B matrix is singular. In order to correct this, we add another row such that

$$\begin{bmatrix} s_1 \\ s_2 \\ \vdots \\ s_M \\ 0 \end{bmatrix} = \begin{bmatrix} & & B & & \\ 1 & 1 & 1 & \cdots & 1 \end{bmatrix} \begin{bmatrix} \phi_1 \\ \phi_2 \\ \vdots \\ \phi_N \end{bmatrix} \tag{6.11}$$

The last row in the new B matrix forces the overall average phase (piston) to be zero across the aperture. This system is overdetermined and the inverse of B cannot be used to determine the phase from the slope. Instead an approximate least squares estimate of the phase can be computed, if B has full rank. The least squares problem can be formulated as

$$\min_{\phi} \ \|\boldsymbol{s} - B\boldsymbol{\phi}\|^2. \tag{6.12}$$

Solving this problem yields the pseudoinverse of B, known as the reconstruction matrix, which is used to compute the phase estimate $\hat{\boldsymbol{\phi}}$

$$\hat{\boldsymbol{\phi}} = \left(B^T B\right)^{-1} B^T \boldsymbol{s}. \tag{6.13}$$

Even with piston mode correction, the B matrix could be poorly conditioned. A practical method for numerically stable inversion is the singular value decomposition (SVD). The pseudoinverse of B can be computed from the SVD

$$B^{\dagger} = V\Sigma^{\dagger} U^T \tag{6.14}$$

where $\Sigma^{\dagger}$ is formed by taking the reciprocal of each nonzero singular value in Σ, transposing the matrix, and padding with zeros as needed to match dimensions.

6.3.3 Modal Reconstruction

In this method, phase is represented in terms of modal coefficients covering the whole aperture. The problem with reconstructing the phase modes from wavefront slopes is similar to that of reconstruction of the phase values at various points as discussed in the zonal methods. The wavefront phase φ is described by a polynomial expansion

$$\phi = \sum_{k=1}^{K} a_k Z_k(x, y) \tag{6.15}$$

where the coefficients are a_k and the polynomial basis functions (modes), such as Zernike polynomials, are $Z_k(x, \ y)$. A Hartmann wavefront sensor measures the slopes of the wavefront at various positions m. There are M total measurements. For example, assume that half of the measurements are slopes in the x direction and half are in the y direction. The set of linear equations is

$$\left.\frac{\partial\phi}{\partial x}\right|_m = \sum_{k=1}^{K} a_k \left.\frac{\partial Z_k(x, y)}{\partial x}\right|_m$$
$$\left.\frac{\partial\phi}{\partial y}\right|_m = \sum_{k=1}^{K} a_k \left.\frac{\partial Z_k(x, y)}{\partial y}\right|_m \qquad (6.16)$$

The subscript m implies evaluation at the coordinates (x_m, y_m). This set can be represented by the familiar matrix equation, $\boldsymbol{s} = B\boldsymbol{a}$, where the vector $\boldsymbol{s}$ is the slope vector and $\boldsymbol{a}$ is the mode coefficient vector, given by

$$\boldsymbol{s} = \left[\left.\frac{\partial\phi}{\partial x}\right|_1 \quad \left.\frac{\partial\phi}{\partial x}\right|_2 \quad \cdots \quad \left.\frac{\partial\phi}{\partial x}\right|_{M/2} \quad \left.\frac{\partial\phi}{\partial y}\right|_1 \quad \left.\frac{\partial\phi}{\partial y}\right|_2 \quad \cdots \quad \left.\frac{\partial\phi}{\partial y}\right|_{M/2}\right]^T,$$

$$\boldsymbol{a} = [a_1 \quad a_2 \quad a_3 \quad \cdots \quad a_K]^T \qquad (6.17)$$

and matrix B is

$$B = \begin{bmatrix} \left.\frac{\partial Z(x, y)_1}{\partial x}\right|_1 & \left.\frac{\partial Z(x, y)_2}{\partial x}\right|_1 & \cdots & \left.\frac{\partial Z(x, y)_K}{\partial x}\right|_1 \\ \left.\frac{\partial Z(x, y)_1}{\partial x}\right|_2 & \left.\frac{\partial Z(x, y)_2}{\partial x}\right|_2 & \cdots & \left.\frac{\partial Z(x, y)_K}{\partial x}\right|_2 \\ \vdots & \vdots & & \vdots \\ \left.\frac{\partial Z(x, y)_1}{\partial x}\right|_{M/2} & \left.\frac{\partial Z(x, y)_2}{\partial x}\right|_{\frac{M}{2}} & \cdots & \left.\frac{\partial Z(x, y)_K}{\partial x}\right|_{M/2} \\ \left.\frac{\partial Z(x, y)_1}{\partial y}\right|_1 & \left.\frac{\partial Z(x, y)_2}{\partial y}\right|_1 & \cdots & \left.\frac{\partial Z(x, y)_2}{\partial y}\right|_1 \\ \vdots & \vdots & & \vdots \\ \left.\frac{\partial Z(x, y)_1}{\partial y}\right|_{M/2} & \left.\frac{\partial Z(x, y)_2}{\partial y}\right|_{\frac{M}{2}} & \cdots & \left.\frac{\partial Z(x, y)_K}{\partial y}\right|_{M/2} \end{bmatrix} \qquad (6.18)$$

The elements are the derivatives of the polynomial basis functions evaluated at the wavefront sensor sub-aperture positions. The solution is found as before by finding the least squares estimate or

$$\hat{\boldsymbol{a}} = (B^T B)^{-1} B^T \boldsymbol{s}. \qquad (6.19)$$

6.4 Fast Steering Mirrors and Deformable Mirrors

6.4.1 Fast Steering Mirrors

Fast steering mirror (FSM) provides two-axis, high bandwidth, precision angular motion for a laser beam. The mirror orientation is controlled by a set of actuators. Typically, voice coil actuators are used for large angular motion and piezoelectric actuators for high resolution and lower angular range. For AO, it is used to correct beam angle error (tip and tilt mode of aberration) which is too large for DM to correct. It is also used for laser communication systems to maintain alignment with a receiver or transmitter in dynamic conditions, imaging systems to stabilize or scan images in cameras or optical sensors, and defense and aerospace in targeting and tracking systems.

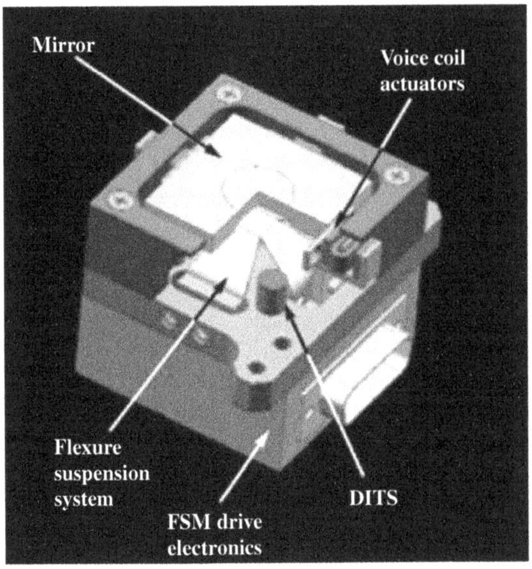

Figure 6.6 Typical voice coil actuated FSM assembly.

6.4.1.1 Voice Coil Actuated FSM

To create a practical FSM, voice coil actuators are used to tilt the mirror substrate (Figure 6.6). Four actuators are mounted behind or to the side of the mirror, one in each quadrant. Two voice coils connected by the diameter of the mirror operate as a push/pull pair, rotating about the axis that bisects them. Therefore, two actuator pairs (four coils) are used to produce two orthogonal rotations (θx, θy). A unique flexure suspension system is used to support the mirror. This flexure allows free rotation about orthogonal axes while constraining side-to-side motion and rotation about the normal (z) axis. In most implementations, the mirror pivots about its front surface, making it a true gimbal mount. Depending upon the application, differential impedance transducers can be included in the assembly to provide absolute position feedback at the 0.1 μrad level, referencing the FSM support structure. Additionally, a reaction-torque cancellation mass can be incorporated into the FSM design. This mass moves in the opposite direction from the mirror and is designed to cancel the recoil motion of the base structure. While this increases the cost of the FSM assembly, it substantially reduces the net vibration caused by rapid mirror motion, thereby simplifying the design of surrounding companion structures.

6.4.1.2 Piezoelectric Actuated Mirrors

The singular outstanding feature of piezo-actuated mirrors is their high resolution; they are capable of producing extremely small and precise motions. Furthermore, they offer high response speeds and are capable of moving fairly large optics. It is also possible to construct piezo-actuated mirror mounts offering two degrees of motion, thus avoiding the displacement jitter inherent in galvanometer scanners.

The major limitation of piezo-actuated mirrors is their severely restricted angular range, typically on the order of a milliradian. Thus, a transient motion could put the beam out of the actuator's angular range will cause the system to lose its alignment lock. As a result, piezo-based beam stabilization systems are best suited to systems that are substantially limited in the range of tilt errors that can be encountered.

6.4.1.3 Identification of the FSM Dynamics

FSMs generally have two operating modes: open-loop and closed-loop. In open-loop mode, FSM simply converts the signal voltage to the current for the voice coil motor controlling the mirror angle. In closed-loop mode, the mirror angle measured by the built-in angle position sensor is fed back to the controller to form the internal feedback control loop. In closed-loop mode, identification of the FSM dynamics is necessary. It is needed to have a plant model of the FSM that includes digital-analog conversion and sampling effect of the command and sensor output. Conducting experiments is a straightforward and accurate way to obtain this information.

Open-loop frequency response can be obtained by the ratio of frequency spectrum of the input and output by the experiment as shown in Figure 6.7.

For closed-loop mode, the feedback control block diagram control is shown in Figure 6.8.

As an example, for OKM FSM, closed-loop response for proportional and integral controller is given in Figure 6.9, for both X and Y axes.

To illustrate the performance characteristics relevant to system identification and control design, Table 6.1. summarizes the typical specifications of a commercial fast steering mirror (FSM), the Newport FSM 300-01.

6.4.2 Deformable Mirrors

A DM is an optical device used to dynamically shape and control the wavefront of light by altering the mirror's surface in real time. It consists of a reflective surface that can be adjusted using actuators (such as piezoelectric elements or micro-electromechanical systems, MEMS) to compensate for distortions in the light path. It has applications in astronomy in AO systems of telescopes; in laser systems in high-power lasers, they optimize beam quality by correcting aberrations in the laser wavefront, vision science to assist in customizing treatment to the patient's optical aberrations, in free-space optical communication to mitigate distortions in signals transmitted through

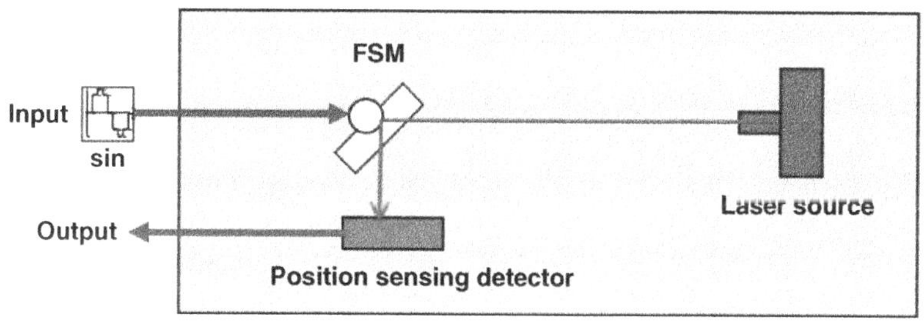

Figure 6.7 Open-loop response.

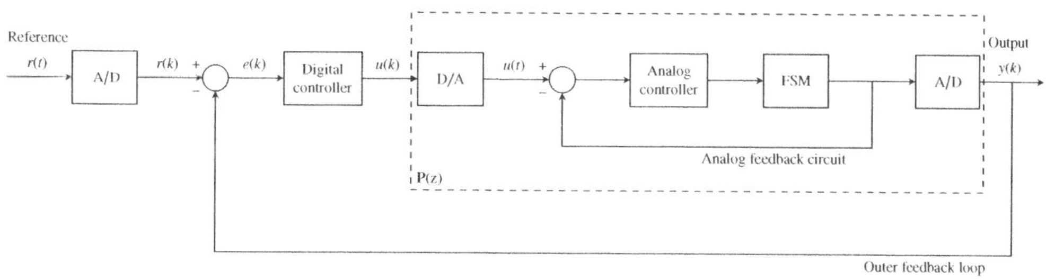

Figure 6.8 Closed-loop response.

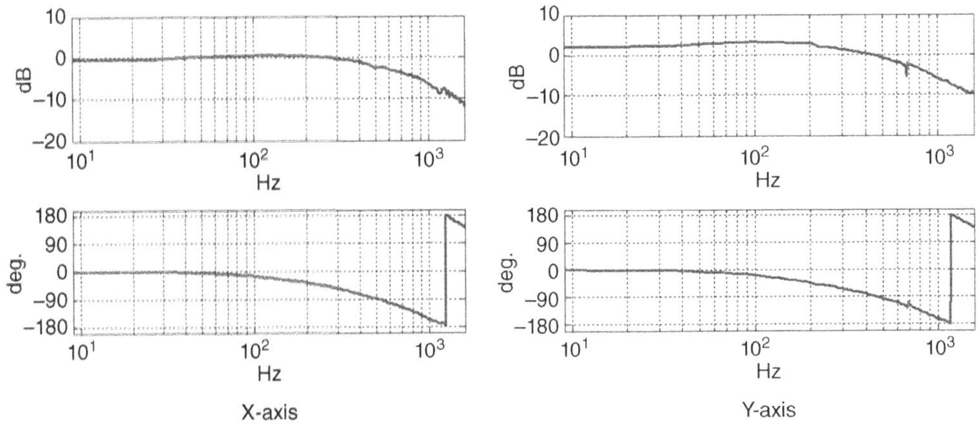

Figure 6.9 Closed-loop response for OKM FSM.

Table 6.1 Typical specifications for Newport FSM 300-01.

Typical specifications	
Newport FSM 300-01	
Mirror	1 inch
Actuators	Voice coil
Angular range	±1.5° (mechanical)
Fast closed-loop control	580 Hz
Resolution	≤1 μrad rms (mechanical)
Accuracy	≤0.262 μrad rms (mechanical)

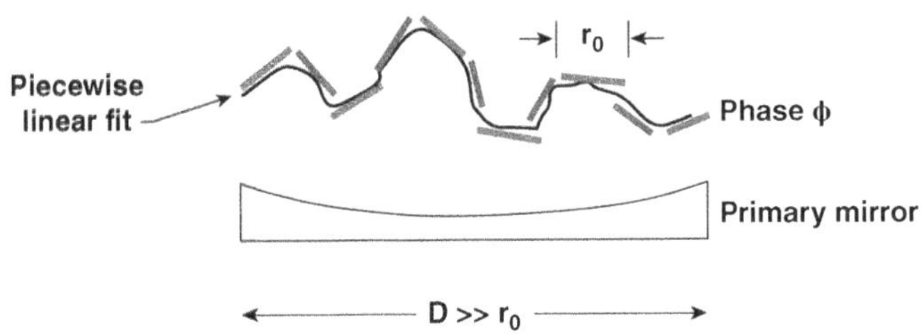

Figure 6.10 Required number of actuators.

turbulent atmospheric conditions, and in military and defense for laser weapons to correct beam aberration due to atmosphere turbulence.

6.4.2.1 DM Characteristics

6.4.2.1.1 Number of Actuators

To calculate required number of actuators, divide the primary mirror into "subapertures" of diameter r_0. Number of subapertures ~$(D/r_0)^2$, where r_0 is evaluated at the desired observing wavelength, as shown in Figure 6.10.

$$r_0 = \left[0.423k^2 \sec\varsigma \int_0^H C_N^2(h)dh\right]^{\left(-\frac{3}{5}\right)} \tag{6.20}$$

where: $k = 2\pi/\lambda$, z = zenith angle, and C_N^2 = refractive index structure parameter, h = height.

6.4.2.1.2 Actuator Stroke

The stroke requirements for the actuators depend upon the strength of the aberrations. For Kolmogorov turbulence, assuming tilt corrected by FSM, variance in units per waves squared) is given by

$$\sigma^2(waves^2) = 0.00357\left(\frac{D}{r_0}\right)^{5/3} \tag{6.21}$$

where D is the aperture diameter, assuming that most of the disturbance will be within $\pm 2.5\sigma$. Also, because of the reflection off the mirror surface, one unit of motion of mirror results in two units of wavefront correction. The total required stroke of the DM will be

$$Stroke(wave) = \frac{1}{2}\left(5\sqrt{0.00357\left(\frac{D}{r_0}\right)^{5/3}}\right) \tag{6.22}$$

6.4.2.1.3 Fitting Error

The fitting error in DMs refers to the discrepancy between the desired wavefront correction and the best approximation that a DM can achieve with its finite number of actuators and specific deformation modes.

For Kolmogorov turbulence, the fitting error is given by

$$\sigma_{\text{fitting}}{}^2 = a_F(d/r_0)^{5/3}\,\text{rad}^2 \tag{6.23}$$

where a_F fitting number depends on the type of DM, d is the diameter of sub-aperture, distance between actuators, and r_0 is a Fried number.

6.4.2.2 Types of DMs

DMs come in various types, each designed for specific applications based on how they achieve surface deformation and the degree of wavefront correction they provide. Here are the main types of DMs:

6.4.2.2.1 Continuous-faceplate DMs

The structure is a single, continuous reflective surface supported by multiple actuators underneath as shown in Figure 6.11.

When you apply voltage to the actuator (PZT, PMN), it expands or contracts in length, thereby pushing or pulling on the mirror. Facesheet thickness must be large enough to maintain flatness during polishing, but thin enough to deflect when pushed or pulled by actuators. Thickness also determines "influence function," response of mirror shape to "push" by 1 actuator. Thick face sheets result in a broad influence function and thin face sheets with a more peaked influence function. Actuators have to be stiff, so they would not bend sideways as the mirror deflects.

To better understand the behavior of deformable mirrors and their response characteristics, it is helpful to examine the properties of common actuator materials used in these systems specifically, PZT (lead zirconate titanate) and PMN (lead magnesium niobate). The characteristics of PZT and PMN) are as follows.

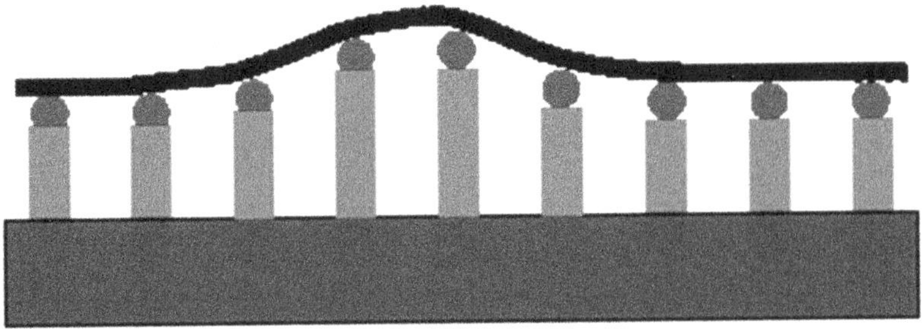

Figure 6.11 Continuous-faceplate DMs.

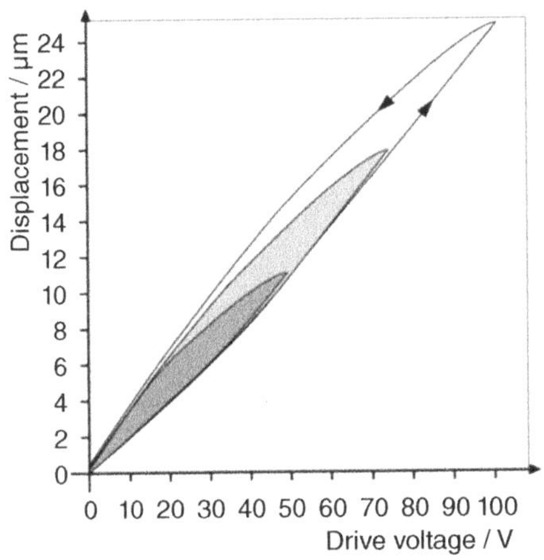

Figure 6.12 Voltage versus displacement characteristic of PZT.

The characteristics of PZT (lead zirconate titanate) and PMN (lead magnesium niobate) are as follows.

6.4.2.2.1.1 PZT (Lead Zirconate Titanate) The displacement is linear in voltage. It has 10–20% hysteresis as shown in Figure 6.12. It consists of a stack of PZT disks with integral electrodes.

6.4.2.2.1.2 PMN (Lead Magnesium Niobate) As shown in Figure 6.13, the displacement is a non-linear, quadratic response. It can "push" and "pull" if a bias is applied. Hysteresis is low, 1–3%; more temperature sensitive than PZT.

Northrop Grumman/AOX developed a 349 channel, 2.5 mm spaced, 2-inch diameter DM for the Naval Postgraduate School. It has a compact configuration with actuators, consisting of 7 × 7 actuator modules assembled in a 3 × 3 array. Actuator stroke is 2.5 µm.

These DMs provide smooth surface deformation, ideal for high-precision applications like AO in telescopes and in high energy lasers, as they can survive large heat. They are, however, expensive, costing about $1000 per channel. Fitting factor for continuous face-sheet DMs is $a_F = 0.28$.

6.4.2.2.2 Segmented DM

The configuration for a DM is shown in Figure 6.14.

Each actuator can move just in piston (one degree of freedom), or in piston plus tip-tilt (three degrees of freedom). For one degree of freedom, $a_F = 1.26$. For three degrees of freedom: $a_F = 0.18$.

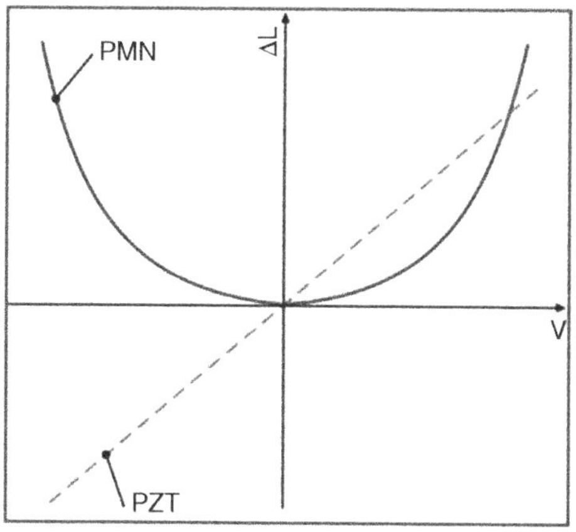

Figure 6.13 Voltage versus displacement characteristic of PMN.

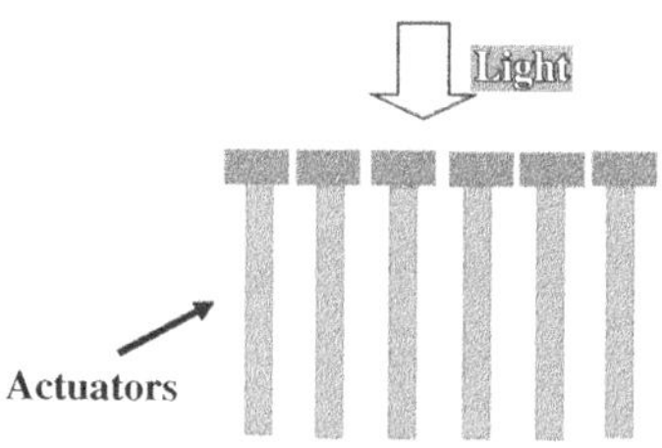

Figure 6.14 Segmented deformable mirror.

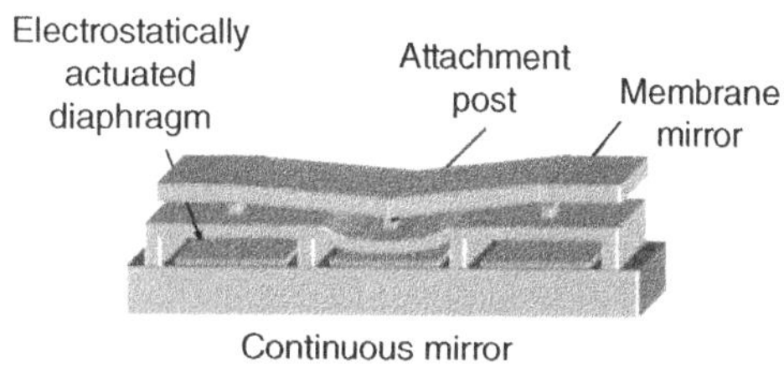

Figure 6.15 MEMS Deformable Mirror.

6.4.2.2.3 Micro-electro-mechanical Systems DM

The structure consists of miniaturized mirrors with a dense array of actuators, fabricated using microfabrication techniques. The DM consists of three layers: an electrode layer, an actuator layer, and a mirror layer. The actuator layer is grounded. When a voltage is put on an electrode, the actuator layer above it is attracted and pulls the mirror layer down. Figure 6.15 shows configuration of a MEMS DM.

It can be made in different configurations. If large, a smooth surface is desired, then the mirror layer is continuous across the entire mirror. If each actuator controls a small region that is independent of the others, then it is a segmented mirror. If a larger segment controlled by multiple actuators is desired, the surface can be divided and one can have tip-tilt and piston motion. These DMs are significantly cheaper than PZT or PMN DMs. However, the mirror size is small and cannot be used for HEL applications.

6.4.2.3 Comparison of DMs

Here is a comparison of continuous-faceplate, segmented, and MEMS deformable mirrors, along with their advantages and disadvantages:

1. Continuous-faceplate DMs
   Advantages: High-quality optical surface suitable for precision imaging. Continuous surface avoids diffraction effects caused by segment edges. Better suited for applications requiring smooth wavefront correction. Can be used for HEL applications.

Disadvantages: Limited stroke compared to segmented mirrors. Generally heavier and less compact. Higher cost due to the need for precise actuator and surface alignment. They are expensive.

2. Segmented DMs
   Advantages: Greater range of motion or "stroke," making them ideal for correcting large aberrations. Modularity simplifies certain repairs and customizations. It is easier to scale for larger aperture systems.
   Disadvantages: Gaps between segments can cause diffraction and reduce optical quality and cannot be used for HEL applications. More complex to control and align due to the need for synchronization between segments. Less suitable for smooth wavefront corrections compared to continuous-faceplate mirrors.
3. MEMS DMs
   Advantages: Highly compact and lightweight, suitable for portable and space-limited systems. Scalable manufacturing allows for lower production costs at high volumes.
   Fast response times due to the small size of actuators.
   Disadvantages: Small mirror size. Limited stroke compared to segmented mirrors. Mechanical limitations and reliability concerns over extended use. Surface roughness is typically higher, which may degrade image quality for high-precision applications.

6.4.2.4 DMs for NPS Multi-conjugate AO Testbed

NPS multi-conjugate AO testbed has three DMS, woofer DM, tweeter DM, and multi-conjugate DM as given in Figure 6.16.

OKO 96ch MMDM (1 inch diameter)				
Stroke max	Refresh rate	Turbulence correction	Power	Hysteresis
19 000 nm (center)	2 kHz	Excellent up to 1.5 m aperture	Up to 150 W with a special coating	Negligible

OKO 109ch PDM (2 inch Diameter)				
Stroke max/inter-actuator	Refresh rate	Turbulence correction	Power	Hysteresis
8000 nm/ 2000 nm	Greater than 2 kHz	Excellent up to 3 m aperture	Up to 1 kW with a special coating	<10%

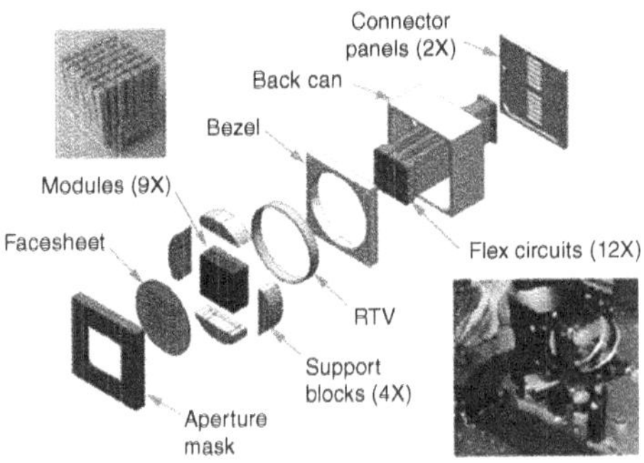

Northrop Grumman Xinetics 349ch (441ch square) PMN 2 inch DM				
Actuator stroke	Full amplitude bandwidth	Coating	Actuator type	Hysteresis
2.5 μm	Greater than 2 kHz	Protected silver	Lead magnesium niobate (PMN)	< 15% at 22C

Figure 6.16 DMs for NPS multi-conjugate AO testbed.

6.5 AO Configurations

6.5.1 Conventional AO System

A conventional AO system refers to an AO system that uses a single wavefront sensor and a single DM. The conventional AO system for imaging an astronomical object is shown in Figure 6.17. The system has a point source reference beacon, which can either be a science target itself or a natural guide star that is nearby with the target. A telescope is used to collect light, reduce diffraction effects, and enhance optical resolution. To compensate for line-of-sight error and optical jitter, the system needs to employ a FSM and a tracker sensor. The FSM is typically positioned at the pupil plane of the telescope, such as the exit pupil of a telescope, to correct the angular position of the entire optical beam. Additionally, a single wavefront sensor, located at the relayed pupil plane, measures the wavefront error across the entire optical beam originating from the reference beacon. Finally, a DM, also situated at the relayed pupil plane, is used to correct the wavefront error of the optical beam originating from both the reference beacon and the science target. The example system shown in Figure 6.17 uses a natural guide star for both target object tracking and wavefront sensing.

In Figure 6.17, anisoplanatic angle θ_0 represents the angular range over which a given guide star can effectively be used to measure and correct for atmospheric distortions of a science target. The effective field of view of a ground telescope with a natural guidestar AO is limited by this anisoplanatic angle, which is as small as 1–2 arcsec in the visible wavelength range and around 2–3 arcsec in the near-infrared range. Using a guide star beam path to correct the other beam paths diviated by the angle θ results in an error represented by the phase variation as

$$\sigma_a^2 = \left(\frac{\theta}{\theta_0}\right)^{\frac{5}{3}} \tag{6.24}$$

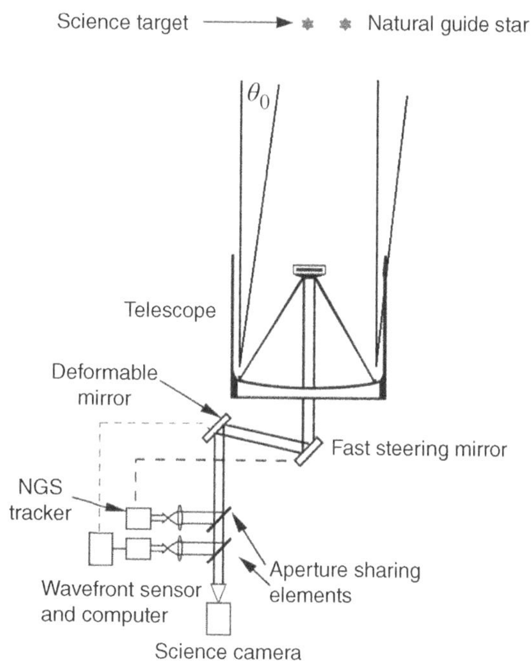

Figure 6.17 Conventional AO system configuration.

6.5.2 Laser Guide Star AO

One of the challenges of conventional AO is that the target object cannot serve as a reference beacon due to its insufficient brightness and/or its extended shape, and a natural guide star that within the anisoplanatic angle θ_0 is not readily available. To address this, a laser guide star is created by projecting a laser beam into the atmosphere. The two common methods for a laser guide star reference beacon are a Rayleigh beacon and a sodium beacon.

Rayleigh beacons use Rayleigh scattering of a laser beam as it interacts with the atmospheric molecules, primarily nitrogen and oxygen, creating a primarily short-wavelength light, such as blue or ultraviolet light. The Rayleigh scattering typically happens in the lower part of the atmosphere about 10–20 km above the earth's surface and a range gate is typically required to produce a reference beacon. A sodium beacon is tuned to a specific wavelength (589 nm) and created by exciting sodium atoms in the mesosphere about 80–100 km above Earth's surface. The light from a sodium beacon is much brighter and more point-like compared to the diffuse light from Rayleigh scattering. However, Rayleigh beacons can be created without relying on specific elements like sodium layer of the atmosphere, making them more versatile in some cases.

Due to the finite altitude of the laser beacons compared to the natural guide star beacons shown in Figure 6.18, laser beacons cause focal anisoplanatism (a.k.a. cone effect), which is the difference between the wavefront distortion from a natural star reference and the laser beacon

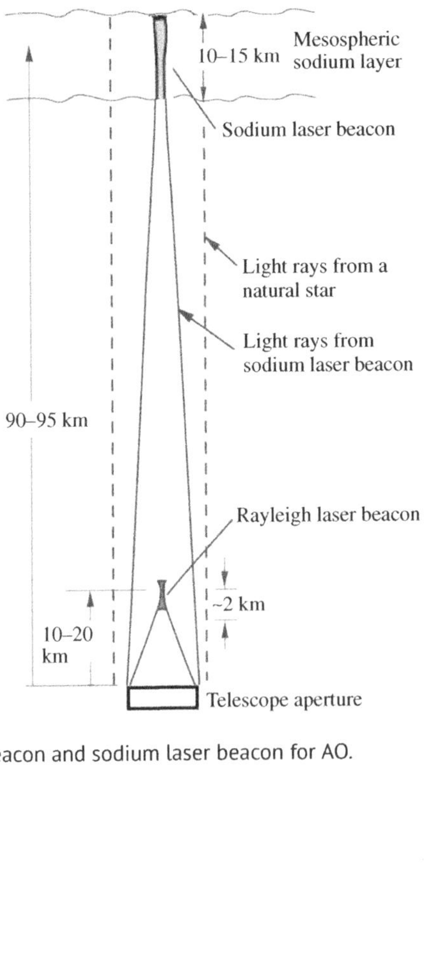

Figure 6.18 Rayleigh laser beacon and sodium laser beacon for AO.

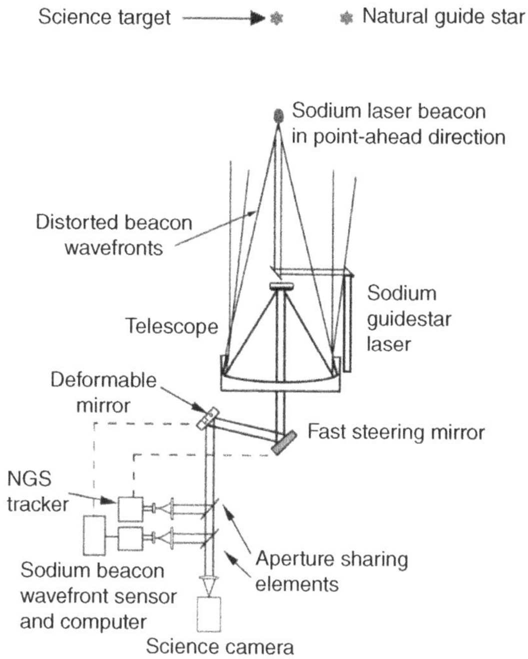

Figure 6.19 Sodium laser guide star AO.

reference. The wavefront phase variance due to focal anisoplanatism is proportional to $D^{5/3}$, such that

$$\sigma_F^2 = \left(\frac{D}{d_0}\right)^{\frac{5}{3}} \tag{6.25}$$

where D is the diameter of the aperture, d_0 is the focal anisoplanatic coherence length determined by the height of the laser beacon, C_n^2 profile, the zenith angle, and the imaging wavelength.

For the correction of line-of-sight error and atmospheric jitter, a natural guide star or a target image is still required, as the laser beacon light travels through the same atmosphere twice in opposite directions, canceling out the tilt information. In Figure 6.19, a natural guide star is used for the line-of-sight tracking and optical jitter mitigation, while the laser beacon is used for wavefront sensing and correction.

The laser guide star AO system configuration is similarly applicable for laser beam control applications such as free-space satellite laser communications, shown in Figure 6.20. Since atmospheric turbulence is concentrated near the earth's surface, downlink laser beam has significantly larger wavefront errors compared to the uplink laser beacon. The downlink optical beam could simultaneously serve as a laser beacon for the line-of-sight tracker, wavefront sensing and correction, and communication data link.

6.5.3 Multi-conjugate AO System

The conventional AO system measures the wavefront error caused by the 3D atmospheric turbulence as a 2D projection of the turbulent volume along the given optical beam path between the

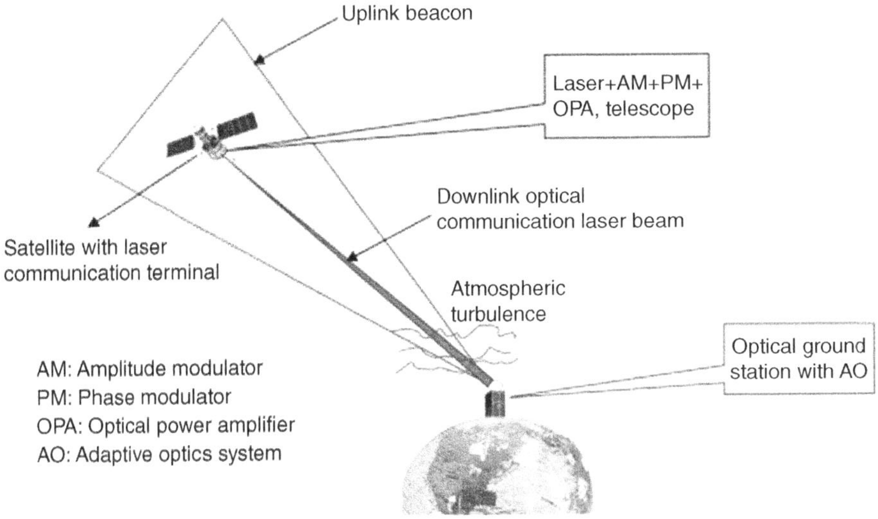

Figure 6.20 AO system for laser communication.

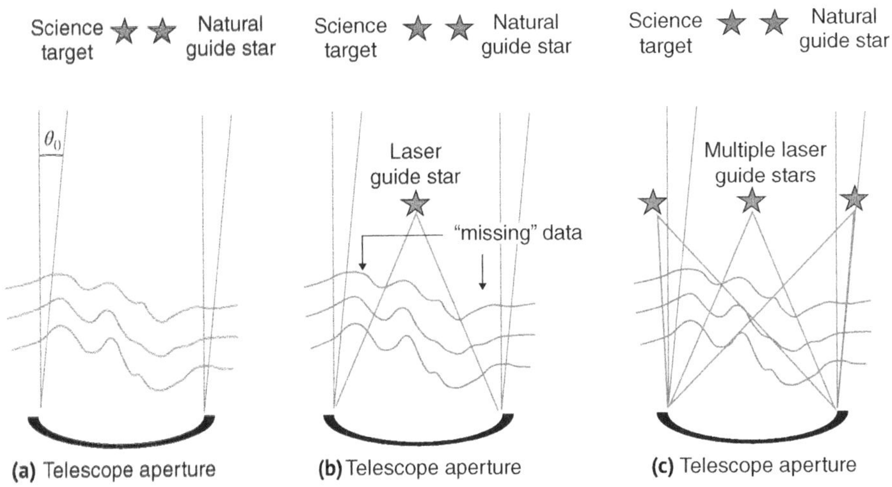

Figure 6.21 (a) Anisoplanatic angle, (b) cone effect of laser guide star, and (c) multiple guide stars for tomographic wavefront sensing.

reference beacon and the telescope aperture. One of the major limitations of conventional AO for imaging of astronomical objects is the limited field-of-view since the optical beam does not cross the same volume of turbulence across the entire field-of-view.

In Figure 6.21a, anisoplanatic angle limits the effective field of view of the AO system. Focal anisoplanatism shown in Figure 6.21b further limits the AO system performance. Mitigating those errors requires that the wavefront aberrations be estimated within the 3D volume of the atmosphere. Multi-conjugate AO for larger field-of-view uses multiple beacons as shown in Figure 6.21c to probe the atmospheric aberration along different lines of sight, followed by computing and applying wavefront corrections at the multiple layers along the beam path.

Figure 6.22 shows two different approaches for MCAO. In the layer-oriented MCAO shown in Figure 6.22a, turbulence at different atmospheric layers is corrected by placing multiple DMs and wavefront sensor pairs conjugated to these layers. Each wavefront sensor is optically

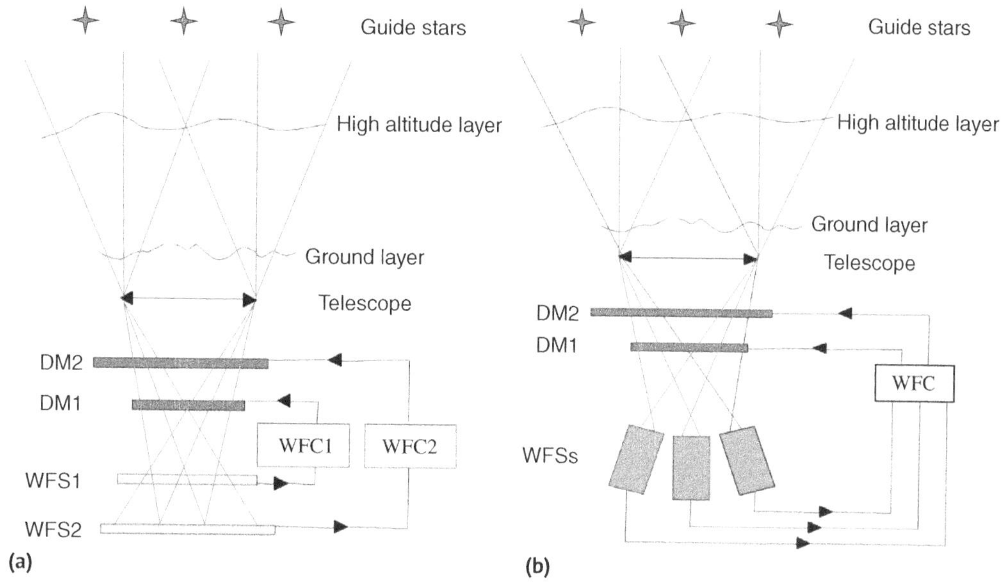

Figure 6.22 (a) Layer-oriented MCAO and (b) star-oriented MCAO.

uncoupled to all the others and co-adding guide stars can improve the SNR of wavefront sensors. The layer-oriented MCAO is suitable for conditions with distinctive atmospheric turbulence layers, and its performance depends on the accurate placement of wavefront sensors and DMs at the correct conjugate layers.

For star-oriented MCAO shown in Figure 6.22b, each wavefront sensor looks at its dedicated guide star. A single global reconstructor is used from multiple wavefront sensors to optimize the correction over a number of line of sights in the field. First, tomographic reconstruction of the atmospheric layers on a set of predefined altitudes is determined. The second is the projection (or least-square fit) of this tomographic information onto the DMs.

Figure 6.23 shows comparison of conventional AO and multi-conjugate AO with improved AO performance.

In the laser beam control applications, multi-conjugate AO could also be used for correction of both amplitude and phase of the coherent optical beam. While AO typically focuses on compensating for phase distortions, optical systems can also experience amplitude distortions due to non-uniform illumination, attenuation, or scintillation caused by atmospheric turbulence. In Figure 6.24, two DMs are placed optically conjugate to the collecting aperture pupil plane and to a finite range from this aperture. Phase shifts introduced by a DM placed at a finite conjugate plane can produce constructive and destructive interference patterns in the pupil plane. By carefully engineering these phase shifts and combining them with those from another DM located in the pupil plane, the MCAO system can effectively modulate both amplitude and phase. Although this approach can be applied to high-contrast imaging, laser beam shaping, and scintillation mitigation, it is less commonly used because amplitude modulation via interference is highly sensitive to spatial variations across the field of view and often requires complex nonlinear optimization to achieve the desired correction.

6.5.4 Woofer–Tweeter AO System

The term woofer–tweeter refers to the system's use of two types of DMs, each serving a different purpose in compensating for different scales of atmospheric distortions. The woofer mirror is

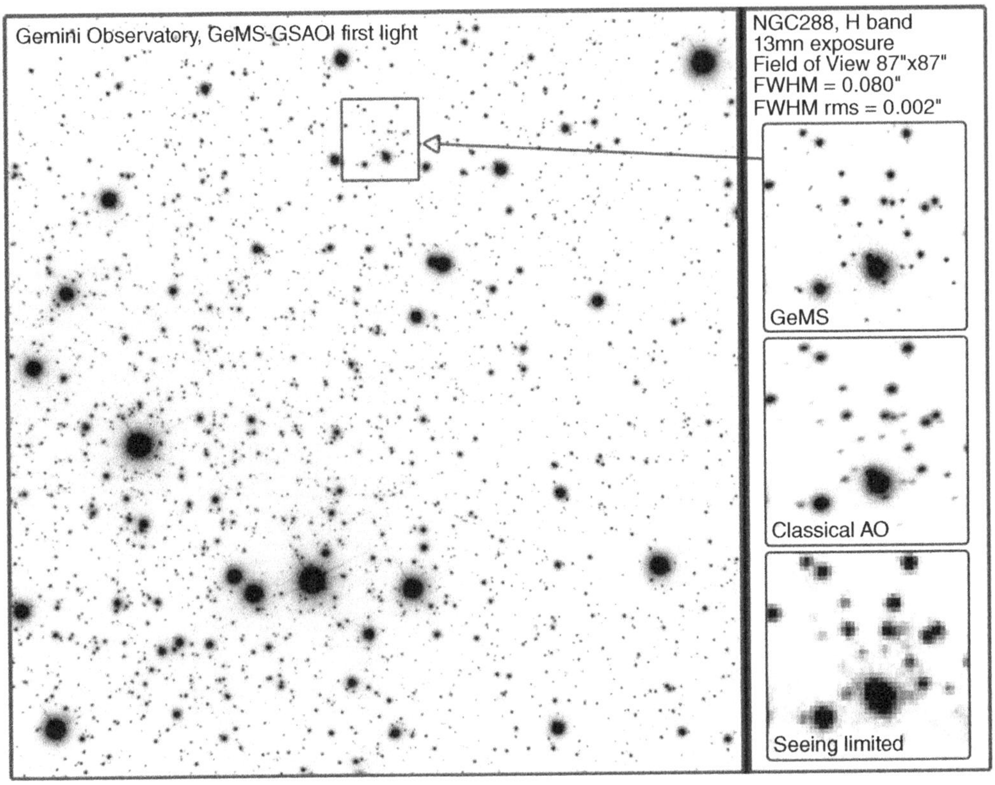

Figure 6.23 Comparison of conventional AO and multi-conjugate AO for Gemini Observatory. *Source*: Gemini Observatory [5].

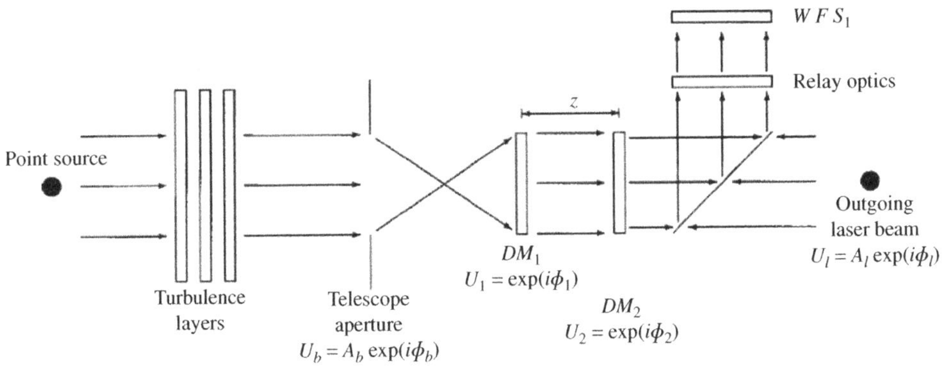

Figure 6.24 Multi-conjugate AO for laser beam control. *Source*: [6].

larger and designed to handle low-frequency distortions that affect the entire telescope's field of view. The tweeter mirror is smaller and is used to correct for high-frequency distortions, which are caused by small-scale atmospheric turbulence. Unlike multi-conjugate AO systems, woofer–tweeter DMs are typically located on an identical planes of the optical system. Woofer–tweeter system can be combined with the multi-conjugate AO system to provide enhanced AO system performance.

6.5.5 Beaconless Target-in-the-loop System

A beaconless target-in-the-loop system is a technique used in AO to correct for wavefront error without requiring a dedicated reference beacon, such as a natural guide star or a laser guide star, and a dedicated wavefront sensor. The beaconless target-in-the-loop approach allows for direct correction of the target itself, which eliminates the need for an external reference source.

Figure 6.25 shows the schematics of target-in-the-loop AO system using the stochastic parallel gradient descent optimization technique. Small random perturbations are applied simultaneously (in parallel) to the actuators of the DM. These perturbations are stochastic in nature, meaning they are randomly chosen. After the perturbation, the result will either improve or worsen the image quality (depending on the nature of the perturbation). The system then calculates the gradient of the image quality cost function with respect to each actuator of a DM. Based on the computed gradient, the DM actuators are iteratively adjusted using the gradient decent until the error is minimized.

Although target-in-the-loop AO allows a beaconless system, iterative nature of the wavefront correction may not be suitable for imaging distant targets and fast changing atmospheric turbulence. Another application of target-in-the-loop system is adaptive phase modulation for coherent beam combination shown in Figure 6.26 where single light detector is used to maximize the

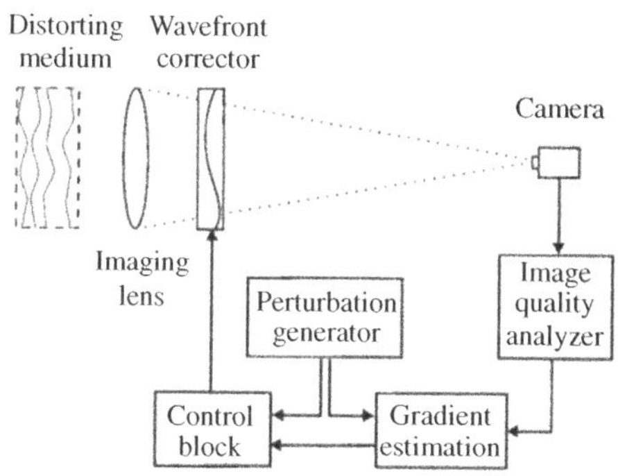

Figure 6.25 AO using stochastic parallel gradient descent optimization. *Source*: [7].

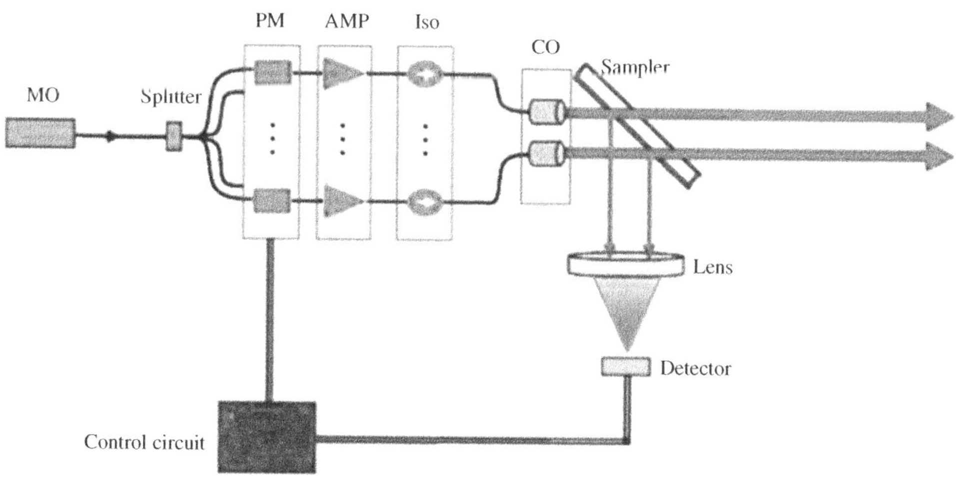

Figure 6.26 Adaptive phase modulation for coherent beam combination using target-in-the-loop approach. *Source*: Zhou et al., (2009) [8]/IEEE.

far-field intensity of the initially incoherent beamlets. The system does not require a dedicated reference beam or a wavefront sensor.

6.6 AO Control

An AO control system is a multi-input, multi-output closed-loop feedback control system. For computational efficiency, the AO control system design often uses a simplified mathematical model of a wavefront sensor and a corrector. A linear reconstructor is also introduced in the control system design to implement decoupled single-input single-output closed-loop feedback control system design.

6.6.1 Influence Function and Influence Matrix

An influence function of an AO system represents a change in the shape of the wavefront when a small perturbation is applied to a single actuator of a wavefront corrector. For the example of a continuous facesheet DM, the influence function could be approximated as a Gaussian function,

$$w(x, y) = \frac{2\pi}{\lambda} \exp\left\{ \frac{\ln b\left[(x - x_c)^2 + (y - y_c)^2\right]}{d^2} \right\} \tag{6.26}$$

where d is the spacing between actuators, x_c and y_c represent the location of the actuator, and b is the Gaussian index determining the spread of the influence function.

Figure 6.27 shows example influence functions represented in 3D and 2D. The influence of wavefront due to all actuator inputs is typically assumed to be a linear combination of individual influence functions such that

$$\phi(x, y) = \sum_{i=1}^{n} u_i w_i(x, y) \tag{6.27}$$

where u_i is the ith actuator input that scales the ith influence function $w_i(x, y)$ and n is the total number of actuators. The equation can be written into a matrix-vector form when the influence function is sampled at a specific AO geometry with l number of actuator points.

$$\boldsymbol{\phi} = \boldsymbol{W}\boldsymbol{u} \tag{6.28}$$

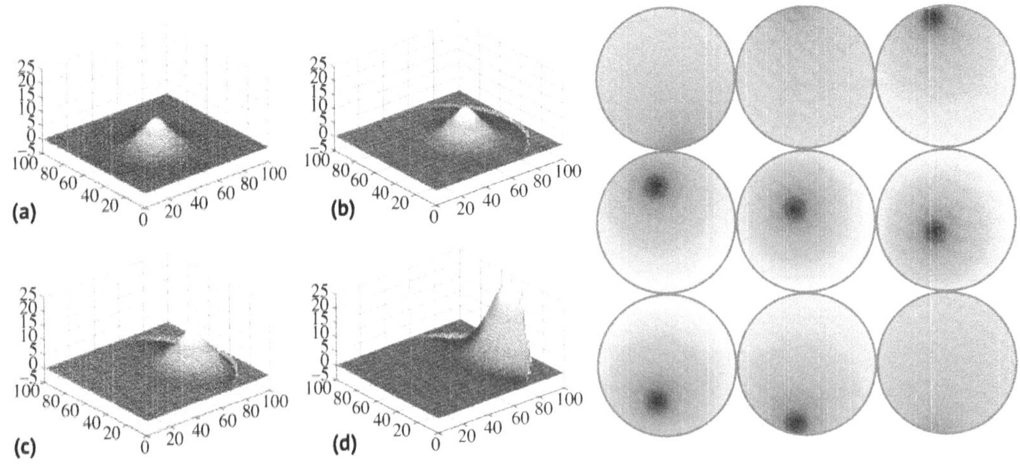

Figure 6.27 Influence function example.

where $\boldsymbol{\phi}$ is a $l \times 1$ vector representing the wavefront at the grid points, $\boldsymbol{u}$ is a $n \times 1$ vector of applied actuator input, $\boldsymbol{W}$ is a $l \times n$ matrix representing the relationship between the actuator input and the wavefront of the corrector. Since the wavefront is measured by a wavefront sensor, the wavefront sensor output with m measurements is related to the actuator input as

$$\boldsymbol{y} = \boldsymbol{B\phi} = \boldsymbol{BWu} = \boldsymbol{\Gamma u} \tag{6.29}$$

where $\boldsymbol{y}$ is a $m \times 1$ vector of wavefront sensor output (either gradient or reconstructed phase), $\boldsymbol{B}$ is a $m \times l$ matrix representing the relationship between wavefront and sensor output (from Eq. 6.10 for zonal wavefront reconstruction and Eqs (6.17–6.19) for modal wavefront reconstruction), and $\boldsymbol{\Gamma} = \boldsymbol{BW}$ is defined as an influence matrix relating the actuator input and the sensor output.

Figure 6.28 illustrates the influence matrix of a conventional AO system consisting of a DM and a SHWS. The influence matrix is often used as a simplified plant model for the AO control system design. This assumes that the dynamics of the DM and the wavefront sensor are much faster than the rest of the dynamics in the AO system such that the actuator input and sensor output are a static relationship represented by the influence matrix. In digital control system design, however, a time-delay is often added in the system plant model to provide a more representative model.

Experimentally determined influence matrix is often used for the control system design since it is often more accurate than the analytically determined influence matrix. The influence matrix is also referred to as a poke matrix since it can be obtained by applying unit input to each channel of the wavefront corrector sequentially and measuring the corresponding wavefront sensor output.

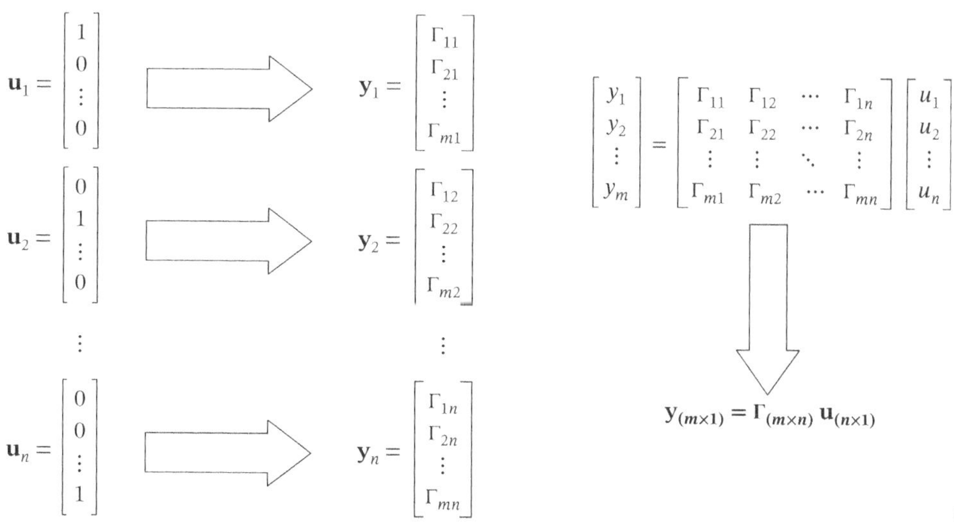

(6.30)

Assuming a plane wavefront of a DM when the actuator input is zero, with unit input on the first actuator, the resulting wavefront sensor output becomes the first column of the influence matrix. With unit input on the second actuator, the resulting wavefront sensor output becomes the second column of the influence matrix. Repeat this poking of each actuator to determine all the columns of the influence matrix as shown in Eq. 6.30. When the wavefront of a DM is not a plane wave when the actuator input is zero (no DM has a perfectly flat surface), both positive unit input and

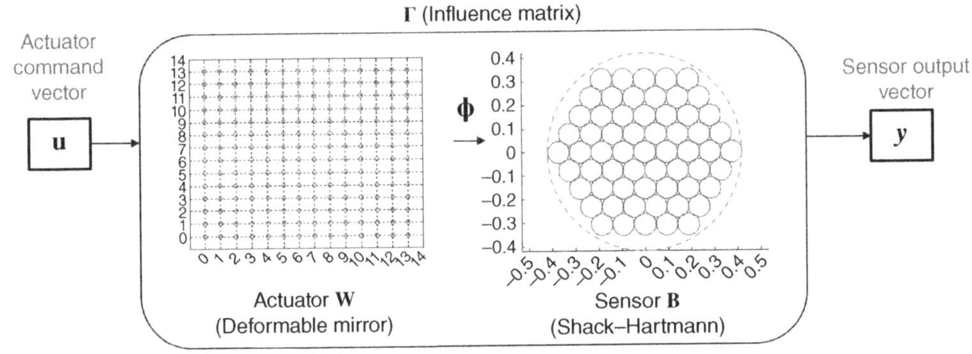

Figure 6.28 Influence matrix of an AO system consists of a DM and a Shack–Hartmann wavefront sensor.

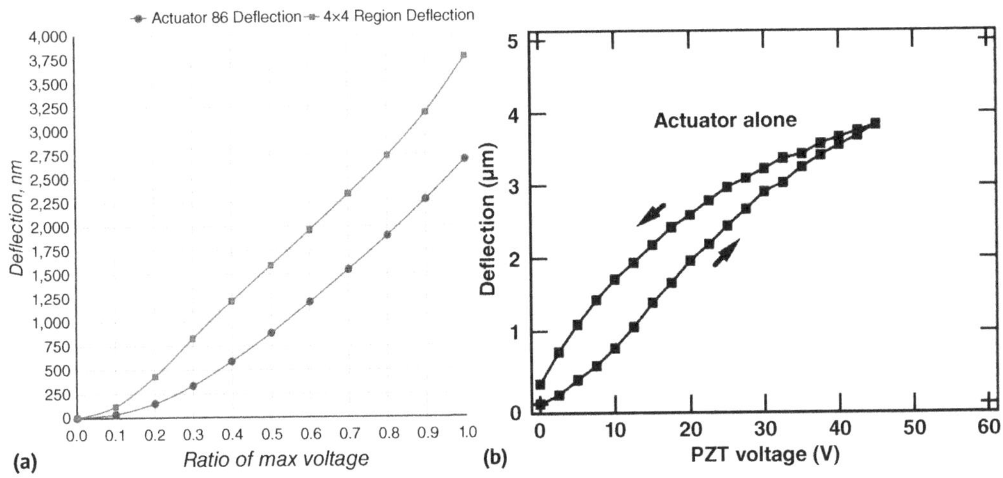

Figure 6.29 Actuator input voltage versus deformable mirror deflection (a) MEMS deformable mirror (b) piezo deformable mirror.

negative unit input are typically applied to each channel and the difference of the two wavefront sensor outputs becomes the corresponding column of the influence matrix.

The determination of influence matrix in this way assumes that the steady-state relationship between the actuator input and wavefront sensor output is linear. For MEMS mirrors, the relationship between voltage applied and its corresponding wavefront deflection is approximately a quartic function as shown in Figure 6.29. Therefore, the actuator input and actual voltage applied need to be adjusted according to the relationship, $u_i = \sqrt{v_i}$. For piezo actuators shown in Figure 6.29, this relationship is approximately linear; however, a nonlinear hysteresis is more significant than MEMS mirrors where the actual deflection of the mirror lags behind the actuator input changes.

6.6.2 Closed-loop Feedback AO Control System Design

Figure 6.30 shows the schematic of a closed-loop AO feedback control. The objective of a control system is to find a controller $\boldsymbol{G}$ which produces the DM command signal $\boldsymbol{u}$ that will provide DM wavefront $\boldsymbol{\phi}_u$ that can compensate wavefront distortion $\boldsymbol{\phi}_d$ using the observed wavefront sensor output $\boldsymbol{y}$.

Figure 6.31 shows the block diagram of the closed-loop AO feedback control system. In the Figure 6.31 block diagram, the plant of the control system including a DM and a wavefront sensor can be modelled with an influence matrix $\boldsymbol{\Gamma} = \boldsymbol{ZW}$. The wavefront aberration $\boldsymbol{\phi}_d$ is acting as a disturbance to the system and $\mathbf{n}$ as a sensor noise. The output of the wavefront sensor $\boldsymbol{y}$ is compared with the reference wavefront sensor output $\boldsymbol{y}_r$, which is typically zero for a plane wavefront. This error signal $\boldsymbol{e}$ is used for a controller to generate a control input.

Since deformation of a DM due to one control channel can affect the shape of the entire wavefront, classical control system design based on single-input single-output (SISO) cannot be directly applied. For computational efficiency and simpler controller design, a reconstructor is introduced in the design of a controller. Figure 6.32 is the block diagram of the AO feedback control system including q number of sample time delays implemented in discrete time domain. The reconstructor $\boldsymbol{F}$ shown in Figure 6.32 is typically chosen as the pseudo-inverse matrix of influence matrix such that $\boldsymbol{F} = \boldsymbol{\Gamma}^{\dagger}$. For example, when the number of sensor output m is greater than the number of control input n, the pseudo-inverse of influence matrix becomes

$$\mathbf{F} = \boldsymbol{\Gamma}^{\dagger} = (\boldsymbol{\Gamma}^T\boldsymbol{\Gamma})^{-1}\boldsymbol{\Gamma}^T \tag{6.31}$$

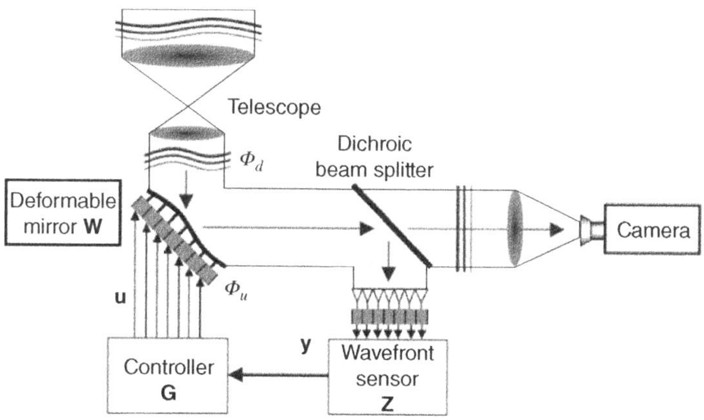

Figure 6.30 Closed-loop feedback control of AO system.

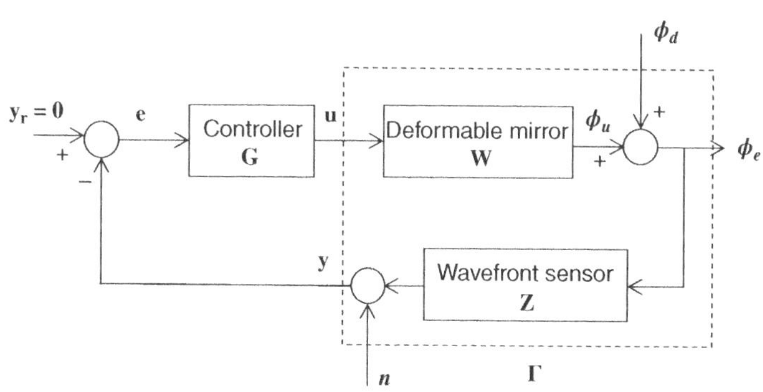

Figure 6.31 Block diagram of closed loop AO feedback control system.

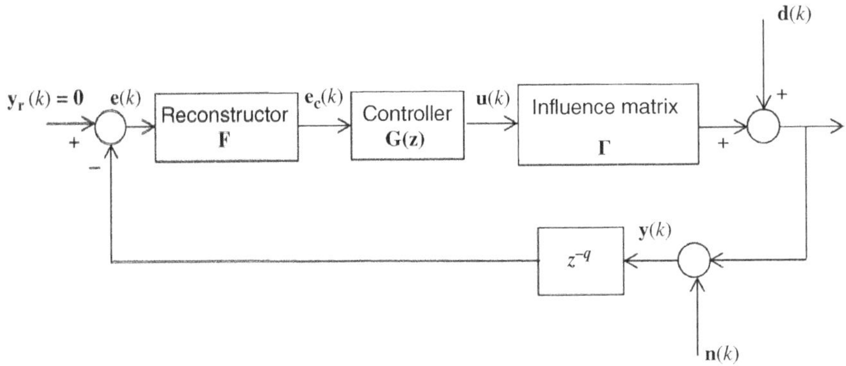

Figure 6.32 Block diagram of equivalent closed-loop AO control system for controller design.

The reconstructed error signal $\boldsymbol{e}_c(k)$ is uncoupled to the control channel $\boldsymbol{u}(k)$ at the steady state since

$$\boldsymbol{e}_c(k) = \mathbf{F}\boldsymbol{e}(k) = -\mathbf{F}\boldsymbol{y}(k) = -\boldsymbol{\Gamma}^{\dagger}\boldsymbol{\Gamma}\mathbf{u}(k) = -\mathbf{I}\boldsymbol{u}(k) \tag{6.32}$$

With the relationship from $\boldsymbol{e}_c(k)$ to $\boldsymbol{u}(k)$ uncoupled, the controller $\boldsymbol{G}(\boldsymbol{z})$ can be determined by applying SISO control design for each individual control input channel.

The most commonly encountered problem when using the reconstructor matrix $\mathbf{F}$ in Eq. (6.31) arises when the matrix $\boldsymbol{\Gamma}^T\boldsymbol{\Gamma}$ is poorly conditioned, making its inverse numerically unstable. This situation corresponds to cases where actuating a certain DM actuator produces little to no effect on the sensor output (e.g. resulting in a near-zero or zero column vector in Eq. 6.30). A practical method for numerically stable inversion is to use singular value decomposition and retain only the modal basis components whose associated singular values (i.e. DM gains) are sufficiently large.

When the SHWS is used, the wavefront sensor output could be either gradient of the wavefront or the reconstructed phase. When the gradient measurements are used, the reference of the control system $\boldsymbol{y}_{\mathbf{r}}(k)$ becomes a zero gradient. This gradient-referenced control is typically used in systems where fast, real-time adjustments are necessary and where phase measurements might be difficult or computationally expensive. While gradient information is easier to acquire, it does not directly provide the phase and it cannot be used for discontinuous wavefront. When the reconstructed phase measurements are used, the reference of the control system $\boldsymbol{y}_{\mathbf{r}}(k)$ becomes a zero phase and the phase-referenced control directly addresses the shape of the wavefront.

Similar to the FSM control system, variation of integral controller is commonly used for each control input. The most frequently used controller is a leaky integrator, whose transfer function is

$$G(z) = \frac{K_i}{1 - cz^{-1}} \tag{6.33}$$

where K_i is the controller gain and c is a constant that is smaller than 1 to ensure stability of the controller. Additional compensators can be designed to provide the desired control bandwidth and disturbance rejection capabilities.

References

1 Claire Max, Adaptive Optics and Its Applications, AO lecture notes, University of California - Santa Cruz, Astronomy 289C Lecture Notes. http://ww.ucolick.org/~max/289C/.
2 Robert Tyson, Principle of Adaptive Optics, Academic Press, 1998.
3 Paul Merritt, Beam Control for Laser Systems, Directed Energy Professional Society, 2012.
4 W. H. Southwell, Wave-front estimation from wave-front slope measurements, J. Opt. Soc. Am. 70 (8), 998–1006, August 1980.
5 Francois Rigaut and Benoit Neichel, Multiconjugate AO for astronomy, arXiv, https://arxiv.org/abs/2003.03097, 2020.
6 Jeffrey D. Barchers and Brent L. Ellerbroek, Improved compensation of turbulence-induced amplitude and phase distortions by means of multiple near-field phase adjustments, J. Opt. Soc. Am. A 18, 399–411, 2001.
7 M. A. Vorontsov and V. P. Sivokon, Stochastic parallel-gradient-descent technique for high-resolution wave-front phase-distortion correction, J. Opt. Soc. Am. A 15, 2745–2758, 1998.
8 Pu Zhou et al., Coherent beam combining of fiber amplifiers using stochastic parallel gradient descent algorithm and its application, IEEE J. Sel. Top. Quantum Electron. 15 (2), 248–256, 2009.

7

Imaging Satellites

7.1 Introduction

Telescopes are optical instruments designed to gather and magnify light, allowing us to observe distant objects. Space telescopes used in imaging satellites are specifically designed for observations outside Earth's atmosphere, which eliminates atmospheric distortions and allows for clearer and more detailed observations. There are two commonly used types of telescopes: reflecting telescopes and refracting telescopes. Reflecting telescopes' primary mirror reflects light to a focus point, where it can be observed or directed to instruments. Refracting telescopes have a primary lens that gathers and refracts light. The eyepiece is used to magnify the focused image. Reflecting telescopes do not suffer from chromatic aberration, a distortion caused by different colors of light focusing on different points. Mirrors reflect all colors equally. Refracting telescopes can suffer from chromatic aberration, where different colors of light focus on slightly different points. This can affect image quality and is mitigated using extra lenses or lens coatings. Reflecting telescopes are easier to manufacture and maintain. Mirrors can be coated to improve reflectivity, and adjustments are often simpler. Refracting telescopes are more complex to manufacture and maintain. Lenses require precise shaping and coating, and misalignment can be more challenging to correct. Reflecting telescopes can be more compact and lighter for a given aperture size. This is advantageous for portable or space-based telescopes. Refracting telescopes tend to be longer and heavier for larger apertures, making them less portable. Reflecting telescopes are generally more cost-effective to manufacture, especially for larger apertures. Refracting telescopes can be more expensive, particularly for larger lenses with high quality optics. Based on these considerations, reflecting telescope design is commonly used for space telescopes. In this chapter, we will cover only reflecting telescopes.

7.2 Telescope Designs

Under reflecting telescopes, three telescope designs are most used: Cassegrain telescopes, Ritchey–Chrétien (RC) telescope, and three-mirror anastigmat. We will discuss the design of these three telescopes. A typical configuration of a reflecting telescope is as given in Figure 7.1.

The primary component of a reflecting telescope is the primary mirror where the light from the object first comes, then it is reflected to the secondary mirror, and after passing through several mirrors, to correct optical aberrations, it goes to the focal plane assembly (FPA) where the collected light is focused to form an image. The principal axis of a space telescope refers to the line passing

Optical Beam Control: Imaging Satellites and Laser Systems, First Edition.
Brij N. Agrawal, Jae Jun Kim and Sachin Agrawal.

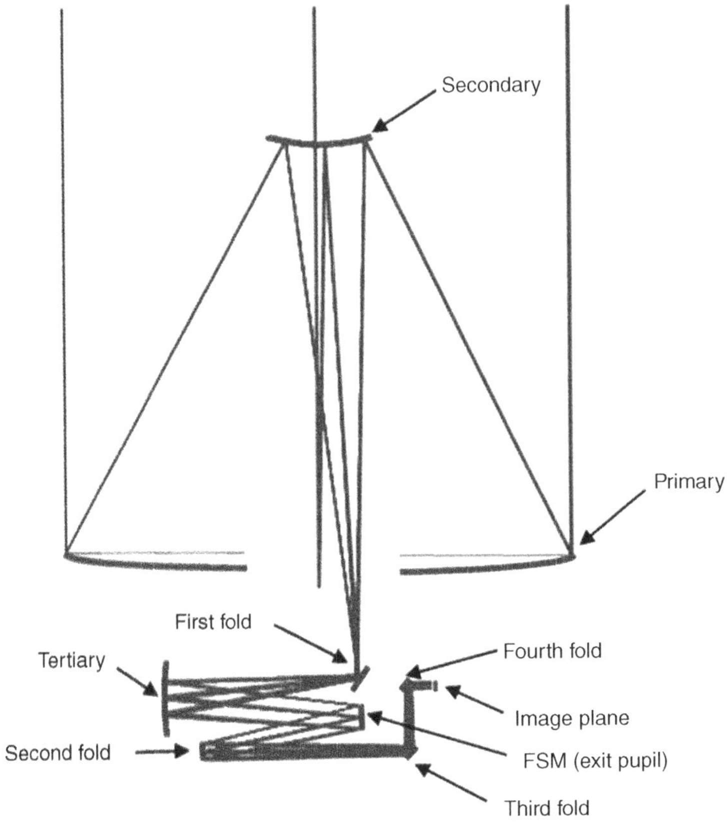

Figure 7.1 A typical configuration of a reflecting telescope.

through the center of the primary mirror and perpendicular to its surface. Before discussing the designs of these telescopes, we will review optical beam aberrations.

7.2.1 Optical Beam Aberrations

7.2.1.1 Spherical Aberration (Third Order)

For telescopes with spherical primary mirrors, as shown in Figure 7.2, along the principal axis of the optic, rays traced from an object point are brought to focus on different image points (along the principal axis) [1]. Each annulus focuses on a different image point due to differences in path length traveled by the rays. This aberration can be removed by using a parabolic mirror.

7.2.1.2 Coma (Fifth Order)

As shown in Figure 7.3, for an object pointing off the principal axis, if spherical aberrations are removed, different annular rings focus rays on different image points (both along and transverse to the principal axis). This aberration is removed in the Ritchey–Chrétien (RC) telescope by using a hyperbolic primary mirror.

7.2.1.3 Astigmatism (Seventh Order)

For an object point off the principal axis, if spherical and coma aberrations are removed, rays traveling on the meridional plane are focused at a different point than those traveling in the

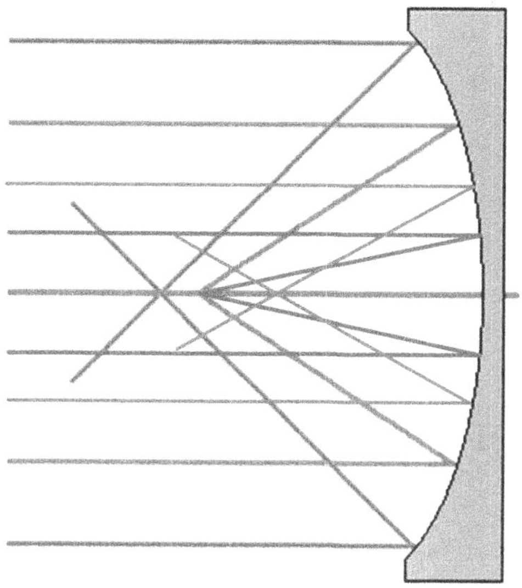

Figure 7.2 Spherical aberrations.

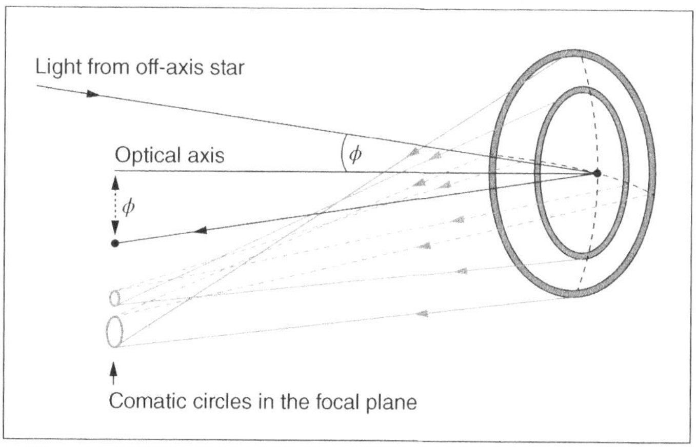

Figure 7.3 Coma aberration.

sagittal plane, as shown in Figure 7.4. The optic is tilted in one of these planes, producing two different focal planes. This aberration is corrected by adding an elliptical mirror in a three-mirror Anastigmat.

7.2.1.4 Curvature of Field (Ninth Order)

If the prior three aberrations are removed, real objects do not focus within a (flat) image plane, but along a curved surface. The curve traversed is the radius of curvature for the optic.

7.2.2 Telescopes

The conic constant is a parameter used in optics to describe the shape of a conic section, which is the curve formed by the intersection of a plane and a cone. In the context of optical surfaces,

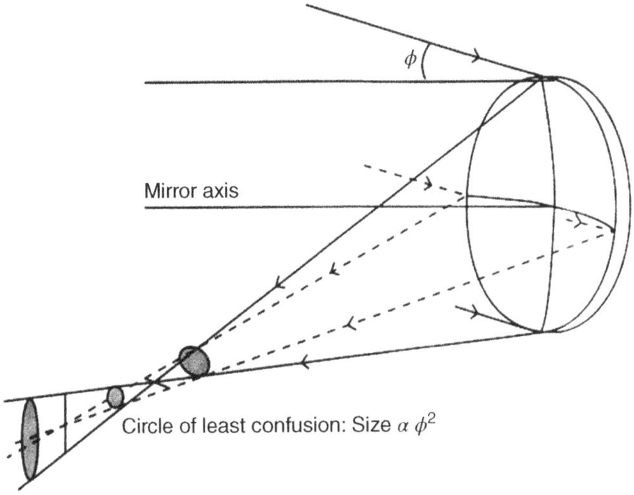

Figure 7.4 Astigmatism aberration.

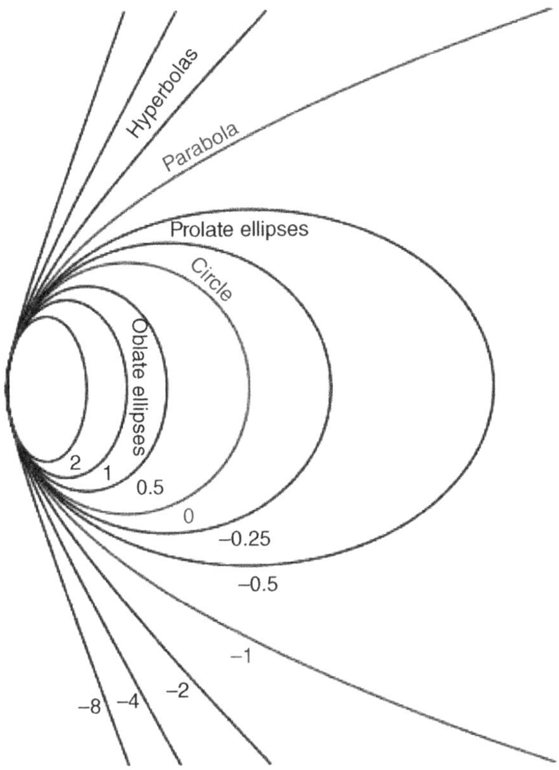

Figure 7.5 Conic sections.

particularly in telescope and lens design, the conic constant is used to characterize the curvature of surfaces such as lenses and mirrors. It helps to determine whether the surface is spherical, parabolic, hyperbolic, or elliptical.

The conic constant is denoted by the letter "K" and can take on different values depending on the type of conic section, as shown in Figure 7.5.

$K = 0$ corresponds to a spherical shape.
$K > 0$ corresponds to an oblate elliptical shape.
$0 > K > -1$ prolate elliptical
$K < 0$ corresponds to a hyperbolic shape.
$K = 1$ corresponds to a parabolic shape.

This parameter is crucial in optical design as it affects properties like focal length, image distortion, and aberrations in optical systems.

Aspheric refers to a surface or lens element that deviates from being purely spherical. In optics, an aspheric surface is one whose curvature changes continuously, rather than following a simple spherical shape.

7.2.3 Cassegrain Telescopes

The Cassegrain reflector, as shown in Figure 7.6, is a combination of a primary concave mirror and a secondary convex mirror, often used in optical telescopes and radio antennas, the main characteristic being that the optical path folds back onto itself, relative to the optical system's primary mirror entrance aperture [2]. This design puts the focal point at a convenient location behind the primary mirror and the convex secondary adds a telephoto effect creating a much longer focal length in a mechanically short system. The classic Cassegrain configuration uses a parabolic reflector as the primary while the secondary mirror is hyperbolic.

A concave parabolic reflector will reflect all incoming light rays parallel to its axis of symmetry to a single point, the focus. A convex hyperbolic secondary reflector has two foci and will reflect all light rays directed at one of its two foci. The mirrors in this type of telescope are designed and positioned so that the primary mirror and the secondary mirror share one focus and so that

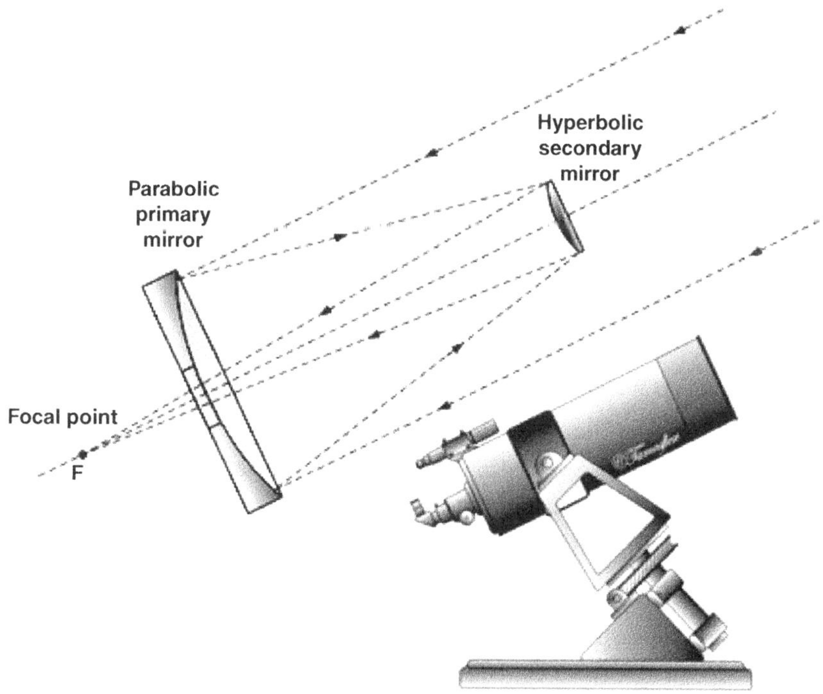

Figure 7.6 Cassegrain telescope.

the second focus of the hyperbolic mirror will be at the same point at which the image is to be observed. In most Cassegrain systems, the secondary mirror blocks a central portion of the aperture. This ring-shaped entrance aperture significantly reduces a portion of the modulation transfer function (MTF) over a range of low spatial frequencies, compared to a full-aperture design such as a refractor or an offset Cassegrain [2]. This MTF notch has the effect of lowering image contrast when imaging broad features. In addition, the support for the secondary (the spider) may introduce diffraction spikes in images.

The radii of curvature of the primary and secondary mirrors, respectively, in the classic configuration are

$$R_1 = -\frac{2DF}{F-B} = -\frac{2F}{M} \tag{7.1}$$

and

$$R_2 = -\frac{2DB}{F-B-D} = -\frac{2B}{M-1} \tag{7.2}$$

where

F is the effective focal length of the system,
B is the back focal length (the distance from the secondary to the focus),
D is the distance between the two mirrors and
$M = (F - B)/D$ is the secondary magnification.

If, instead of B and D, the known quantities are the focal length of the primary mirror, and the distance to the focus behind the primary mirror, b, then.

$$D = f_1(F,-,b)/(F,+,f_1) \tag{7.3}$$

and

$$B = D + b. \tag{7.4}$$

The conic constant of the primary mirror is that of a parabola, $K_1 = -1$. so that there is no spherical aberration introduced by the primary mirror. For the secondary mirror, K_2 is given by.

$$K_2 = -1 - \alpha - \sqrt{\alpha(\alpha+2)},$$

where

$$\alpha = \frac{1}{2}\left[\frac{4DBM}{(F+BM-DM)(F-B-D)}\right]^2. \tag{7.5}$$

7.2.4 Ritchey–Chrétien Telescope

The Ritchey–Chrétien telescope was invented in the early 1910s by American astronomer George Willis Ritchey and French astronomer Henri Chrétien. A Ritchey–Chrétien telescope (RCT or simply RC) is a specialized variant of the Cassegrain telescope that has a hyperbolic primary mirror, and a hyperbolic secondary mirror designed to eliminate off-axis optical errors (coma) [3]. RCT has a wider field of view free of optical errors compared to a more traditional reflecting telescope configuration. Since the mid-20th century, most large professional research telescopes have been Ritchey–Chrétien configurations; some well-known examples are the Hubble Space telescope, the Keck telescope, and the ESO very large telescope. The RCT offers good off-axis optical performance, but its mirrors require sophisticated techniques to manufacture and test. Hence, the Ritchey–Chrétien configuration is mostly found on high-performance professional telescopes. The basic Ritchey–Chrétien two-surface design is free of third-order coma and spherical aberration [1]. However, the two-surface design does suffer from fifth-order coma, severe large-angle astigmatism, and comparatively severe field curvature.

The design equations for the RC telescope are as follows.

The radii of curvature of the primary and secondary mirrors, respectively, in a two-mirror Cassegrain configuration are:

$$R_1 = -\frac{2DF}{F - B} = -\frac{2F}{M} \tag{7.6}$$

and

$$R_2 = -\frac{2DB}{F - B - D} = -\frac{2B}{M - 1}, \tag{7.7}$$

where

F is the effective focal length of the system,
B is the back focal length (the distance from the secondary to the focus),
D is the distance between the two mirrors and
$M = (F - B)/D$ is the secondary magnification [3].

If, instead of B and D, the known quantities are the focal length of the primary mirror, f_1, and the distance to the focus behind the primary mirror, b, then

$$D = f_1(F - b)/(F + f_1) \text{ and } B = D + b. \tag{7.8}$$

For a Ritchey–Chrétien system, the conic constants K_1 and K_2 of the two mirrors are chosen so as to eliminate third-order spherical aberration and coma; the solution is:

$$K_1 = -1 - \frac{2}{M^3} \cdot \frac{B}{D} \tag{7.9}$$

and

$$K_2 = -1 - \frac{2}{(M - 1)^3}\left[M(2M - 1) + \frac{B}{D}\right]. \tag{7.10}$$

Note that K_1 and K_2 are less than -1 (since $M > 1$), so both mirrors are hyperbolic (the primary mirror is typically quite close to being parabolic, however.)

7.2.5 Three-mirror Anastigmat Telescope

A three-mirror anastigmat is an anastigmat telescope built with three curved mirrors, enabling it to minimize all three main optical aberrations – spherical aberration, coma, and astigmatism. This is primarily used to enable wide fields of view, much larger than possible with telescopes with just one or two curved surfaces. A telescope with only one curved mirror, such as a Newtonian telescope, will always have aberrations. If the mirror is spherical, it will suffer from spherical aberration. If the mirror is made parabolic, to correct the spherical aberration, then it must necessarily suffer from coma and off-axis astigmatism. With two curved mirrors, such as the Ritchey–Chrétien telescope, coma can be minimized as well. This allows a larger useful field of view, and the remaining astigmatism is symmetrical around the distorted objects, allowing astrometry across the wide field of view. However, astigmatism can be reduced by including a third curved optical element. When this element is a mirror, the result is a three-mirror anastigmat. In practice, the design may also include any number of flat-fold mirrors, used to bend the optical path into more convenient configurations. The James Webb Space Telescope is a three-mirror anastigmat featuring ellipsoidal primary, hyperboloidal secondary, and ellipsoidal tertiary. The Segmented Mirror Telescope at the Naval Postgraduate School is a three-mirror anastigmat incorporating Ritchey–Chretien Cassegrain telescope and an ellipsoid tertiary. These designs are ideal for wide-field observations. The ellipsoidal tertiary removes astigmatism and increases the wide field of view beyond the two-mirror Ritchey–Chretien design. This design is referred to as a "Paul-Baker" three-mirror anastigmat. The mathematical relationships for these three mirrors are more complex than Ritchey–Chretien two mirror design.

7.3 Telescope Performance

Before we talk about telescope performance, let us review some basic optics fundamentals.

7.3.1 Diffraction

Diffraction is a phenomenon in optics that occurs when light encounters an obstacle or aperture and bends around it, spreading out in different directions. This bending of light is a consequence of the wave nature of light. When light waves encounter an obstacle or aperture that is comparable in size to the wavelength of the light, the waves are forced to bend around it, leading to the spreading out of light in various directions. This phenomenon is described by Huygen's principle, which states that every point on a wavefront can be considered a source of secondary spherical wavelets. The interference of these wavelets results in the observed diffraction pattern. Geometric optics does not predict diffraction. Geometric optics is a branch of optics that deals with the behavior of light as rays, focusing on the paths that light rays take and the formation of images through lenses and mirrors. Geometric optics simplifies the study of light by treating it as rays and ignoring the wave nature of light.

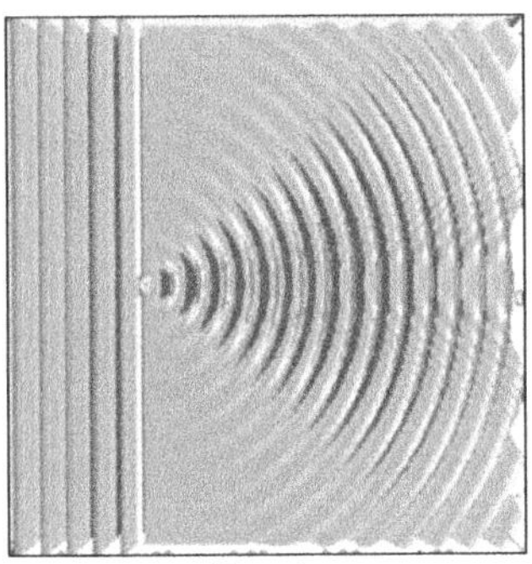

$D_O \approx \lambda$.

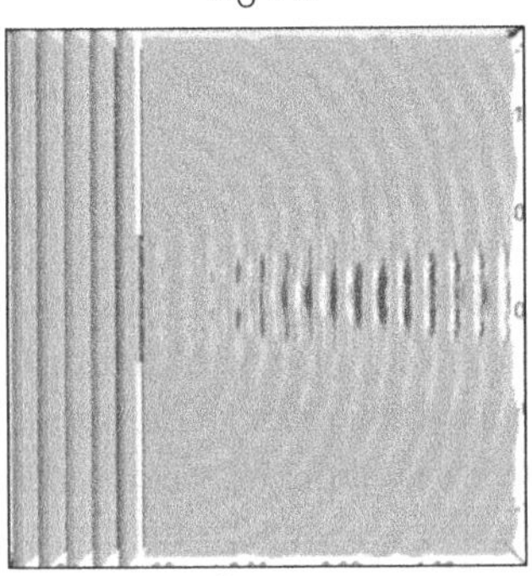

$D_O > \lambda$.

Figure 7.7 Diffraction.

The extent of diffraction depends on the size of the obstacle or aperture relative to the wavelength of the light. If an obstacle or aperture is on the order of the wavelength of the light, significant diffraction occurs. If the obstacle is much larger than the wavelength, then diffraction is less, as shown in Figure 7.7.

Fresnel and Fraunhofer diffraction are two types of diffraction patterns that occur when light waves encounter obstacles or apertures. The main difference between them lies in the distance between the diffracting object and the observation screen. Fresnel diffraction occurs when the diffracting object and the observation screen are at a close or intermediate distance from each other. The light waves are diverging or converging when they reach the diffracting object. In Fresnel diffraction, the curvature of the wavefronts cannot be ignored, and the wave nature of light is more pronounced. This occurs when the distance between the diffracting object and the observation screen is comparable to the dimensions of the diffracting object.

The mathematical analysis of Fresnel diffraction is more complex than Fraunhofer diffraction, and it often involves integrals. Fraunhofer diffraction is a "far-field" effect where the distance between the source, the obstacle/aperture, and the screen is large enough to treat wavefronts as planar, resulting in a Fourier Transform relationship and a simpler analysis. In contrast, Fresnel diffraction is a "near-field" effect, occurring at finite distances where wavefronts are spherical or cylindrical, and their curvature must be accounted for, making the analysis more complex and the pattern dependent on the screen's distance.

7.3.1.1 Beam Propagation

For a circular aperture, propagating a collimated beam in a vacuum, with no jitter and distortion, the intensity diffraction pattern is shown in Figure 7.8.

The intensity pattern is Airy disc as shown in Figures 7.9 and 7.10 is given by the equation.

$$I(u) = I_o\left(\frac{2J_1(u)}{u}\right)^2 \text{ where } u = \pi\left(\frac{\theta}{\lambda/D}\right) \tag{7.11}$$

Half-angle beam spread to the first minimum, θ_{12}, is

$$\theta_{12} = \frac{1.22\lambda}{D} \tag{7.12}$$

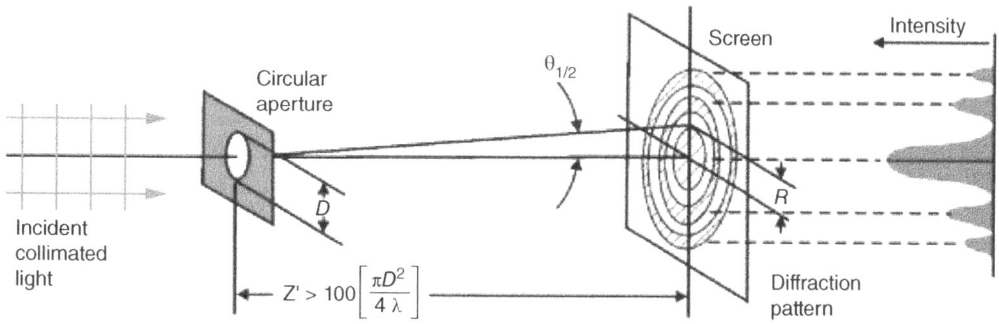

Figure 7.8 Diffraction intensity pattern of a distant point source for a circular aperture.

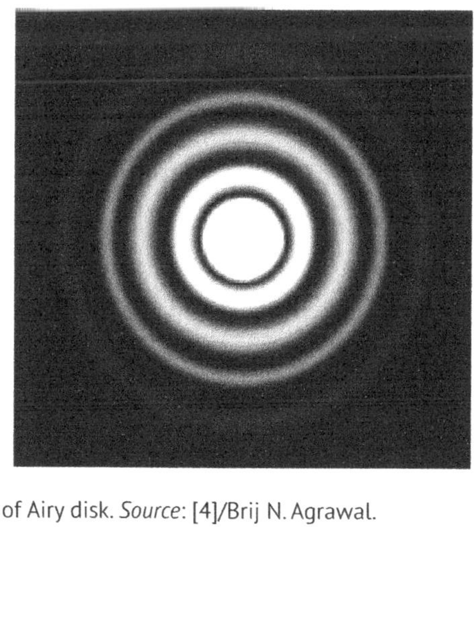

Figure 7.9 Intensity pattern of Airy disk. *Source*: [4]/Brij N. Agrawal.

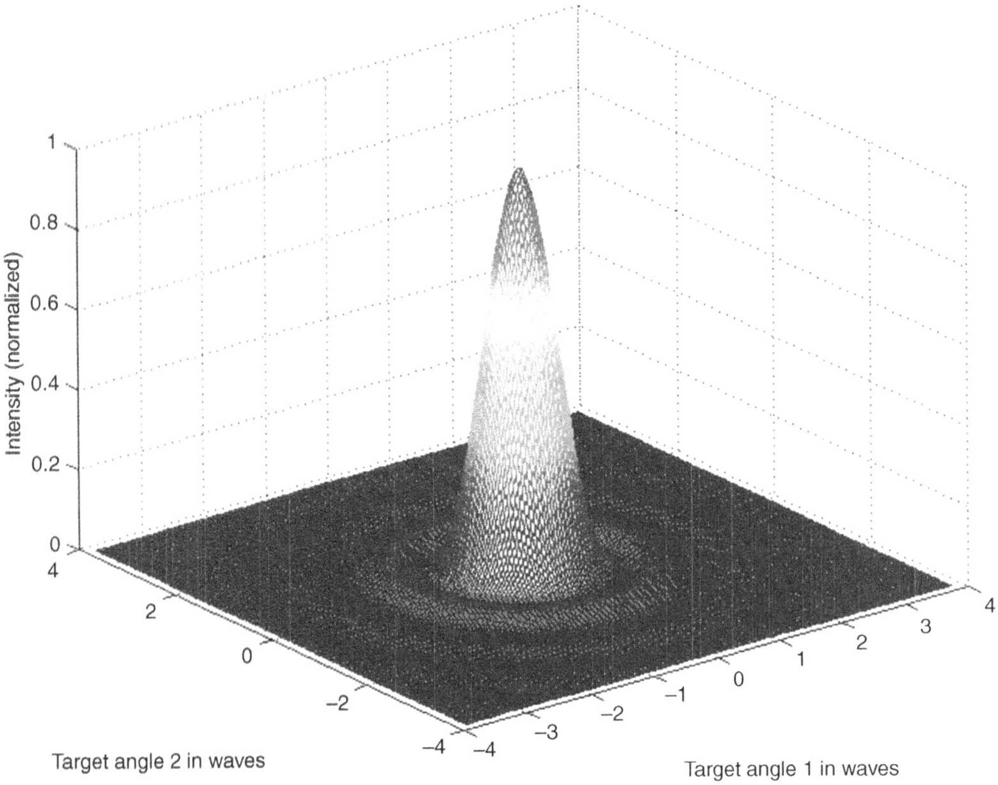

Figure 7.10 Airy pattern.

Radius of central bright disk (Airy disk), R, is:

$$R = \frac{1.22\lambda Z'}{D} \tag{7.13}$$

where λ = wavelength of light,
D = diameter of aperture, and
Z' = aperture-to-screen distance
J_1 = first order Bessel function

For almost all the performance analysis done for tracking and pointing, we assume we have a Gaussian beam. This is very close to the Airy disk, as shown in Figure 7.11. It is given by.

$$I(u) \approx I_0 e^{\frac{-u^2}{4}} \tag{7.14}$$

The focused light at the focal point of a telescope can be mathematically related to the Fourier transform of the light at the primary mirror. This connection arises from the principles of optics and wave optics. The relationship involves the concept of Fourier optics, which states that the image formed by an optical system is related to the Fourier transform of the object (or light distribution) in the entrance pupil of the system. The intensity distribution and spatial characteristics of the focused light at the focal point are related to the Fourier transform of the light distribution at the primary mirror.

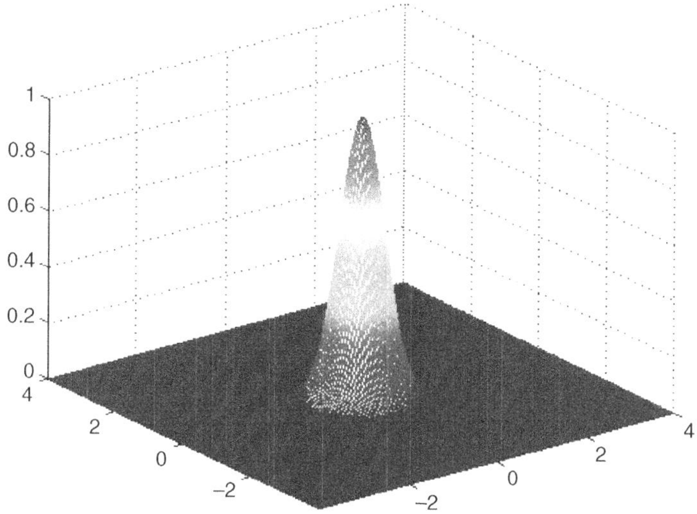

Figure 7.11 Gaussian pattern.

7.3.1.2 Modulation Transfer Function

The point spread function (PSF) describes the response of an imaging system to a point source (impulse response). For example, the PSF of a point source at infinity, imaged by an aberration-free optical system with a finite aperture is an Airy disc. The image I(x, 'y') of a complex object can be seen as a convolution of object O(x, y) and PSF.

$$\mathbf{I(x, 'y') = O(x, y) \otimes PSF} \tag{7.15}$$

The optical transfer function (OTF) is defined as the Fourier transform of the PSF,

$$\mathbf{OTF}(\upsilon) = \mathbf{F(PSF)} \tag{7.16}$$

The MTF is given by the absolute value of the optical transfer function,

$$\mathbf{OTF}(\upsilon) = \mathbf{MTF}(\upsilon)\, e^{iPhTF(\upsilon)} \tag{7.17}$$

Figure 7.12 shows different relationships of optical parameters.

A telescope acts like a linear filter in the spatial frequency domain with a cutoff frequency that depends on the aperture diameter (diffraction limit in the Fourier domain).

Figure 7.13 shows scene and image modulation.

Another way to look at resolving power is with the MTF, as shown Figure 7.14.

MTF expresses the output modulation depth of an image given a perfectly modulated object at a given spatial frequency.

MTF is unity at 0 spatial frequency (DC) and goes to zero at the resolution limit of the system.

In summary, for a telescope, the PSF describes the response of the telescope to a point source of light, typically a star. It characterizes how the light from a point source in the sky is spread out in the image formed by the telescope. It represents the spatial distribution of light in the image formed by the system in response to a point source in the object space. In the context of astronomy and telescopes, the PSF is often affected by factors such as diffraction, aberrations in the telescope optics, and atmospheric turbulence.

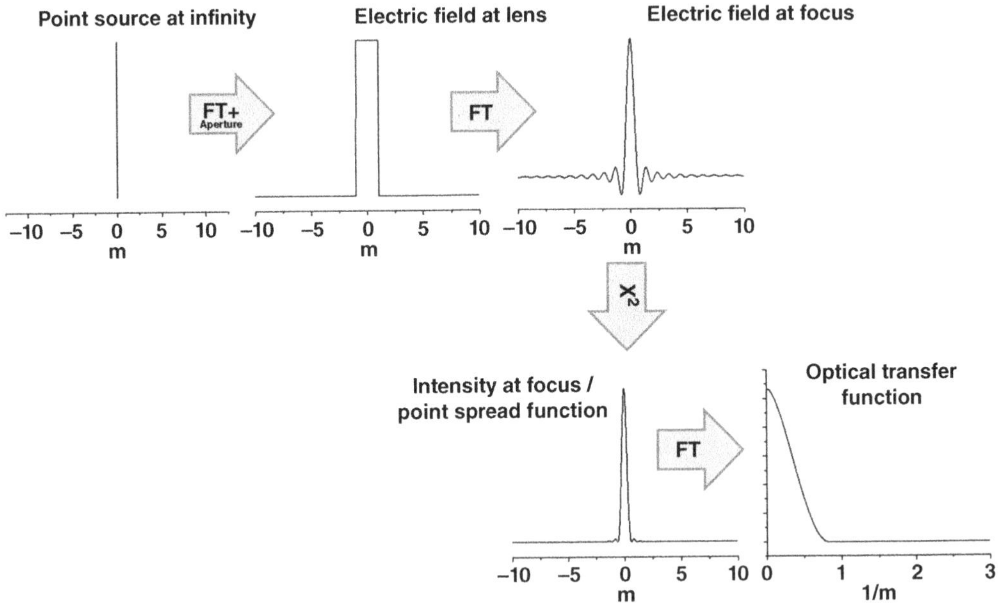

Figure 7.12 Optical properties.

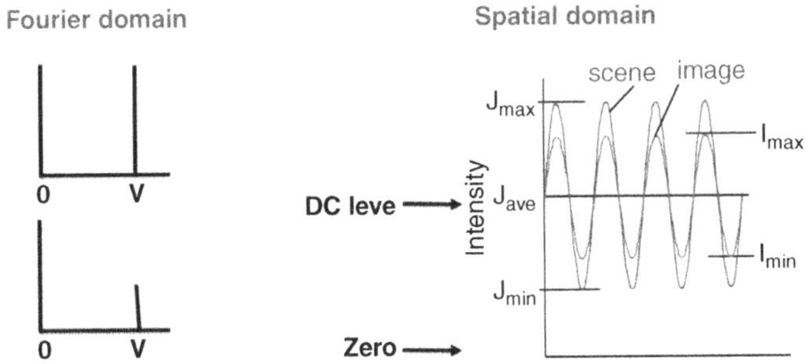

Figure 7.13 Scene and image modulation [4].

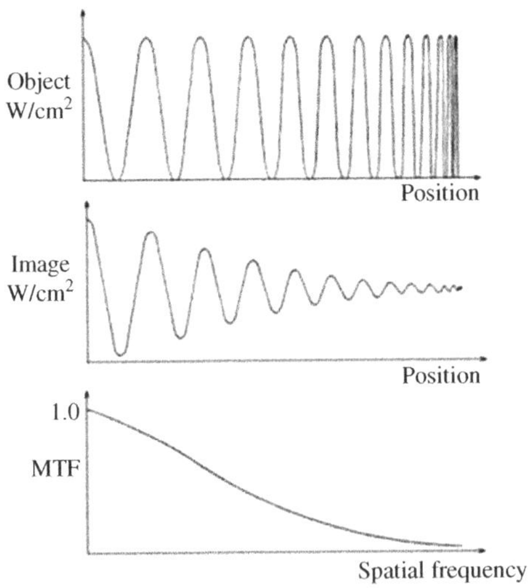

Figure 7.14 Representation of telescope performance by MTF [5].

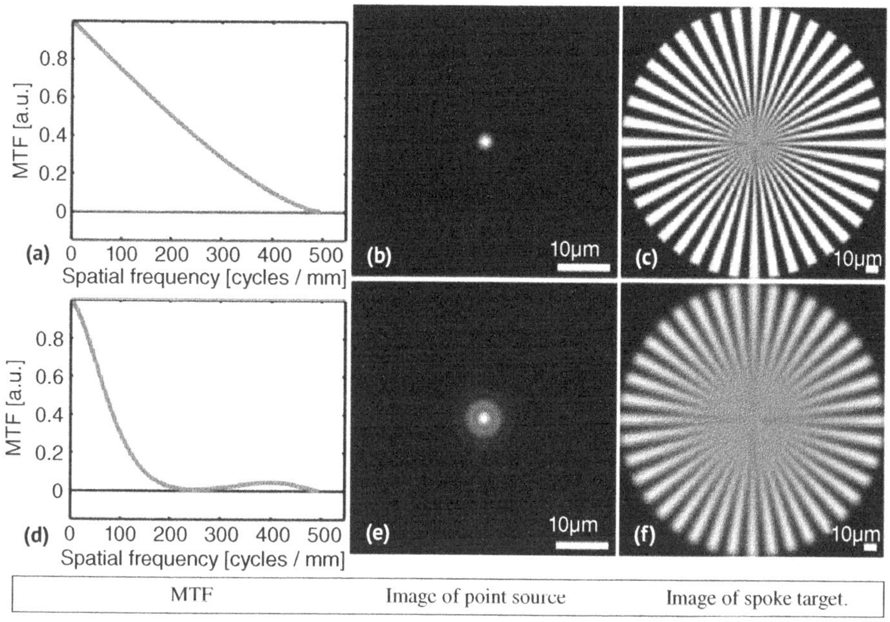

Figure 7.15 Impact of MTF on PSF and image of spoke target. *Source*: [5]/Brij N. Agrawal.

7.3.1.3 Optical Transfer Function

Optical transfer function is the Fourier transform of PSF. In mathematical terms, the MTF(f) is the absolute value of the Fourier Transform of the PSF (x, y), where f represents spatial frequency and (x, y) are coordinates in the image plane.

PSF describes how a point source is spread out in the image, and the MTF quantifies the system's ability to reproduce different spatial frequencies, providing information about image contrast at different levels of detail.

From Figure 7.15, for out of focus optical system has a lower spatial cutoff frequency, broader PSF, and lower resolution.

7.3.1.4 Resolution

The image of a scene that we obtain from a telescope is the collection of point images formed from the collection of point sources in the scene. The image of a point source created by a circular aperture is the Airy disk. The width of the Airy disk (diameter of the first minimum ring) is 2.44 λ/d, where d is the diameter of the aperture and λ is wavelength. The degree of detail visible in an image is dependent on the size of the Airy disk – the smaller the disk, the more visible details, and resolution.

The resolution of an imaging system is its ability to discriminate two closely spaced objects. Figure 7.16 marked as clearly resolved shows two images at a distance of 2.44 λ/d, first minimum of one image falls on the first minimum of another image. Baron Rayleigh formulated the Rayleigh criterion for resolution: objects can be distinguished if they are separated by an angle of 1.22 λ/d, so that the maximum of one image falls on the first minimum of the other image, as shown in Figure 7.16 marked as Raleigh. For practical applications, the resolution of an image is often approximated as λ/d in the figure marked as Sparrow. For two objects with an angle less than λ/d, it is not resolved as shown in Figure 7.16 marked not resolved. It should be noted that this resolution

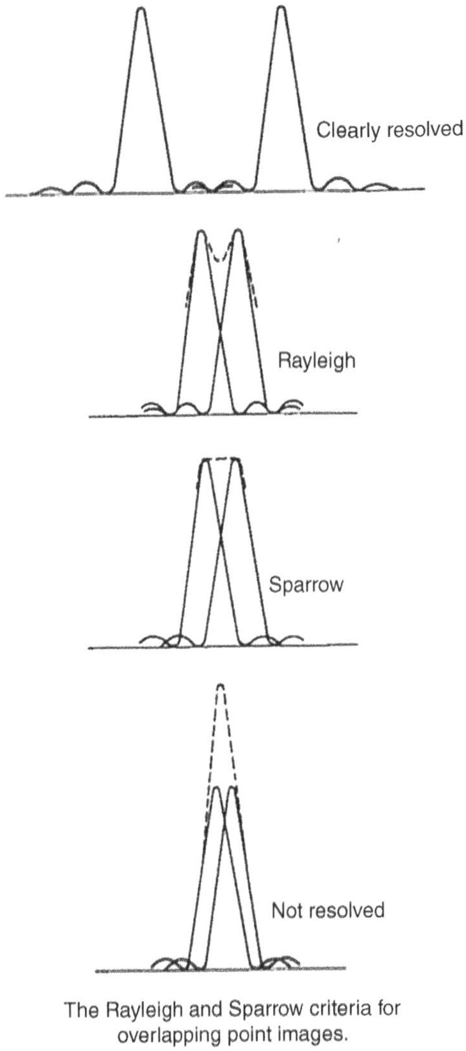

Figure 7.16 Different image resolution definitions [1].

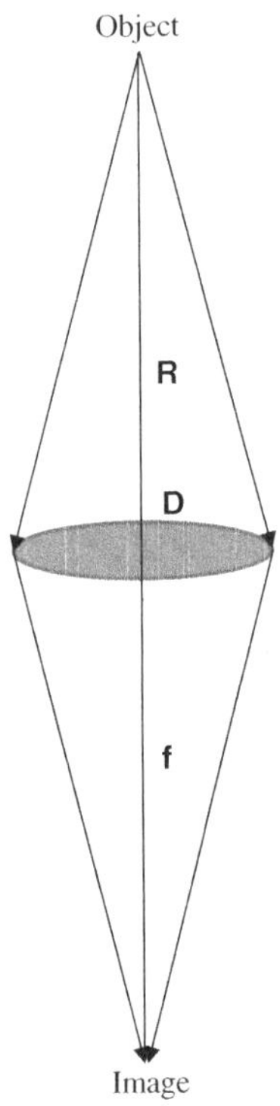

Figure 7.17 Object image generation in a telescope [4].

is due to a diffraction for perfect optics. It is called diffraction-limited resolution. In general, the resolution will be lower than this due to many factors including optic imperfections.

7.3.2 Object Image Generation

In a simplified form the focal telescope can be represented by Figure 7.17 consisting of aperture, object, and image.

where R = range (distance to object), f = focal length (distance to image), and D = aperture diameter, and λ = wavelength

7.3.2.1 Digital Focal Plane

A typical telescope digital focal plane is shown in Figure 7.18.

Figure 7.18 Telescope digital plane. *Source*: [5]/Brij N. Agrawal.

Focal plane parameters

p = pixel size (assume x and y are the same)
N = number of pixels in a row (assume square array)

Derived Parameters

$$\text{IFOV} = \text{instantaneous field of view, pixel field of view} = p/f$$
$$\text{FOV} = \text{focal plane field of view} = \sqrt{2}\text{N} \times \text{IFOV(diagonal)}$$

7.3.2.2 Ground Sampling Distance

Ground sampling distance (GSD) is given by the following equation.

$$\text{GSD} = \text{IFOV} \cdot R = p/f \cdot R \tag{7.18}$$

It represents the projection of the pixel field of view on the ground.

7.3.2.3 Quality Factor

λ/D as discussed earlier is a measure of the angular size of the PSF or the resolution of the optical system. Q is the ratio of the "angular resolution" and the angular size of a pixel (IFOV).

Q is proportional to the number of pixels across the central lobe of the PSF (Airy disk) as shown in Figure 7.19. It is a measure of how finely the detector samples the image plane with respect to the diffraction-limited PSF of the optical system.

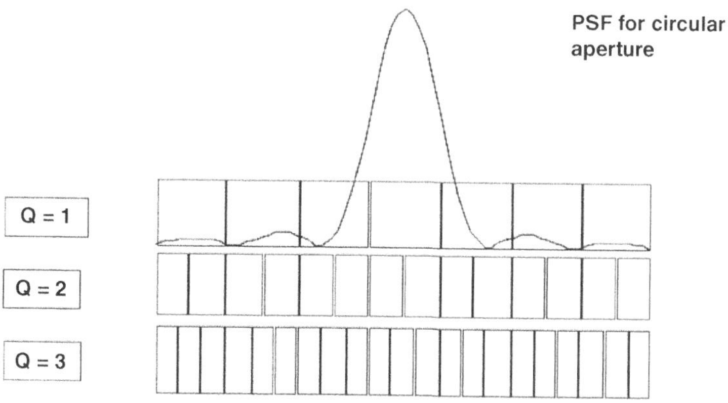

Figure 7.19 Different optical values of Q [5].

There is a value of Q that provides the maximum resolution (limited by diffraction) with the minimum number of pixels; reducing Q below that value reduces the resolution due to under-sampling (aliasing); increasing Q above that value uses more pixels to cover the same field of view due to oversampling. Q is normally less than 2. An increase in Q will result in a decrease in GSD, resulting in higher image quality. On the other hand, it increases integration time resulting in more jitter, which degrades image quality. A decrease in Q increases GSD resulting in lower image quality. It will reduce integration time resulting in less jitter which improves image quality (limited by focal plane read-out time/shitter time). Q typically has been between 1 and 1.5 for heritage earth imaging systems.

7.3.2.4 Ground Resolution Distance

Ground resolution distance (GRD) is determined by the Airy resolution angle λ/D. It is given by.

$$\mathrm{GRD} = \lambda/D \cdot \mathrm{R} \tag{7.19}$$

$$\frac{p}{f} = \frac{1}{Q} \cdot \frac{\lambda}{D} \tag{7.20}$$

So

$$GSD = \frac{1}{Q} \cdot \frac{\lambda}{D} \cdot R = \frac{1}{Q}, GRD \tag{7.21}$$

In Figures 7.20 and 7.21, these parameters are shown.

7.3.2.5 Jitter Requirements

Jitter should not be more than one-third of the pixel, so the maximum value should be one-third IFOV.

7.3.2.6 Pointing Accuracy

Pointing accuracy depends on geolocation requirements. Pointing accuracy (radians) = Geolocation requirement(m)/height of the satellite (m).

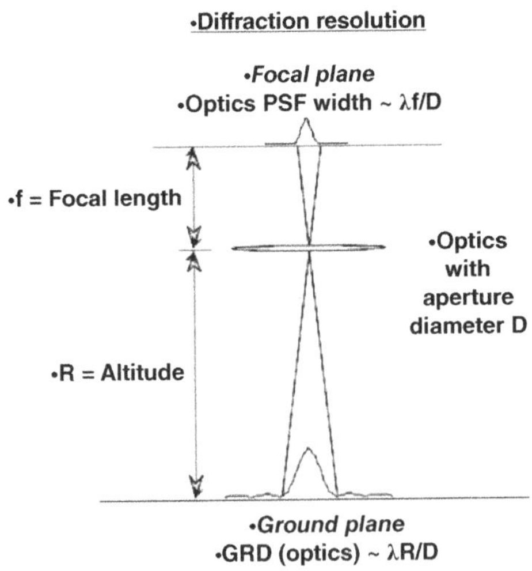

Figure 7.20 Telescope diffraction resolution.

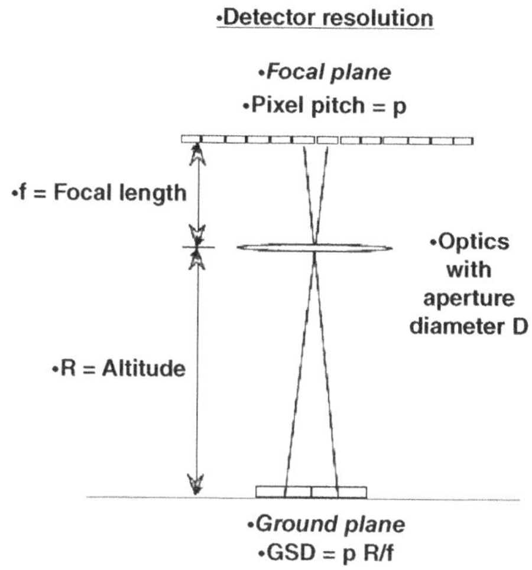

Figure 7.21 Telescope detector resolution [5].

7.3.3 Example

Given:

Wavelength = 650 nm; Orbit altitude = 575 km; Aperture = 50 cm; Focal plane array = 8000 × 8000 pixels; Pixel size = 5 micro m; GSD = 0.5 m, and Geolocation = 10 m.

Find

Focal length, IFOV, ground resolution distance. Q, focal field of view (Diagonal), Jitter requirements, and pointing requirement

1. Focal length

$$\text{GSD} = \frac{p}{f} \cdot R$$

Or

$$f = \frac{p}{\text{GSD}} R = \frac{5 \times 10^{-6}}{0.5} \times 575 \times 10^{3}$$
$$= 5.75 \text{ m}$$

2. $\text{IFOV} = \frac{p}{f} = \frac{5 \times 10^{-6}}{5.75} = 8.69 \times 10^{-7}\, radians$

On ground

$$\text{GSD} = \frac{p}{f} R = 8.69 \times 10^{-7} \times 575 \times 10^{3}$$
$$= 0.5\,\text{m}$$

3. Ground resolution distance

$$\text{GRD} = \frac{\lambda}{D} \cdot R$$
$$= \frac{650 \times 10^{-9} \times 575 \times 10^{3}}{0.5}$$
$$= 0.7475\,\text{m}$$

4. Q

$$Q = \frac{\text{GRD}}{\text{GSD}} = \frac{0.7475}{0.5} = 1.495$$

5. Focal field of view (Diagonal)

$$\begin{aligned}\text{FFOV} &= \sqrt{2} \times 8000 \times \text{IFOV} \\ &= \sqrt{2} \times 8000 \times 8.69 \times 10^{-7} \\ &= 9.83 \times 10^{-3}\,\text{radians}\end{aligned}$$

On ground FFOV

$$\begin{aligned}&= 8000 \times .5 = 4000 \\ &= 4\,\text{km}\end{aligned}$$

6. Jitter requirements

$$\begin{aligned}&= \frac{1}{3}\text{IFO} \\ &= \frac{1}{3} \times 8.69 \times 10^{-7} \\ &= 0.289\,\text{microradians}\end{aligned}$$

7. Pointing Requirements

$$\begin{aligned}&= \text{geolocation/orbit latitude} \\ &= \frac{10}{575 \times 10^3}\ \text{radians} \\ &= 17.4 \times 10^{-6}\ \text{radians} \\ &= 17.4\,\text{microradians} \\ &= .996 \times 10^{-3}\,\text{deg}\end{aligned}$$

7.3.4 Sampling MTF

The sampling cutoff frequency depends on its pixel spacing p and the telescope focal length f as follows.

$$\chi_{\text{sampling}} = \frac{f}{p} = \frac{1}{\text{IFOV}} \tag{7.22}$$

The sampling MTF is given by

$$\text{MTF}_{\text{sampling}} = \frac{\sin(\pi\chi)}{\pi\chi} \tag{7.23}$$

where χ is the frequency normalized on sampling. This function is plotted in Figure 7.22.

7.3.5 Aliasing

When the scene contains spatial frequencies higher than half the sampling frequency, artifacts, called aliasing errors, appear in the image, which is not present in the scene. This phenomenon is peculiar to periodic sample data systems and occurs because the scene frequencies beat with the sampling frequency. The insidious nature of aliasing is that, without a "truth" image to compare, there is no way to tell if aliasing is present in an image.

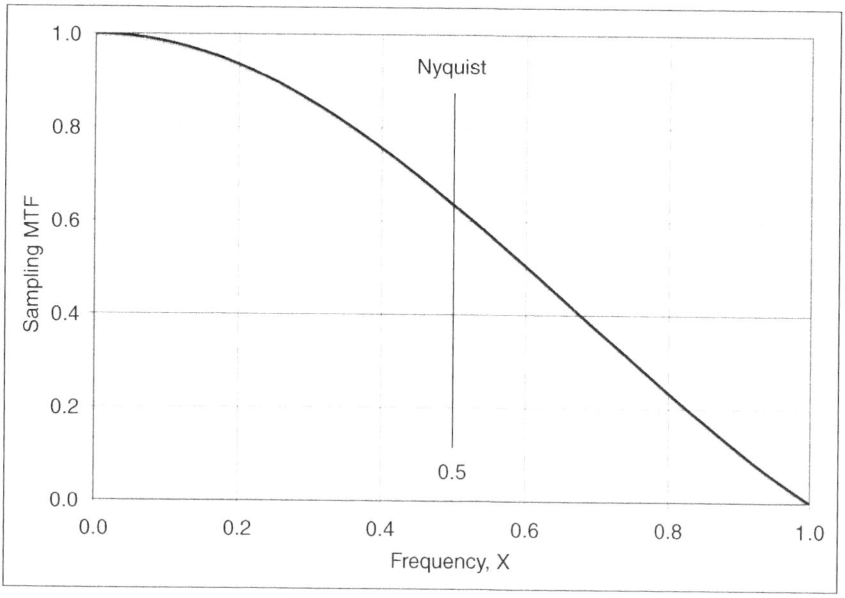

Figure 7.22 Sampling MTF (normalized on the cutoff frequency 1/IFOV).

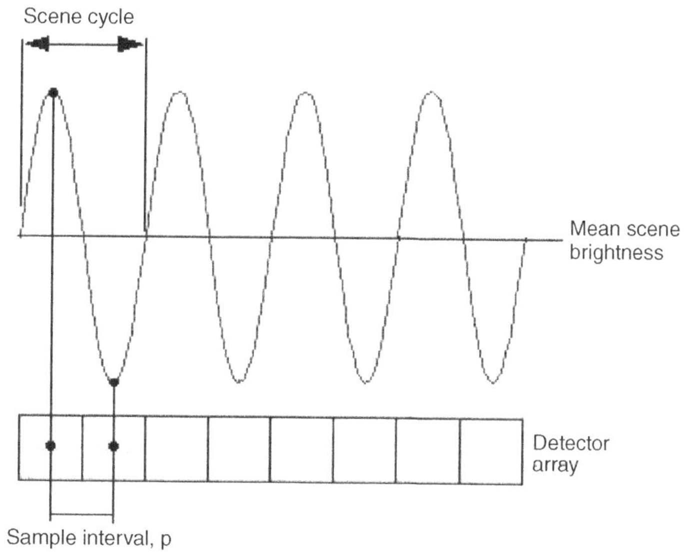

Figure 7.23 Sampling at the Nyquist frequency. *Source*: [6]/R. R. Auelmann.

To avoid aliasing either the system modulation transfer must be zero at frequencies above the Nyquist frequency (defined as half the sampling frequency) or the scene must contain no frequencies above the Nyquist frequency. This is the so-called Nyquist sampling criterion. Figure 7.23 illustrates the situation when the scene frequency is exactly equal to the Nyquist frequency; that is, there are two samples per scene cycle.

The sample interval is half the scene cycle interval. What happens when this criterion is not satisfied? Consider the situation in Figure 7.24. There are only six samples within the four-cycle

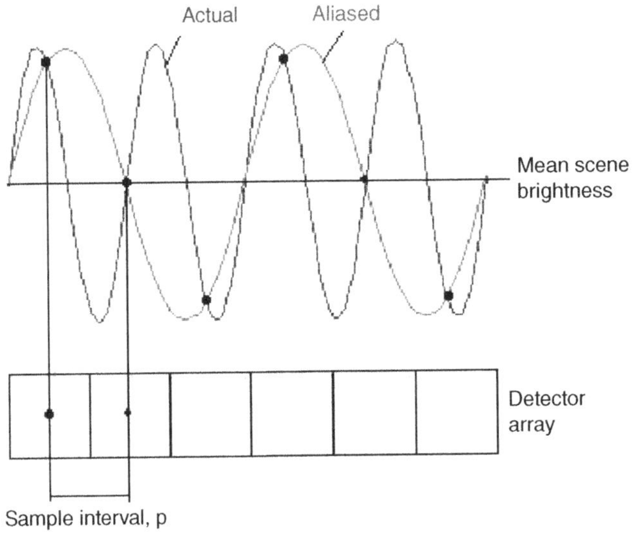

Figure 7.24 Example of aliasing. *Source*: [6]/R. R. Auelmann.

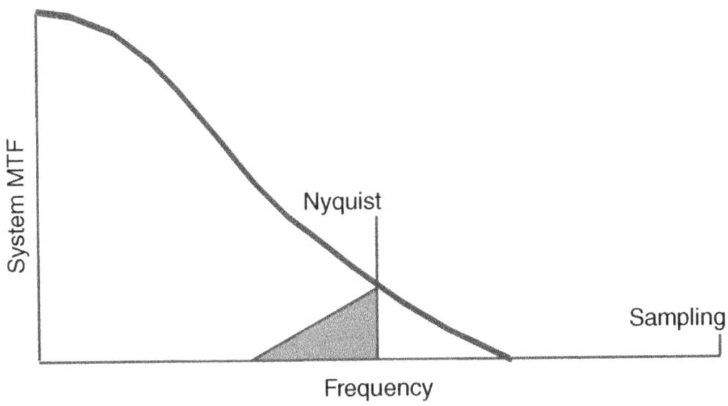

Figure 7.25 Aliasing folds the MTF back about the Nyquist frequency. *Source*: [6]/R. R. Auelmann.

scene, which means that the scene frequency is 3/2 times the Nyquist frequency. The aliased image is the lowest frequency pattern consistent with the sampling. In this case, it is at half the frequency of the real scene. The aliased image appears at a frequency as far below the Nyquist frequency as the scene frequency is above the Nyquist frequency.

It is as though the portion of the MTF above the Nyquist frequency flips these high-frequency scene inputs into lower-frequency image outputs as depicted in Figure 7.25 showing the total system MTF with a cutoff past the Nyquist frequency.

7.4 Optical Components

7.4.1 Focal Plane

7.4.1.1 Focal Plane Technology

Focal plane assembly (FPA) is where the collected light is focused to form an image. It typically consists of an array of sensors, such as charge-coupled device (CCD) or complementary metal-oxide-semiconductor (CMOS) detectors. CCD and CMOS detectors are two types of semiconductor-based image sensors used in digital cameras, including those found in imaging satellites. While both serve the same fundamental purpose of capturing light and converting it into digital signals, they differ in their underlying technology and characteristics.

7.4.1.1.1 Technology

CCD: CCDs use a specialized manufacturing process where charge is transferred through a series of capacitors to produce an electrical signal. Each pixel in a CCD sensor is connected to an adjacent pixel by a shift register, which moves the charge from one pixel to the next until it reaches an amplifier.

CMOS: CMOS sensors, on the other hand, utilize a different approach. Each pixel in a CMOS sensor has its own amplifier and readout circuitry integrated directly into the sensor chip. This allows for parallel readout of each pixel's charge, providing faster readout speeds compared to CCDs.

7.4.1.1.2 Power Consumption

CCD: CCDs typically consume more power compared to CMOS sensors, primarily due to the need for external circuitry to read out the charge from each pixel sequentially.

CMOS: CMOS sensors have lower power consumption because they can selectively activate individual pixels for readout, rather than reading out the entire sensor array at once.

7.4.1.1.3 Image Quality

CCD: CCD sensors traditionally have been favored for their superior image quality, particularly in terms of sensitivity to light and dynamic range. They tend to produce images with lower noise levels and better color accuracy.

CMOS: While CMOS sensors historically lagged CCDs in terms of image quality, advancements in CMOS technology have narrowed the gap. Modern CMOS sensors can offer comparable image quality to CCDs while providing other advantages such as faster readout speeds and lower power consumption.

7.4.1.1.4 Applications

CCD: CCD sensors are still commonly used in applications where high image quality is paramount, such as professional photography, astronomy, and certain scientific imaging applications.

CMOS: CMOS sensors have become increasingly popular due to their versatility, lower power consumption, and faster readout speeds. They are widely used in consumer electronics, mobile devices, and many industrial and scientific imaging systems, including those used in imaging satellites.

In summary, while CCD and CMOS sensors have different underlying technologies and characteristics, both can capture high-quality images and are used in various imaging applications, including those in satellite systems. The choice between CCD and CMOS often depends on factors such as desired image quality, power consumption, and cost considerations.

7.4.1.2 Focal Plane Devices

7.4.1.2.1 Scanning Sensors

The linear array, as shown in Figure 7.26, is scanned across a target area (push broom scan). Time delay integration (TDI) is a technique used in imaging satellites and other remote sensing systems

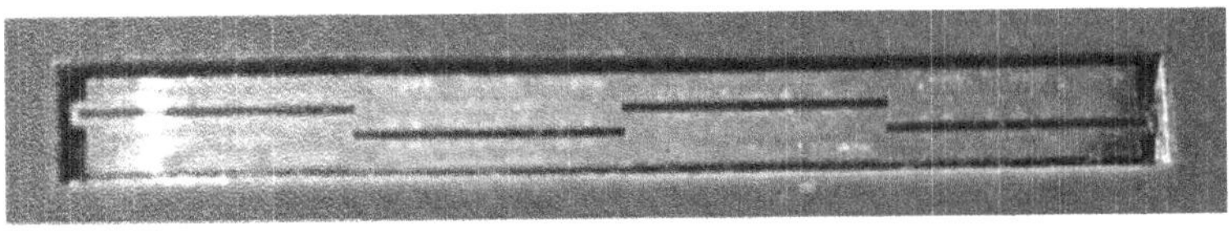

Figure 7.26 Line scanning array. *Source*: [5]/Davis.

to enhance the signal-to-noise ratio and improve image quality, especially in scenarios where the target is moving relative to the sensor. TDI is particularly useful in scenarios like satellite imaging where the satellite is moving over the Earth's surface while capturing images. In TDI operation, the sensor accumulates the signal from multiple adjacent pixels along the direction of motion over multiple exposure times. This accumulation effectively increases the exposure time for each pixel, allowing the sensor to collect more photons and improve the signal-to-noise ratio. Each exposure captures a strip of the scene, and the strips are then combined to form a complete image.

TDI is especially beneficial in satellite imaging applications because it allows for better image quality despite the motion of the satellite relative to the Earth's surface. It helps reduce motion blur and increases the ability to detect and resolve features in the imagery, even in low-light conditions.

Overall, TDI technology improves the performance of imaging satellites by enhancing image quality, enabling better detection and analysis of features on the Earth's surface, and contributing to more accurate and valuable satellite imagery.

7.4.1.2.2 Staring Sensors

Arrays of square pixels that take snapshots of an area. Areas are mosaicked in a step-stare fashion.

7.4.1.2.3 Visible Focal Planes

Silicon PIN photodetectors hybridized with CMOS readout-integrated circuit (ROIC).

7.4.1.2.4 Infrared Focal Plan

InSb, HgCdTe photodetectors hybridized with CMOS ROIC. InSb, HgCdTe photodetectors hybridized with CMOS ROIC.

Requires cooled operation (~77–170°K).

Table 7.1 gives an example of specifying a focal plane.

7.4.2 Operation Concepts

Most satellite sensors operate in the visible and near-infrared (VNIR) portion of the spectrum roughly 0.4–0.9 μm and employ a high-resolution panchromatic (pan) band with imagery portrayed in grey levels. For some satellites (Worldview 1, for example) this is the only band. But more often there are lower resolution multispectral bands, three of which are blue-green-red and sometimes a fourth near-infrared, as shown in Figure 7.27. The lower-resolution multispectral bands are also useful for other purposes such as determining the health of crops and vegetation. Gray images use only one spectral channel, allowing the satellite sensor to dedicate all its pixels to capturing fine details in that single channel. In contrast, color images require capturing multiple channels (usually red, green, and blue), which divides the sensor resolution across these channels.

Figure 7.28 illustrates the operation of pushbroom scanner, whiskbroom scanner, and step-stare scanner. Mechanisms advantages, disadvantages, and applications for these scanners are given as follows.

Table 7.1 Specifying a focal plane.

Parameter	NFOV	WFOV	Comments
Pixels	2048 × 2048 sq	1024 × 1024 sq	Minimum value for NFOV WFOV is approximate and can be bigger.
Frame rate	≥30 Hz max	~1 Hz	Full frame, short integration time. Slower maximum frame rates will be considered for both.
Technology	CCD / CMOS / Hybrid	Beggars cannot be choosers.	
Integration Time	<100 μs to >1s	<100 μs to >1s	Must be adjustable, shorter integration times may be desired.
QE * Fill Factor	>20%	>30%	Solar weighted. Higher QE is better for WFOV.
Noise	<30 e-	<30 e-	All sources except quantization and image shot noise.
Digitizer Resolution	≥12 bits	≥10 bits	Number of taps is not specified.
Environment	1–3 yrs.	1–3 yrs.	In GEO or LEO
Data Interface	LVDS	LVDS	Preferred, camera link and space wire are acceptable. Please indicate preference.
Com / Tel Interface	RS422 / mil-1553	RS422 / mil-1553	Preferred, others acceptable. Please specify.
Spectral Band [nm]	700–350	Silicon Pan	
Power	<15 W	<10 W	Estimated power, includes cooling, power estimate is a target allocation; higher numbers will be considered.
Mass	<10 kg	<5 kg	Mass of FPA including power and electronics. This is a target allocation, indicating your design mass.
TRL	>6	>6	Flight proven preferred. Repackaging to fit new optics is allowed. New electronics design is not preferred.

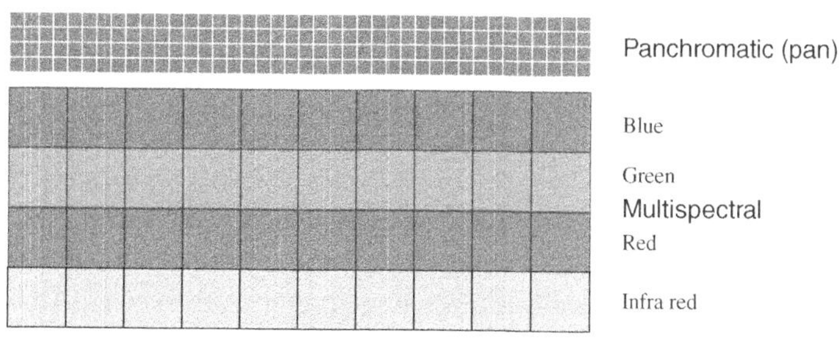

Figure 7.27 High-resolution pan band with low-resolution multispectral bands. *Source*: [7]/R. R. Auelmann.

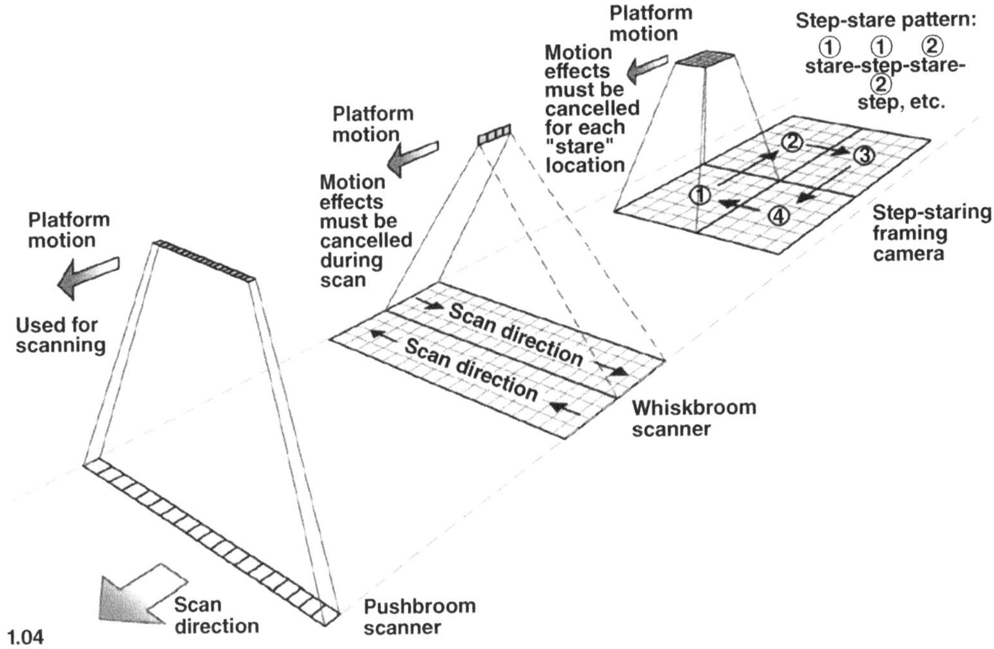

Figure 7.28 Illustration of push broom/whisk broom scanner and step starer [7].

7.4.2.1 Pushbroom Scanner

The mechanism is to use a linear array of detectors aligned perpendicular to the satellite's motion. As the satellite moves forward, the detectors capture data line-by-line, building an image. Pushbroom imaging typically employs linear sensor arrays, such as CCD or CMOS detectors. Advantages are high radiometric sensitivity due to continuous observation of each ground point, no moving parts, leading to increased reliability and reduced wear, and the ability to achieve high spatial resolution. The disadvantages are that it requires precise calibration of all detectors in the array and is susceptible to variations in detector performance (stripping artifacts). Applications are that it is used in modern satellites like Landsat 8 (OLI) and Sentinel-2.

7.4.2.2 Whiskbroom Scanner

The mechanism uses a single or small set of detectors to scan across the satellite's swath width using a rotating or oscillating mirror. The mirror sweeps side-to-side (perpendicular to the satellite's path) while the satellite moves forward. The advantages are that each ground point is viewed by the same detector, reducing calibration complexity. Well-suited for older and simpler sensor designs. Disadvantages are moving parts can wear out over time, reducing reliability and lower dwell time on each ground point, leading to reduced radiometric sensitivity. Applications are used in older satellite systems like Landsat 1-7 (TM and ETM+ sensors).

7.4.2.3 Step-stare Scanner

The Mechanism is to capture images of specific areas sequentially by "staring" at one scene, stepping to the next, and repeating. It often employs detectors or mirrors to shift the field of view between scenes. Staring imaging typically utilizes area-array sensors, such as CCD or CMOS detectors, capable of capturing images of the entire target area simultaneously. Advantages are high dwell time on each scene allows for better radiometric and spectral resolution. It is suitable for

Table 7.2 Comparison of push broom and starring imaging.

	Pushbroom imaging	Staring imaging
Integration time	0.1–3 ms (TDI)	Up to 1000 ms
Orbital constraints	Sun-synchronous (10 AM–2 PM)	Highly relaxed
Focal plane technology	1D linear arrays (TDI)	2D focal plane arrays
Multispectral implementation	Multiple linear arrays w/filters	Filter wheel, Dichroics, Bayer-Filter...
Imaging Q	0.25–1.0+ (optimized for int time)	Up to 2.0 (Optimized for resolution)
Aperture diameter	Large to overcome low-Q	Significantly smaller (matched to Q)
Frames per pass	1–10's	100–1000's
FOV	Large (array + sweep limited)	Small (FPA limited)
Targets per orbit	Many 10's	2–10
Persistence on target	Very limited	4–10 min
Data quality issues	2D image not rigid (sweep/LOS errors) Aliasing in Imagery from low-Q	Time-separation between bands "Grain" in images from high-Q

high-resolution or hyperspectral imaging. Disadvantages are limited to imaging specific targets and not suitable for continuous, wide-area mapping. It has slower data acquisition compared to pushbroom and whiskbroom scanners. Applications are typically used in high-resolution reconnaissance or targeted hyperspectral imaging.

7.4.2.4 Comparison of push broom and starring imaging

A comparison between push broom and staring imaging is given in Table 7.2.

7.5 Image Aberration

7.5.1 Pointing, Jitter, and Smear

Consider a least square linear fit to the pointing error over an integration period, as shown in Figure 7.29. Pointing error is constant. Smear is the slope of a linear fit. Jitter is residual error relative to linear fit.

Jitter variance in kth integration time window

$$\eta_{i,k}^2 = \frac{1}{T_{INT}} \int_{t_k}^{t_k+T_{INT}} [x_i(t) - s_i(t) - \mu_{i,k}]^2 dt \tag{7.24}$$

$$\mu_{i,k} = \frac{1}{T_{INT}} \int_{t_k}^{t_k+T_{INT}} [x_i(t) - s_i(t)] dt \tag{7.25}$$

$$\text{steady state jitter} = \sqrt{\frac{1}{N} \sum_{k=1}^{N} \eta_{i,k}^2} \tag{7.26}$$

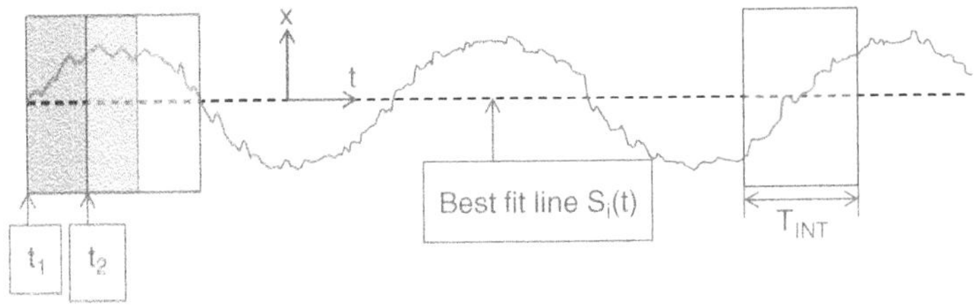

Figure 7.29 Pointing, jitter, and smear representation [7].

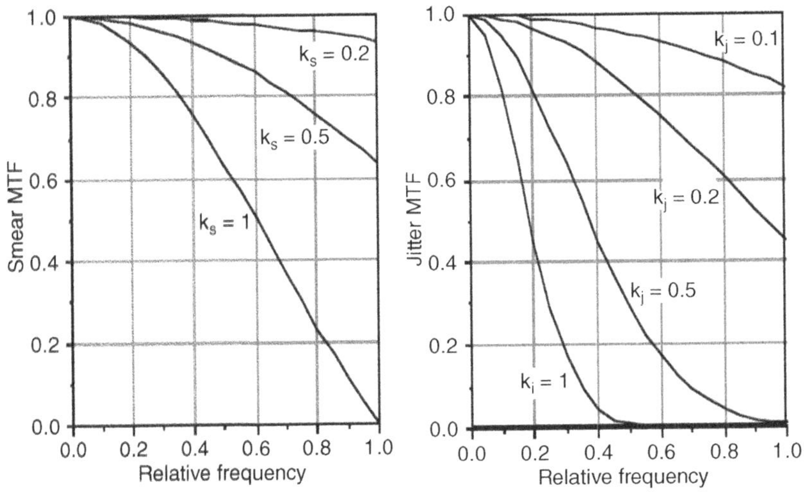

Figure 7.30 Smear and Jittr MTF normalized to sampling frequency. *Source*: [6]/R. R. Auelmann.

If the camera LOS translates at a nearly constant rate across the scene during the integration time, the motion is termed smear. If, on the other hand, it oscillates during the integration time, the motion is termed jitter. Both effects can be expressed in terms of their MTFs. The MTF for smear is.

$$\mathrm{MTF}_{\mathrm{smear}} = \frac{\sin(\pi k_s \chi)}{\pi k_s \chi}, \quad k_s = \frac{\dot{\theta}_{LOS} t_{\mathrm{int}}}{\mathrm{IFOV}} \tag{7.27}$$

where χ is normalized on the sampling frequency. The constant k_s is the change in LOS direction during the integration interval tint normalized on the IFOV. Thus, $k_s = 1$ corresponds to one pixel of smear. The smear MTF is a sinc function, which for $k_s = 1$ is identical to the sampling MTF. Moreover, for scanners, $k_s = 1$ corresponds to the smear in the along-scan direction produced by the scan process itself.

The MTF for jitter is.

$$\mathrm{MTF}_{\mathrm{jitter}} = \exp\left[-2(\pi k_j \chi)^2\right], \quad k_j = \frac{\sigma_j}{\mathrm{IFOV}} \tag{7.28}$$

Here, σ_j is the rms jitter amplitude. The MTFs for smear and jitter are depicted in Figure 7.30.

The MTF is two to three times more sensitive to jitter than to smear (i.e. comparing the sensitivity to k_s and k_j). This is fortunate because for short exposures (say a few milliseconds), most disturbances appear as smear rather than jitter.

7.5.2 National Image Interpretability Rating Scale

National Image Interpretability Rating Scale (NIIRS) is used to qualitatively describe the quality of an image. The rating relates to the observability of details in an image as shown in Table 7.3.

NIIRS is often used synonymously with GRD (or GSD).

7.5.2.1 Signal-to-noise Ratio

Signal-to-noise ratio (SNR) is another important characteristic of imaging satellite payload. SNR can be thought of as the number of electrons generated in a pixel due to photons received from a point on the target divided by the variance in the number of electrons generated in the pixel from all other sources. Signal electrons are lost due to atmospheric absorption or scattering, optics transmissivity, and detector quantum efficiency. Noise electrons are generated by background photons from sources other than the target and noise processes in the detector from shot noise, readout, and dark current.

7.5.2.2 General Image Quality Equation

The general image quality equation (GIQE) has been developed, which makes it possible to compute the NIIRS as a function of three main factors: (1) resolution (expressed in terms of the ground sample distance (GSD), (2) the relative edge response (RER), which is indirectly related to the PSF, and (3) the signal-to-noise ratio (SNR). The form of the GIQE is

$$\mathrm{NIIRS} = c_0 + c_1 \ \log_{10}(\mathrm{GSD}) + c_2 \ \log_{10}(\mathrm{RER}) + c_3 \frac{\mathrm{G}}{\mathrm{SNR}} + c_4 \ \mathrm{H}, \tag{7.29}$$

$$\mathrm{NIIRS} = \mathrm{A_{sys}} - \mathrm{a} \ \log_{10} \mathrm{GSD_{GM}} + \mathrm{b} \ \log_{10} \ \mathrm{RER_{GM}} - 0.656\mathrm{H_{GM}} - 0.344 \ (\mathrm{G/SNR}) \tag{7.30}$$

where $\mathrm{A_{sys}} = 10.251$ for the visible, and $\mathrm{A_{sys}} = 10.751$ for the infrared images. $\mathrm{GSD_{GM}}$ is the geometric mean (GM) of ground-sampled distance (GSD) in inches, $\mathrm{RER_{GM}}$ is the geometric mean of the relative edge response (RER), HGM is the geometric height of overshoot due to post-processing image sharpening, G is the noise gain due to image enhancement, and SNR is the signal-to-noise ratio. The coefficients $a = 3.32,\ \ b = 1.559$ if $\mathrm{RER} \geq 0.9$, and $a = 3.16,\ \ b = 2.817$ if $\mathrm{RER} < 0.9$.

Table 7.3 NIIRS for images.

NIIRS	GRD (m)	Nominal capability (EO)
1	>9.0	Detect medium-sized port
2	4.5–9.0	Detect large buildings
3	2.5–4.5	Detect trains on tracks
4	1.2–2.5	Identify railroad tracks
5	0.75–1.2	Identify theater ballistic missile
6	0.40–0.75	Identify spare tyre on truck
7	0.20–0.40	Identify individual rail ties
8	0.10–0.20	Identify windshield wiper
9	<0.1	Identify individual rail spikes

With $Q = 2$, GRD = 2XGSD

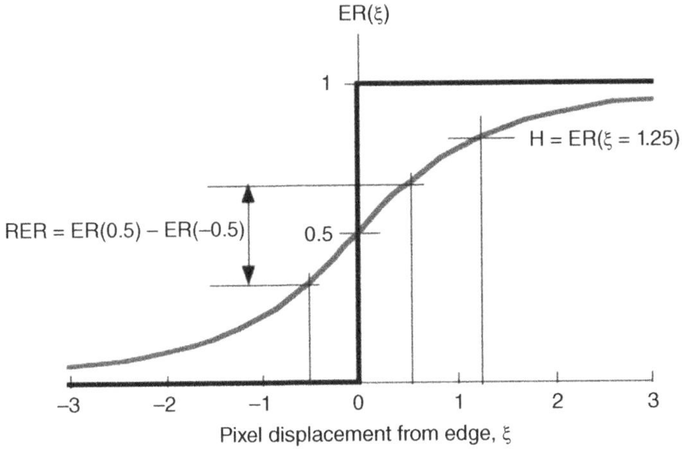

Figure 7.31 ER and RER. *Source*: [6]/Brij N. Agrawal.

7.5.2.3 Edge Response Function

Edge response (ER) is a measure of how well an imager can reproduce sharp edges in a scene. The scene edge has zero value to the left of center, and a unity value to the right of center. It is defined by the equation.

$$ER(\xi) = 0.5 + \frac{1}{\pi}\int_0^{1/Q}\left(\frac{\mathrm{MTF}(\chi)}{\chi}\sin 2\pi\chi\xi\right)d\chi \tag{7.31}$$

where ξ is the number of pixels offset measured from the edge, and χ is the spatial frequency normalized on the sampling cutoff frequency 1/IFOV.

The relative edge response (RER) and the overshot parameter H are key parameters, derived from the ER, which are used to assess image quality. The RER is the difference in ER measured at half a pixel on both sides of the edge and is given by

$$\begin{aligned} RER &= ER(0.5) - ER(-0.5) \\ &= 2\int_0^{1/Q}\left(MTF(\chi)\frac{\sin\pi\chi}{\pi\chi}\right)d\chi \end{aligned} \tag{7.32}$$

The overshoot parameter H is also determined from the ER function. It is the peak value of the ER over the range $\xi = 1 - 3$. However, if the ER monotonically increases (over this range), H is the value of the ER at $\xi = 1.25$. Figure 7.31 shows ER and RER.

7.6 Space Telescopes

7.6.1 Segmented Mirror Telescope

In this section, we will discuss the segmented mirror telescope (SMT) at the Naval Postgraduate School. SMT was developed by Lockheed Martin Corporation and ITT for National Reconnaissance Office under a research program. The SMT at the Naval Postgraduate School is a three-mirror anastigmat incorporating Ritchey–Chretien Cassegrain telescope, a hyperboloids primary mirror, an ellipsoid secondary mirror, and an ellipsoid tertiary. These designs are ideal for wide-field

Figure 7.32 Segmented mirror telescope.

observations. The ellipsoidal tertiary removes astigmatism and increases the wide FOV beyond the two-mirror Ritchey–Chretien design.

SMT, as shown in Figure 7.32, is a 3 m aperture space telescope consisting of six hexagonal 1 m mirror segments. The segments are constructed using a nano-laminate foil mirror surface bonded to a silicon carbide substrate. Built into the substrate of each segment, are 156 piezo face sheet actuators (FSAs). The FSAs are used to actively change the surface of each mirror segment. The mirror segments are mounted at three points to fine control actuators (FCAs). Each FCA is mounted to two coarse control actuators (CCAs) in a hexapod arrangement. The FCAs and CCAs provide six degrees of freedom (DOF) motion for alignment and phasing. SMT has a Shack–Hartmann wavefront sensor for surface control. It has a Phase diversity sensor for segment alignment. Fast steering mirror (FSM) is for jitter control and to utilize the entire field of regard (FOR) available for viewing. The FSM is set at the exit pupil to avoid residual image motion during jitter correction.

7.6.1.1 Optical Configuration

The optical configuration of SMT is shown in Figure 7.33. Mirror parameters are given in Table 7.4.

7.6.1.2 Analytical Model

The analytical models of imaging satellites, as shown in Figure 7.34, are complex and multidisciplinary. It involves optical models, CAD models, structural models, thermal models, and control models and their integration. The block diagram of SMT analytical modes is as follows.

The first block in Figure 7.34 determines both the layout of the physical structure and the optical details of the telescope, each of which is modeled separately. The mission requirements of Block 1 drive the design of the telescope.

Block 2 drives optical model. Popular optical modeling software packages include Zemax, Optics Software for Layout and Optimization (OSLO), and Code V. Optical software packages trace simulated light fields, or wavefronts, through the system to either the exit pupil or to the image plane. This process typically employs geometric ray tracing to propagate the wavefront from one component to another; however, when appropriate, it will use more complex propagations that accurately model Fraunhofer and Fresnel diffraction.

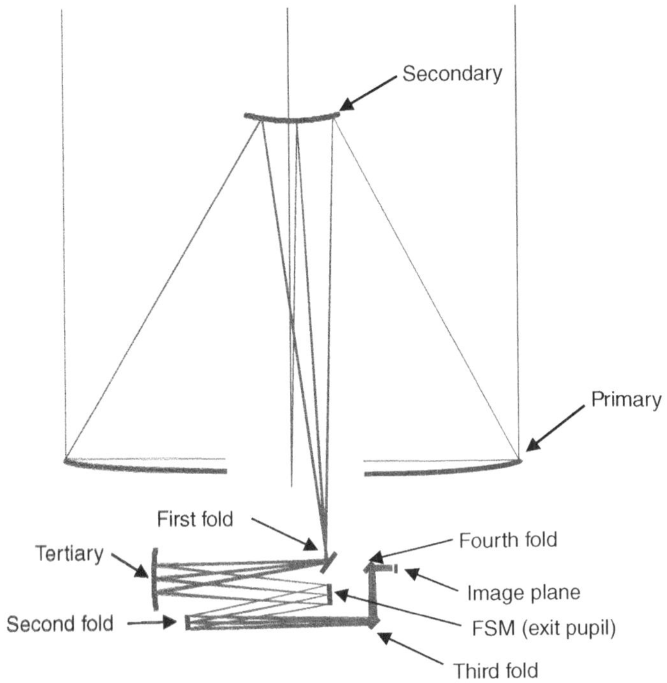

Figure 7.33 Optical configuration of SMT.

Table 7.4 Mirror parameters of SMT.

Surface	ROC (mm)	Clear aperture (CA) (mm)	Aspheric	Conic constant
Primary	6031.1	3090.00	1.02E-14	−1.014
Secondary	−1092.9	551.6	0.0	−1.824
1st fold mirror	Flat	298.6	0.00	0.000
Tertiary	1844.8	454.7	0.00	−0.672
FSM	Flat	154.8	0.00	0.000
2nd fold mirror	Flat	101.8	0.00	0.000
3rd fold mirror	Flat	75.9	0.00	0.000
4th fold mirror	Flat	57.0	0.00	0.000

The computer-aided drafting (CAD) model of Block 3 defines the physical layout of all the components in the system and is used to generate the thermal and structural models shown in Blocks 4 and 5, respectively. Although the nominal positions of optical elements in the CAD model are dictated by the optical model, the CAD model details the supporting structure.

The thermal model in Block 4 is used to generate thermal profiles that the system will be exposed to during operation, such as those caused by orbital transitions into and out of direct sunlight. These profiles are used as disturbances in the structural model to predict how temperature gradients will cause the structure to expand or contract. Thermal Desktop and SINDA are industry-accepted thermal packages.

To create the structural model, also called a finite element model (FEM), the CAD model is discretized by meshing the components into small elements defined by points in Euclidean space,

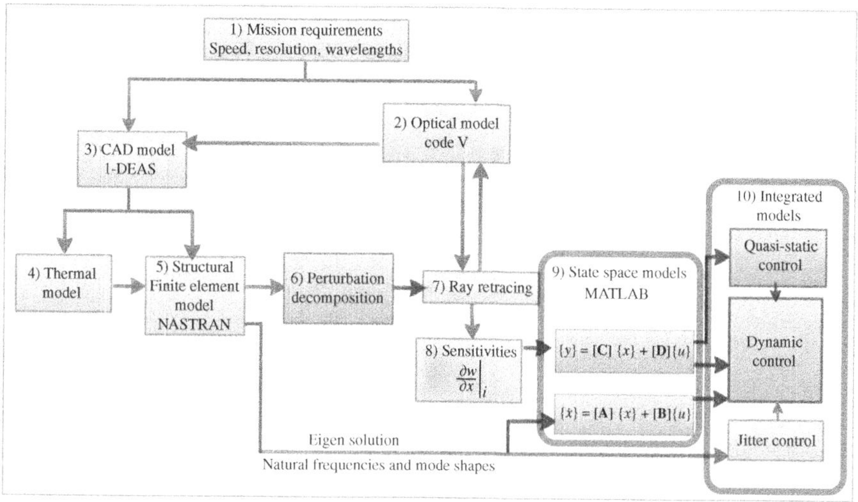

Figure 7.34 The integrated design process for segmented mirror systems [8].

a geometric space that represents physical space. These points are called nodes. Material properties, boundary conditions, and disturbances are attributed to either the elements or the grid defining the elements to approximate the real system. NASA STRuctural Analysis (NASTRAN) is used exclusively to solve various static and dynamic solutions.

The basic equation of motion for any finite element model is of the form

$$[\mathbf{M}]\{\ddot{x}\} + [\mathbf{K}]\{x\} = [\mathbf{G}]\{u\}, \tag{7.33}$$

Note that this equation is a balance of forces, where $[\mathbf{M}]\{\ddot{x}\}$ constitutes inertial forces, $[\mathbf{K}]\{x\}$ constitutes elastic forces, and $[\mathbf{G}]\{u\}$ constitutes externally applied forces.

Free vibration analysis results in

$$[\mathbf{M}]\ddot{x}\} + [\mathbf{K}]\{x\} = \{0\}. \tag{7.34}$$

The eigenvalue solution in NASTRAN finds the eigenvalues λ_i and eigen- vectors φ_i using a form like all eigenvalue solutions,

$$\left[[\mathbf{K}] - \lambda_i \cdot [\mathbf{M}]\right] \cdot \varphi_i = \{0\}. \tag{7.35}$$

The eigenvalues λ_i are related to the natural frequencies f_i of the structure by

$$_i\ \lambda_i = \omega^2 = (2\pi f_i)^2 \text{ for } i = 1 \text{ to } m \text{ modes},$$

where ω_i is in units of rad/s, and f_i is in units of Hertz (Hz).

The modal transformation matrix $[\mathbf{\Phi}]$ is the eigenvector φ_i given by

$$[\mathbf{\Phi}] = [\phi_1 \quad \phi_i \quad \cdots \quad \phi_m] = \left[\begin{bmatrix} x_1 \\ x_2 \\ \vdots \\ x_{6n} \end{bmatrix}_1 \begin{bmatrix} x_1 \\ x_2 \\ \vdots \\ x_{6n} \end{bmatrix}_2 \cdots \begin{bmatrix} x_1 \\ x_2 \\ \vdots \\ x_{6n} \end{bmatrix}_m\right] \tag{7.36}$$

for m modes and $6n$ DOF,

where each mode shape φ_i describes the displacement shape at frequency f_i.

An important property of $[\mathbf{\Phi}]$ is that each mode shape is orthogonal to the others, which will be leveraged to develop a modal-based state-space model.

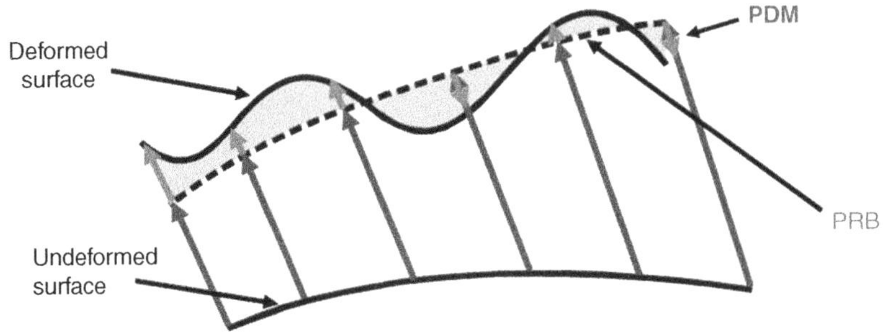

Figure 7.35 Perturbed surface includes a PRB and PDM [8].

To ensure continuity when passing information between the FEM and the optical model, it is important to use coordinate systems that are consistent between the two models.

As shown in Block 6, structural perturbations must be decomposed into two components before they can be used for optical analysis: rigid body perturbations (PRB) and deformable mirror perturbations (PDM). These perturbations are described in Figure 7.35.

The rigid body perturbations in Figure 7.35, annotated with long arrows, start at the undeformed surface and extend to the best-fit plane. The best-fit plane is the least-squares solution to the deformed surface and is represented with a dashed line. The deformable mirror perturbations are the short arrows. They start from the best-fit plane and extend to the deformed surface. Note the rigid body perturbations are much larger in magnitude than the deformed perturbations. Also, note that only the deformable component would change the optical prescription of an optical element.

In the optical model, the PRB component is applied to the centroid of its respective optical element, thus perturbing the optic rigidly. Here rigidly infers that the optical prescription is not changing. On the other hand, the PDM component describes the curvature of the optical surface and affects the prescription. Historically, this decomposition was accomplished by writing in-house code and was prone to inaccuracies. Moreover, this decomposition often limits the speed of the design cycle. Fortunately, a new software package, called Sigfit [9], is available to perform these operations.

To develop sensitivities, the relation between structural motion and optical performance, the NASTRAN perturbations must be decomposed into an optically useful basis, e.g. Zernike polynomials. After decomposing the NASTRAN perturbations, Sigfit generates input interferometry files (*.int) that are usable by optical packages like Code V.

7.6.1.3 First-order Modeling to Develop Sensitivities

With the initial modeling complete, the task becomes to relate structural perturbation states {x} to optical performance. This is accomplished as shown in Figure 7.36.

In Figure 7.36, the sensitivity ∂w is equivalent to the wavefront $w(\partial x)$ because the initial states are assumed to be zero, i.e. the WFE is zero for zero perturbation,

$$\frac{\partial w}{\partial x} = \frac{w(\partial x) - w(0)}{\partial x - 0} = \frac{w(\partial x)}{\partial x}. \tag{7.37}$$

For this work, the sensitivities to modal coordinates $\{q\}$ were desired and are related to $\{x\}$ by the modal transformation matrix,

$$\{x\} = [\Phi]\{q\}. \tag{7.38}$$

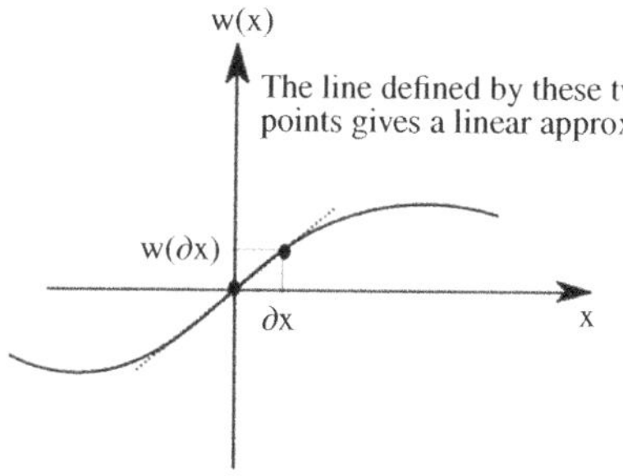

Figure 7.36 Sensitiveness of a wavefront to structural perturbation [9].

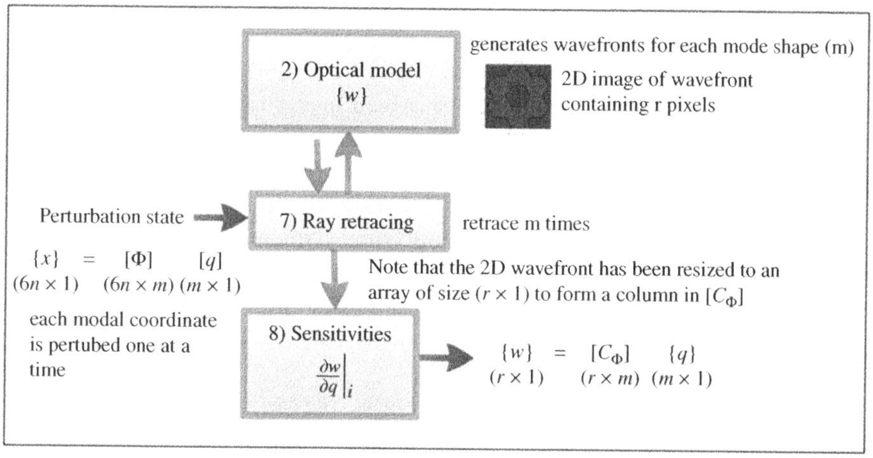

Figure 7.37 System sensitivities to perturbations are derived by re-tracing the wavefront for each perturbed degree of freedom [8].

The sensitivities are created in blocks 7 and 8 of Figure 7.34, the process of which is annotated with more detail in Figure 7.37.

For this work, the sensitivity matrix was annotated with a subscript of Φ to indicate that it was developed using mode shape perturbations. With $[C_\Phi]$ written in terms of its partial derivatives, the relationship of the wavefront to the modal coordinates is.

$$\underset{(r\times 1)}{\{w\}} = \underset{(r\times m)}{\begin{bmatrix} \dfrac{\partial w_1}{\partial q_1} & \cdots & \dfrac{\partial w_1}{\partial q_m} \\ \vdots & \dfrac{\partial w_j}{\partial q_i} & \vdots \\ \dfrac{\partial w_r}{\partial q_1} & \cdots & \dfrac{\partial w_r}{\partial q_m} \end{bmatrix}} \underset{(m\times 1)}{\{q\}} \text{ for } i = 1 \text{ to } m \text{ modes and } j = 1 \text{ to } r \text{ pixels,} \tag{7.39}$$

where m is the number of states in $\{q\}$, r is the number of pixels in $\{w\}$, and the partials $\dfrac{\partial w_j}{\partial q_i}$ are constant values.

As annotated in Figure 7.37, each ith perturbation is applied in optical software, and a wave-front of r pixels is generated using geometric raytracing. Each ray re-tracing in Block 7 produces a new set of sensitivities in Block 8, which constitutes a column in the [C] matrix. This technique is like the method Modeling and Analysis for Controlled Optical Systems (MACOS) employs. MACOS was used to develop sensitivities for the PRB and PDM perturbations discussed earlier.

There are two software packages currently available for creating sensitivities; they are MACOS and CVtoolkit. MACOS is a powerful software package created by JPL. Currently, it requires that definitions from the original optical model in Block 2 be used to manually create a MACOS input file. It then performs ray re-tracing itself and does not use the original optical modeling software directly. A newer MATLAB toolkit, called CVtoolkit, was developed by NASA Goddard and leverages the existing optical model and design tools of Code V to generate sensitivities. It consists of a series of MATLAB scripts designed to call Code V and store variables in MATLAB. Although not as capable as MACOS, CVtoolkit itself does not require the creation of a separate optical model.

7.6.1.4 Experimental Verification

To evaluate techniques for closed surface control of all segments using FSAs and a 4D interferometer, the block diagram of the testbed is shown in Figure 7.38.

4D laser interferometer and the null corrector (which makes the primary mirror look like a spherical mirror) are mounted at the primary mirror center-of-curvature. A laser beam from the interferometer passes through the null corrector at the center-of-curvature and out to the primary mirror segments. The reflected beam passes back through the null corrector and then into the interferometer. The interferometer measures the wavefront error and can be used to drive the face sheet actuators to minimize the aberrations present in the segments.

Next, we generate an influence coefficient matrix by actuating one actuator at a time. The relationship between the actuator input u and interferometer output y, representing wavefront is in a linear equation of the form.

$$\{y\} = [\phi]\{u\} \tag{7.40}$$

ith column of the influence coefficient matrix represents the influence function of the ith actuator as shown below.

$$[\phi] = \begin{bmatrix} \phi_{1,1} & \phi_{1,2} & \phi_{1,3} & \phi_{1,4} & \cdots & \phi_{1,n} \\ \phi_{2,1} & \phi_{2,2} & \phi_{2,3} & \phi_{2,4} & \cdots & \phi_{2,n} \\ \phi_{3,1} & \phi_{2,2} & \phi_{3,3} & \phi_{3,4} & \cdots & \phi_{3,n} \\ \phi_{4,1} & \phi_{3,2} & \phi_{4,3} & \phi_{4,4} & \cdots & \phi_{4,n} \\ \vdots & \vdots & \vdots & \vdots & \ddots & \vdots \\ \phi_{m,1} & \phi_{m,2} & \phi_{m,3} & \phi_{m,4} & \cdots & \phi_{m,n} \end{bmatrix} \tag{7.41}$$

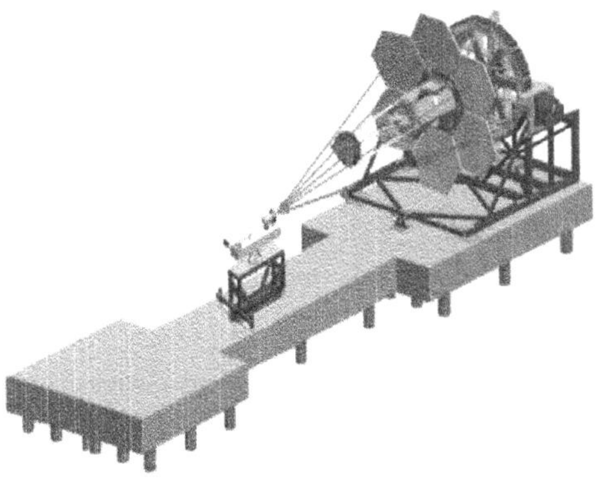

Figure 7.38 SMT test configuration.

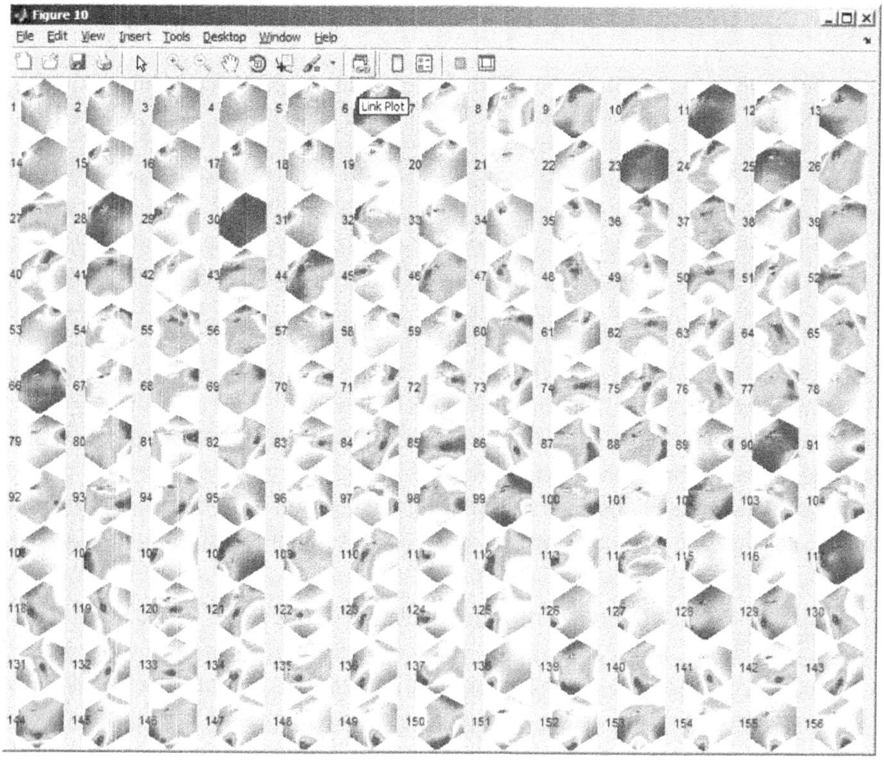

Figure 7.39 Influence function for SMT. *Source*: [10]/Naval Postgraduate School.

As an example, typical influence functions determined from the experiments are shown in Figure 7.39.

The simplest control command is the least squares solution is

$$\mathbf{u} = \mathbf{\Phi}^{\dagger}\mathbf{y} \tag{7.42}$$

where $\mathbf{\Phi}^{\dagger}$ is pseudo inverse of $\mathbf{\Phi}$

Control commands can also be found by constrained optimization.

$$\min_{x} \frac{1}{2}\|\mathbf{\Phi} \cdot \mathbf{u} - y\|_2^2 \text{ subject to LB} \leq \mathbf{u} \leq \text{UB} \tag{7.43}$$

LB is the lower control bound and UB is the upper control bound.

Iterative update of control command yields better performance.

$$\boldsymbol{u}(k+1) = \boldsymbol{u}(k) - \mu\mathbf{\Phi}^{\dagger}\boldsymbol{y}(k) \text{ (unconstrained)} \tag{7.44}$$

$$\boldsymbol{u}(k+1) = \boldsymbol{u}(k) - \mu\boldsymbol{\beta}(k) \text{ (Constrained)} \tag{7.45}$$

where μ is updated gain, and $\boldsymbol{\beta}(k)$ is constrained optimization solution.

DM command voltage relationship to control command

$$\mathbf{v}(i) = \mathbf{u}(i)/c + \mathbf{v}_0 \tag{7.46}$$

where v(i) is applied voltage, u(i) is desired actuator force, c is calibration constant between force and voltage, and v_0 is biased voltage.

7.7 Telescope Design Example

This telescope design was performed by the space system engineering students at the Naval Postgraduate School under course AE 4871 Spacecraft Design and Integration in 2017 [10].

7.7.1 Misson

Natural disasters cannot be predicted with significant certainty beyond theories or confidence ranges. After the disaster strikes, relief agencies begin gathering information, supplies, and establishing plans and strategies for humanitarian efforts. The Affordable Persistent Collection System (APCS) program is a constellation of satellites forming a Disaster Relief Assessment Global Observation Network (DRAGON). This program fills critical gaps in current collection capabilities by providing persistent imagery and video products for forensic analysis of ongoing events. In a time of challenging budget constraints, the design and implementation of this program must minimize cost and schedule while maximizing coverage.

This spacecraft design was performed by the space engineering students as part of their Master's Astronautical Engineering requirements. In this section, we will focus on the telescope payload design.

7.7.1.1 Requirements

1. Coverage: The APCS shall be capable of imaging a contiguous area of at least 2500 square kilometers per access
2. Latitude: The APCS shall provide coverage of targets between 60° north latitude and 60° south latitude.
3. Revisit: The APCS constellation shall provide access to targets within the specified latitude band at least once every 48 hours.
4. Duration: The APCS constellation shall provide access to regional targets for a minimum of one hour per access.
5. Maximum Gap during Access: For the APCS constellation gaps in coverage during a target, access shall not exceed 5 minutes.
6. Instantaneous Field of Regard: The instantaneous field of regard for targets with target elevation angles (TEA) greater than 20° shall be at least 1 080 000 km^2.
7. Spatial Resolution: The APCS shall provide panchromatic imagery with a ground sample distance (GSD) no greater than 0.5 m for targets at Nadir. Multispectral bands (near-infrared, red, green, blue) imagery shall be provided with a GSD no greater than 1 m.
8. Spectral Content: The APCS shall collect data in the following five spectral bands, as shown in Table 7.5.

Table 7.5 Optical bands.

Band	Wavelength (nanometers)
Panchromatic	400–900
Near Infrared	750–900
Red	600–750
Green	500–600
Blue	400–500

9. Geolocation: The APCS shall provide knowledge of the satellite ephemeris and attitude adequate to enable geolocation of an object on the ground to within 5 m linear error in each orthogonal axis.
10. Derived Requirement: GPS sensor for enhanced geolocation accuracy.
11. Imaging Time: The APCS shall be capable of imaging at least 400 minutes per day.
12. Contiguous Area: The APCS shall be capable of imaging a contiguous area of at least 2500 km^2 per access.

7.7.2 Trade Space

A Ritchey–Chretien Cassegrain telescope design with a tertiary mirror, a three-mirror anastigmat telescope, was chosen. This design eliminates spherical, coma, and astigmatism wave aberration.

The Affordable Persistent Collection System (APCS) requirements specify a maximum wet mass (launch mass) of 300 kg. The mass budget for the payload was finally 59 Kg. Considering the allocated mass, the primary mirror diameter was selected to be 50 cm. The material for the optical mirrors selected was Zerodur. Zerodur has a very low coefficient of thermal expansion and a density of 2.54 g/cm^3. Using this material, a 50 cm primary mirror with 75% lightning has a mass of approximately 10 kg.

A filter wheel (FW) was chosen to provide a single beam path with minimal mass and aberrations. A single beam path with a filter wheel provides the various pan and multispectral images required. A trade study was conducted between a 1° Field of View (FOV) and a 2° FOV telescope. Based on the target area of 2500 km^2 for a 50 × 50 km^2 target, the larger FOV provided significantly reduced ADCS requirements, a larger area for video capture, and better coverage persistence. Typical telescopes have a 1° FOV. There were no standard 2° FOV designs available that fit all the requirements to prove feasibility. The Spacecraft Research and Development Center (SRDC) at NPS conducted extensive design and simulation tests using Zemax to develop a feasible 2° FOV design for the satellite. This design provided required mirrors and sizes, which were then sourced to manufacturers.

To provide for the large target area, the FSM is used to scan the sensor field of view (SFOV) within the telescope FOV. The FSM is also used to reduce jitter from satellite vibration (jitter control) and smear due to satellite motion.

A single focal plane array sensor is used for all image collection. For the sensor, CMOS was chosen as opposed to CCD. Based on the state of the industry, CMOS provides better cost and performance while still meeting space-rated standards. The V59M, a space-rated Teledyne CMOS focal plane sensor, with a pixel pitch of 5.0 μm, was chosen as an industry-standard COTS product with a 7680 × 7680-pixel focal array for a total of 59 megapixels (Mp). The staring sensor is necessary (as opposed to a line array sensor) because of the requirement for video capability.

Design calculations were made with the 50 cm mirror, a mean panchromatic wavelength (cutoff frequency) of 650 nm, and the requirement for GSD to be no more than 0.5 m. A Q of 1.5 was used based on consultations with subject matter experts to determine the focal length (f).

7.7.3 Optical Telescope Design

Figure 7.40 shows a notional optical telescope design, incorporating a 2° FOV Ritchey–Chretien telescope with tertiary mirror, FSM, 6-position multicolor filter wheel, and focal plane array sensor.

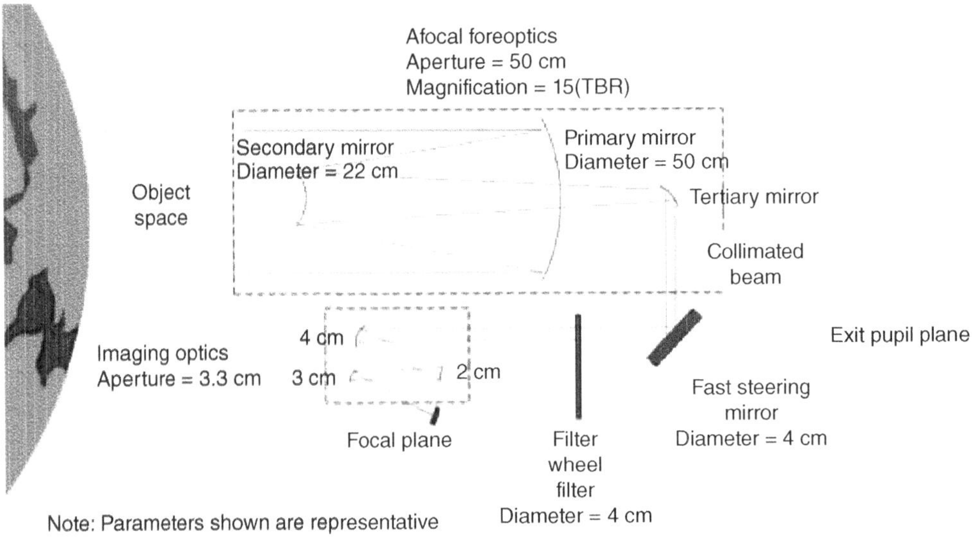

Figure 7.40 Telescope optical configuration. *Source*: [10]/Naval Postgraduate School.

Design calculations were made with the 50 cm mirror, a mean panchromatic wavelength (cutoff frequency) of 650 nm, and the requirement for GSD to be no more than 0.5 m. A quality factor (Q) 1.5 was selected based on consultation with subject matter experts. The equation for calculating focal length is as follows:

$$Q = \frac{\lambda f}{dp} \tag{7.47}$$

where Q is the quality factor, λ is the wavelength, f is the focal length, d is the diameter of primary mirror, and p is the pixel size.

For our design, $Q = 1.5$, $\lambda = 650$ nm, $p = 5.0\ \mu$m, and $d = 0.5$ m, resulting in the focal length $f = 5.77$ m.

Next, we determine the satellite altitude to meet GSD requirement of 0.5 m by using the equation

$$GSD = \frac{pR}{f} \tag{7.48}$$

where GSD is ground sampling distance, p is pixel pitch, R is the altitude of the satellite, and f is focal length.

For our design, GSD = 0.5 m, $p = 5.0\ \mu$m, $f = 5.7$ m, resulting in $R = 575$ km. This is the maximum altitude to meet GSD requirements with the chosen optical parameters for Q, λ, p, and d.

7.7.3.1 Image Collection

A basic calculation of the diameter of the telescope field of view (TFOV) considering a flat Earth, a good approximation for LEO for this view cone angle at nadir is calculated by:

$$\text{D(Diameter)} = R * \tan(1.0°) * 2 = 19\,\text{km} \tag{7.49}$$

Telescope Pixel FOV (TPF) also known as instantaneous FOV (IFOV) can be determined from the equation

$$TPF = GSD = \frac{pR}{f} \tag{7.50}$$

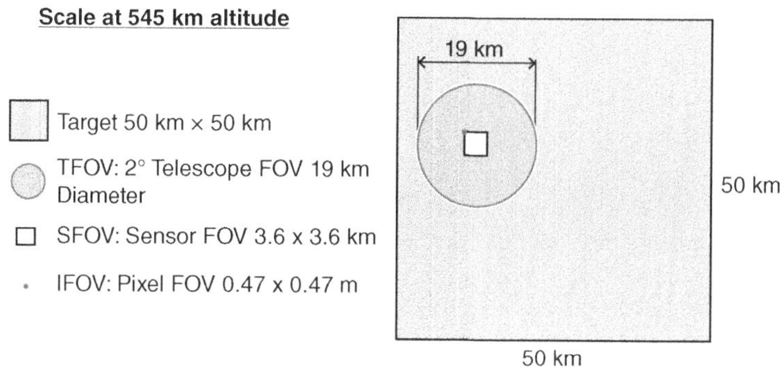

Figure 7.41 Field of views for telescope, focal plane, and pixel. *Source*: [10]/Naval Postgraduate School.

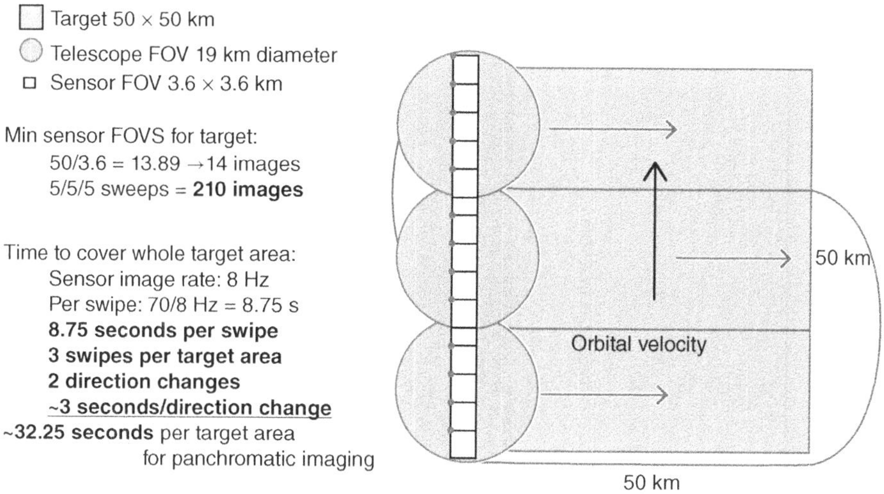

Figure 7.42 Panchromatic scanning mode. *Source*: [10]/Naval Postgraduate School.

The FPA sensor FOV (SFOV), nadir pointing, can be calculated as:

$$\text{SFOV} = \text{TPF} \times \text{number of pixel} = 0.47 \times 7680 = 3.6\,\text{km}$$

SFOF is 3.6 km × 3.6 km.

Figure 7.41 summarizes all fields of view.

In mode 1, covering the total target area, $50 \times 50\,km^2$, as shown in Figure 7.42, it needs continuous bus slewing and three sweeps for a nadir pass. Since the SFOV is $3.6 \times 3.6\,\text{km}^2$, it will require 14 images across and 15 images high with a total of 210 images for panchromatic. MSI requires four times the number of images to capture all 4 spectral bands. The figure shows an example of the TFOV following a back-and-forth sweeping scan by a bus while the FSM scans images (yellow squares) in TFOV.

An image scanning collection rate of 8 Hz was used. This comes from the Teledyne specifications for the V59M having a frame rate of 8 Hz. The FSM slew and settle time and the sensor integration time are included in 8 Hz. Time per sweep will be $70/8 = 8.75$ seconds. There will be three sweeps. There will be two directional changes with 3 seconds each. Therefore, the total time for panchromatic imaging of the target area will be 32.5 seconds.

Table 7.6 Parts list for size, mass, and power allocation.

Component	Mass (kg)	Dimensions (mm)	Power (W)	Voltage (V)	Heat load (W)	Reliability	TRL	Comments
Primary mirror	10	500 × 150 disk	0			1	9	
Secondary mirror	1.5	220 × 37 disk	0			1	9	
Tertiary mirror 6	1.9	200 × 34 disk	0			1	9	
FSM	0.027	40 × 8 cylinder	30	24	6	1	9	
FSM driver	2.5	51 × 102 × 156 box	(included in FSM)			1	9	
Telescope mirror 3 (fold)	1.5	220 × 37 disk	0			1	9	
Telescope mirror 4 (fold)	1.7	220 × 37 disk	0			1	9	
Telescope mirror 5 (fold)	1.9	220 × 37 disk	0			1	9	
Telescope mirror 7	0.0216	40 × 6.8 disk	0			1	9	
Telescope mirror 8	0.0027	20 × 3.4 disk	0			1	9	
Telescope mirror 9	0.0091	30 × 5 disk	0			1	9	
Filter wheel	1.5	305 × 38 disk	160	24	32	1	5	
Filter wheel motor	1	50 × 40 × 70 box	(included in FW)			1	5	
Sensor	0.25	67 × 55 × 10 plate	3			1	6	
Dust cover	0.18	550 × 5 disk	0			1	9	Cyanate
Payload structure	33.09	See NX Model	0			1	9	Cyanate
Dust cover hinge	0.5	50 × 50 × 50	10	24	6	1	9	
Total	57.5804		203	24				
With 10% margin	63.3384							

For video mode, TFOV is held fixed on the target while the FSM either remains in place or scans a pattern to produce video at up to 8 Hz. The frame rate is limited by the image capture rate of 8 Hz.

7.7.4 Telescope Components

Table 7.6 shows the telescope components with projected mass, power, and thermal heat load. Additionally, 10% margin was allocated for small-scale components not accounted for in the preliminary design, such as mounting brackets, wires, and other hardware.

References

1 J. Bagnasco, Segmented mirror telescope optics, Lecture for spacecraft design class at Naval Postgraduate School, 2017.
2 https://en.wikipedia.org/wiki/Cassegrain reflector.
3 Ritchey–Chrétien telescope – Wikipedia, https://en.wikipedia.org/wiki/Ritchey.
4 A. L. Duncan, Optical imaging payload design, Lecture for spacecraft design class at Naval Postgraduate School, 2017.
5 Davis, Electro-optical space payload design, Lecture for spacecraft design class at Naval Postgraduate School, 2017.
6 R. R. Auelmann, Image quality metrics, Lecture material for spacecraft design class at Naval Postgraduate School, 2017.
7 A. Scott, AE 4830 spacecraft systems I, space mission and payloads, Class lectures, Naval Postgraduate School.
8 A.J. Yingling, Integrated structures optics and controls of the segmented mirror telescope, Ph.D. Dissertation, Naval Postgraduate School, September 2012.
9 Sigmadyne, Sigfit reference manual, version 2010r2d, Report, Sigmadyne Inc., Rochester NY, May 2011.
10 L. Atwood, A. R. Bradstreet, A M. Roush, J. L. Eddy, L. C. Ozeck, M. R. Roncoroni, N. D. Pandya, N. R. Koeppen, S. M. Sundey, and T. A. Lippman, AE4871 design project 2017, Naval Postgraduate School, 2017.

8

Laser Systems

Laser systems have a wide range of applications across various fields due to their ability to produce coherent, focused, and high-intensity light. The applications of laser systems include laser cutting and manufacturing, laser lithography, laser spectroscopy, laser-based sensing and metrology, free-space optical communications, and high-energy laser systems. Laser beam control technologies are a collection of multidisciplinary technical areas such as laser devices, geometric optics, Fourier optics, atmospheric optics, structures, thermal systems, sensors, actuators, and control theory. In general, the goal of laser beam control system is to precisely determine a target location and direct the laser beam to the desired aimpoint with minimal loss. This chapter introduces laser beam propagation as well as hardware and control methods commonly used for laser systems. Example laser systems for free space optical communications and high-energy laser systems are also introduced.

8.1 Laser Fundamentals

8.1.1 Stimulated Emission

A laser (light amplification by stimulated emission of radiation) is a light source created by the stimulated emission process by which an incoming photon of a specific frequency interacts with an excited atomic electron of the lasing medium, causing it to drop to a lower energy level. The liberated energy transfers to the electromagnetic field, creating a new photon with identical frequency, polarization, and direction of travel to the photons of the incident wave. The energy of the photon emitted during the stimulated emission is given by

$$E_{\text{photon}} = E_{\text{upper}} - E_{\text{lower}} = h\nu$$

where h is Planck's constant $\left(6.62607015 \times 10^{-34}\ \text{J Hz}^{-1}\right)$ and ν is the frequency of the photon.

If the system is in thermal equilibrium, the two-level energy system shown in Figure 8.1a will have more atomic electrons in the lower energy state E_1 than in the upper energy state E_2. Since the probability of emission is lower than the probability of absorption with the two-level energy system, any photon created during the stimulated emission is quickly absorbed and jumps back to a higher energy state. Population inversion refers to a condition where more electrons are in the upper energy state E_2 than in a lower energy state E_1, which is a necessary condition for sustained simulated emission to occur. Instead of a two-level energy system, lasers typically use either a

Optical Beam Control: Imaging Satellites and Laser Systems, First Edition.
Brij N. Agrawal, Jae Jun Kim and Sachin Agrawal.

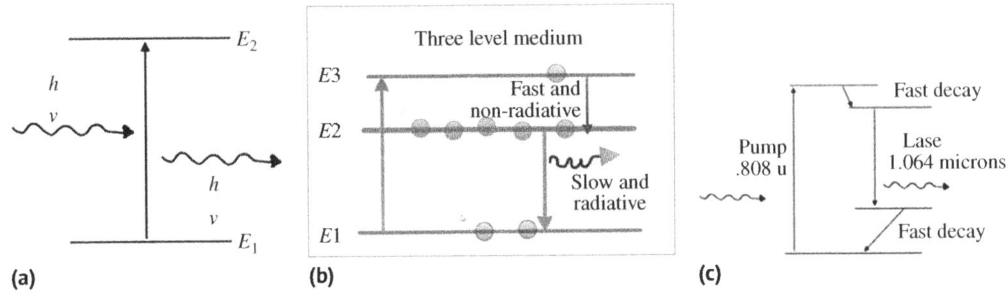

Figure 8.1 (a) Two-level energy system, (b) three-level energy system, and (c) four-level energy system in the lasing medium.

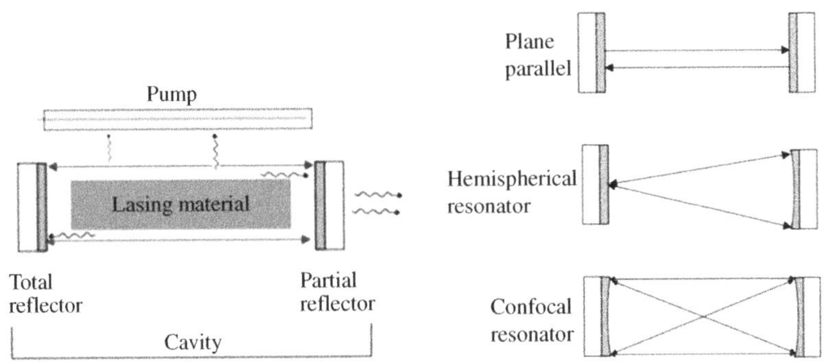

Figure 8.2 Generic laser and common resonant cavity designs.

three-level or four-level energy system in the lasing medium as shown in Figure 8.1b,c. Due to the variations in the energy decay rate between discrete energy levels, population inversion between two energy levels can occur. To sustain the stimulated emission, energy is supplied to the system through electrical, optical, or chemical means (pumping) to ensure stimulated emission occurs more frequently than absorption, leading to light amplification.

8.1.2 Resonant Cavities

A generic laser shown in Figure 8.2, pumps energy into a lasing medium, also called a gain medium, to sustain the stimulated emission. To use the energy most efficiently, the gain medium is contained in a reflective resonant cavity (optical resonator) that will redirect most of the pump energy into the lasing direction. If the applied pump power is too small, the gain will never be sufficient to overcome the cavity losses, and laser light will not be produced. The minimum pump power needed to begin laser action is called the lasing threshold.

To provide light amplification, the waves need to be propagating between the reflective mirrors at the resonator ends through the gain medium multiple times. The length between reflective mirrors is designed as an integer multiple of the laser wavelength to support standing electromagnetic waves. One of the mirrors in the cavity partially transmits light, allowing the laser output to escape the cavity. Common resonant cavity design is shown in Figure 8.2, which can influence the beam quality and stability depending on their optical properties and alignment accuracies.

8.1.3 Laser Characteristics

8.1.3.1 Monochromaticity

Unlike other light sources, laser light is composed of light waves with a very narrow range of wavelengths. The monochromaticity of a laser light is represented by the laser linewidth, which refers to the range of frequencies (or wavelengths) over which the laser emits light. The laser linewidth of a typical laser is on the order of kHz to GHz range. Many factors contribute to the laser linewidth including the lasing medium material, optical cavity design and alignment, and environmental disturbances such as temperature fluctuations and mechanical vibrations.

8.1.3.2 Coherence

Coherence refers to both temporal coherence and spatial coherence, which is a measure of laser beam's ability to interfere with time-shifted and position-shifted samples of itself. The coherence length is a measure of coherence, which is closely related to the spectral properties of the laser light, particularly its linewidth. The coherence length is defined as

$$L_c = \frac{c}{\Delta f} = \frac{\lambda_0^2}{\Delta\lambda} \tag{8.1}$$

where λ_0 is the wavelength of the laser and $\Delta\lambda$ is the laser linewidth in wavelengths. Coherence length of different types of lasers varies significantly ranging from millimeters to tens of kilometers.

8.1.3.3 Polarization

Lasers can be designed to have various polarization states and often linearly polarized in a specific direction out of the box. Polarization is an important consideration for beam control since it could affect how light interacts with optical components, materials, and atmosphere. Polarization is also important for certain applications such as optical communication, where it can be used to increase the data transmission rate by utilizing multiple polarization modes.

8.1.3.4 Mode Shapes

The mode shape of a laser refers to the spatial distribution of the electromagnetic field within the laser cavity. Various transverse electromagnetic wave modes shown in Figure 8.3 illustrate how the light's intensity varies in space across the laser beam.

The general expression for the transverse electric field in a beam is given by the Hermite–Gaussian mode function,

$$HG_{p,q}(x, y) = H_p\left(\frac{\sqrt{2}\,x}{w(z)}\right) H_q\left(\frac{\sqrt{2}\,x}{w(z)}\right) e^{-(x^2+y^2)/w^2} \tag{8.2}$$

where p and q are integer number starting from zero, H_p and H_q are the Hermite polynomials of order p and q, and w is the minimum spot size of the beam inside the cavity. Typically, TEM_{00} is the most used mode shape for laser systems.

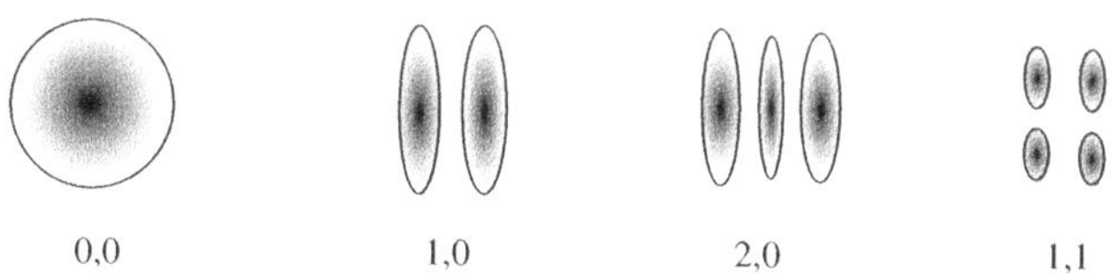

Figure 8.3 Transverse electromagnetic (TEM) wave modes. *Source:* Brij N. Agrawal.

8.1.4 Type of Lasers

Lasers are commonly classified based on the type of lasing medium used within the optical cavity. The various types of lasers, categorized according to their gain medium and their specific characteristics, are discussed in the following sections. Additionally, a summary of commercially available lasers, including details on their gain medium, wavelength, power, and energy, is provided in Chapter 2, Table 2.1.

8.1.4.1 Gas Lasers

Gas lasers use gas as a gain medium. The gas medium can be a mixture of noble gases like helium–neon (HeNe) or a molecular gas like carbon dioxide (CO_2). Gas lasers excite the electrons to higher energy states using the electrical current passing through the gas. Gas lasers typically have a narrow linewidth, usually around 1–10 MHz and highly monochromatic.

8.1.4.2 Diode Lasers

Diode lasers, or semiconductor lasers, use the semiconductor material as a gain medium. Diode lasers generally have broader linewidths than gas lasers. The linewidth for a single-mode diode laser can range from about 1 MHz to several GHz, depending on the laser's design, temperature, and operating conditions. Specialized diode lasers, such as distributed feedback lasers, can have linewidths as narrow as 10 kHz–1 MHz, making them suitable for high-precision applications like optical communication and spectroscopy.

8.1.4.3 Dye Lasers

Dye lasers work by using organic dye molecules as an active medium. Dye lasers can be tuned to emit a broad range of wavelengths depending on the dye used, making them highly versatile for a range of scientific and industrial applications.

8.1.4.4 Solid-state Lasers

Solid-state lasers use a solid gain medium such as a glass or crystal, rather than a liquid as in dye lasers or a gas as in gas lasers. Solid-state lasers are widely used in optical beam control applications such as a sodium beacon laser that is generated by means of sum frequency generation with Nd:YAG solid-state lasers at 1064 and 1319 nm.

8.1.4.5 Fiber Laser

A fiber laser is a laser with an optical fiber as the active gain medium. The core of the optical fiber is doped with a rare-earth element, such as ytterbium (Yb), erbium (Er), neodymium (Nd), or thulium (Tm). These dopants are responsible for the laser's ability to amplify light when pumped with an external light source (such as a diode laser). Fiber lasers offer advantages such as high output power, compactness, and flexibility, making them ideal for high-energy laser applications. Their long active regions allow high optical gain and efficient cooling due to the fiber's surface area to volume ratio. Additionally, the fiber's waveguide properties reduce thermal distortion, producing a high-quality optical beam that can be effectively coupled with laser beam control systems. Fiber lasers are typically smaller than solid-state or gas lasers with similar power.

8.1.4.6 Chemical Lasers

Chemical lasers such as CO_2 lasers, chemical oxygen iodine lasers (COIL), and nitrogen lasers use exothermic chemical reactions to generate population inversion. Because energy sources come from a chemical reaction rather than an electrical current or external optical pump, chemical lasers can produce high-power beams suitable for specific applications.

8.1.4.7 Free-electron Lasers

In a free-electron laser, an electron beam moving at a speed close to the speed of light is made to pass through the magnetic field generated by a periodic structure. The stimulated emission process comes about through the interaction of the electromagnetic field of the laser beam with these relativistic electrons moving in the periodic magnetic structure. As in any other laser, two end mirrors are used to provide feedback for laser oscillation. The electron beam is injected into the laser cavity and then deflected out the cavity using suitable bending magnets. FELs can produce a very broad range of wavelengths, from X-rays to microwaves, by adjusting the energy of the electrons and the magnetic field in the undulator.

8.1.4.8 Continuous Wave Lasers

Continuous wave (CW) lasers emit a continuous, uninterrupted beam of light without any pulsing or modulation. In a CW laser, the balance of pump power against gain saturation and cavity losses produces an equilibrium value of the laser power inside the cavity. As the laser is continuously pumped, the active gain medium is excited to a steady state, allowing for a stable and uninterrupted laser beam.

8.1.4.9 Pulsed Lasers

A pulsed laser refers to any laser that is not classified as a CW laser, where the optical power is emitted in discrete pulses, each with a specific duration and repetition rate. Pulsed lasers play a crucial role in various beam control applications, including laser beam modulation for optical communication systems and time-of-flight measurement for light detection and ranging (LIDAR). Pulsed lasers are often employed in laser beacon systems for beam control, as their range-gate capability helps mitigate the impact of stray light in light sensing applications. The following discusses common methods for generating pulsed lasers.

8.1.4.9.1 Gain Switching

Gain switching is a method to modulate the output of a laser by rapidly varying the optical gain of the laser medium, typically by pulsing the pump source, creating a population inversion that is then rapidly released as a pulse. Gain switching is relatively easier to implement and can produce high-speed pulsed lasers ideal for applications needing fast data transmission.

8.1.4.9.2 Q-switching

Q-switching is a widely used technique for generating high-intensity, short-duration laser pulses, typically in the range of nanoseconds to microseconds. In this process, the laser cavity is initially maintained in a closed state (high-Q condition), which allows energy to accumulate in the laser medium. An optical modulator, such as an electro-optic device, rapidly alters the cavity conditions, effectively opening the cavity. This sudden change allows the stored energy to be released in a brief, concentrated burst, producing a high-intensity laser pulse. As the entire pulse energy is emitted in a single event, the peak power of the pulse is significantly higher than that of a CW laser, with pulse durations potentially as short as nanoseconds.

8.1.4.9.3 Mode-locking

In a mode-locked laser, the laser cavity is constructed so that all the different modes of the laser combine coherently, leading to constructive interference at regular intervals. This causes the emission of a series of pulses, with each pulse corresponding to the time interval between modes. Mode-locking can be used for generating extremely short pulses, down to femtosecond scales.

8.2 Laser Beam Profile

The laser beam profile refers to the spatial distribution of light intensity or irradiance within the cross-section of the laser beam, typically in the plane perpendicular to its direction of propagation. The laser beam profile provides a snapshot of the beam's characteristics at a specific point in time and space, serving as a crucial tool for analyzing and predicting the performance of laser systems.

The beam profile at the source is influenced by various factors, including the properties of the lasing medium, the design of the resonant cavity, and other elements that impact the stimulated emission process. The two most common laser beam intensity profiles are the Gaussian beam and the Top-hat beam, as illustrated in Figure 8.4.

A Gaussian beam is one of the most common laser beam profiles, characterized by a Gaussian intensity distribution that is most intense at the center and decays symmetrically as you move away from the center. The intensity distribution of an ideal circular Gaussian beam with a radial distance r from the center of the beam at position z along the propagation axis is defined as

$$I(r, z) = I_0 e^{-2\left(\frac{r}{w(z)}\right)^2} \tag{8.3}$$

where I_0 is the peak intensity at the beam center and $w(z)$ is the beam radius at position z. Since the beam intensity profile represented by the Gaussian function is infinite in extent, the radius of the beam $w(z)$ in Eq. 8.4 is defined as the radius when the irradiance is $1/e^2$ (13.5%) of peak intensity I_0. This is related to the full width at half maximum (FWHM) of the intensity distribution such that

$$w(z) = \frac{\text{FWHM}(z)}{\sqrt{2\ln 2}} \tag{8.4}$$

A top-hat beam, also known as a uniform intensity beam, has a flat intensity profile across the beam's cross-section within a certain radius, with a sharp drop-off outside of that radius. The width of a top-hat beam with equal total power and peak irradiance of a Gaussian beam (Figure 8.4a) with radius w is

$$2w_T = \sqrt{2}\,w \tag{8.5}$$

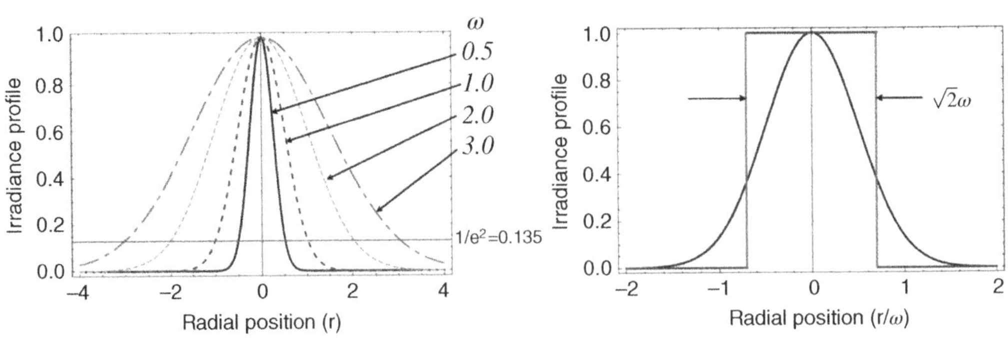

Figure 8.4 (a) Gaussian beam and (b) top-hat beam.

which is shown in Figure 8.4(b). Therefore, the intensity profile of a top-hat beam with equal total power and peak irradiance of a Gaussian beam with radius w is written as

$$I(r) = \begin{cases} I_0 & \text{for } r \leq w_T \\ 0 & \text{for } r > w_T \end{cases} \tag{8.6}$$

where w_T is an equivalent diameter of a top hat beam.

Instead of the intensity profile of a beam, the irradiance profile of a beam in terms of power per unit area is also interchangeably used to represent laser beam profiles. For a Gaussian beam, the total power $P_0(W)$ of the beam is computed by integrating the intensity over the entire cross-sectional area of the beam such that

$$P_0 = \int_0^{2\pi} \int_0^{\infty} I_0 e^{-2\left(\frac{r}{w(z)}\right)^2} r dr\, d\theta = 2\pi r \int_0^{\infty} I_0 e^{-2\left(\frac{r}{w(z)}\right)^2} dr = \frac{\pi w(z)^2}{2} I_0 \tag{8.7}$$

Therefore, the irradiance profile $(\mathrm{W/m^2})$ becomes

$$H(r, z) = \frac{2P_0}{\pi w(z)^2} e^{-2\left(\frac{r}{w(z)}\right)^2} \tag{8.8}$$

In practical applications, Gaussian beams are often truncated at a certain radius due to apertures, lenses, or mirrors with finite dimensions. As a result, the truncated Gaussian beam exhibits reduced power compared to the ideal Gaussian beam. Additionally, the propagation of the beam introduces diffraction effects that cause deviations in the irradiance profile from the ideal Gaussian distribution. Figure 8.5 illustrates the changes in the radius of the truncated Gaussian beam as it propagates along the z-axis in space. Both focused and collimated laser beams will eventually diverge due to diffraction. The beam waist (w_0), as shown in Figure 8.5, refers to the radius of the narrowest point of a propagating truncated Gaussian beam, where the beam has the smallest cross-sectional diameter. For a focusing beam, the beam waist is typically located at the focal point of the optical system, with its size constrained by the ratio λ/D. For a collimated beam, the beam waist is generally situated at the exiting aperture, as illustrated in Figure 8.5.

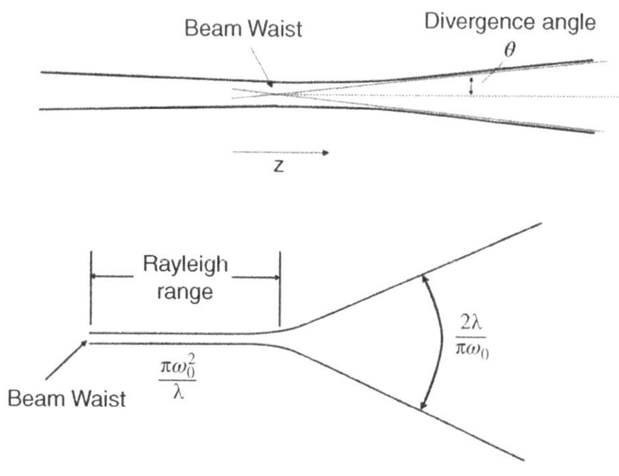

Figure 8.5 Truncated Gaussian beam propagation – focusing beam (top) and collimated beam (bottom).

The Rayleigh range (z_0) provides a measure of how far the laser beam can travel from the beam waist before the radius of the beam increases significantly due to divergence. The threshold for Rayleigh range is given by

$$w(z_0) = \sqrt{2}\, w_0 \tag{8.9}$$

Beyond the Rayleigh range, the beam starts to diverge significantly, and the radius of the beam increases according to the beam divergence angle. The Rayleigh range is given by

$$z_0 = \frac{\pi w_0^2}{\lambda} \tag{8.10}$$

and the beam divergence angle (θ_D) is

$$\theta_D = \frac{\lambda}{\pi w_0} \tag{8.11}$$

The radius of the beam ($w(z)$) and the wavefront radius of curvature ($R(z)$) of the diverging beam becomes

$$w(z) = w_0\sqrt{1 + \frac{z^2}{z_0^2}} = w_0\sqrt{1 + \left(\frac{\lambda z}{\pi w_0^2}\right)^2} \tag{8.12}$$

$$R(z) = z + \frac{z_0^2}{z} \tag{8.13}$$

where $R(z) \approx z$ when the propagation distance z becomes much larger than z_0 (approaching to a plane wavefront).

Figure 8.6 presents a simulation of the Rayleigh range for a collimated Gaussian beam. As shown, the beam remains relatively collimated and does not experience significant divergence until it reaches the Rayleigh range z_0. To extend the propagation distance of a collimated beam, the diameter of the beam at the exiting aperture must be increased for a given wavelength. The same principle applies to a focusing beam, where the focal distance must be shorter than the Rayleigh range for a given beam diameter at the exiting aperture. This increase in beam diameter is typically

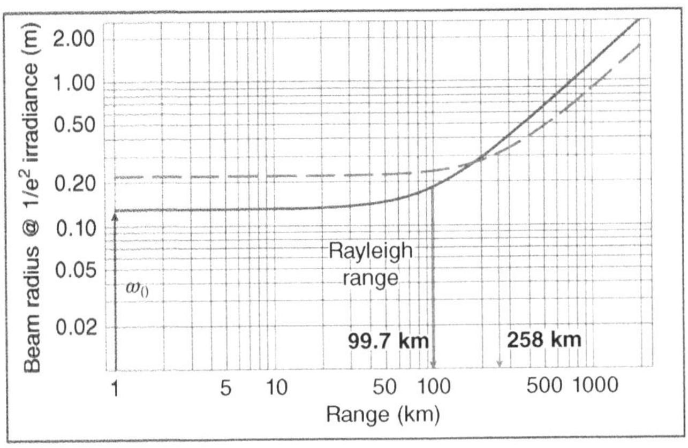

Figure 8.6 Rayleigh range of a collimated Gaussian beam (beam waist located at origin) with two different wavelengths and beam waist values (589 nm with $w_0 = 22$ cm shown in dotted line, and 532 nm with $w_0 = 13$ cm shown in solid line).

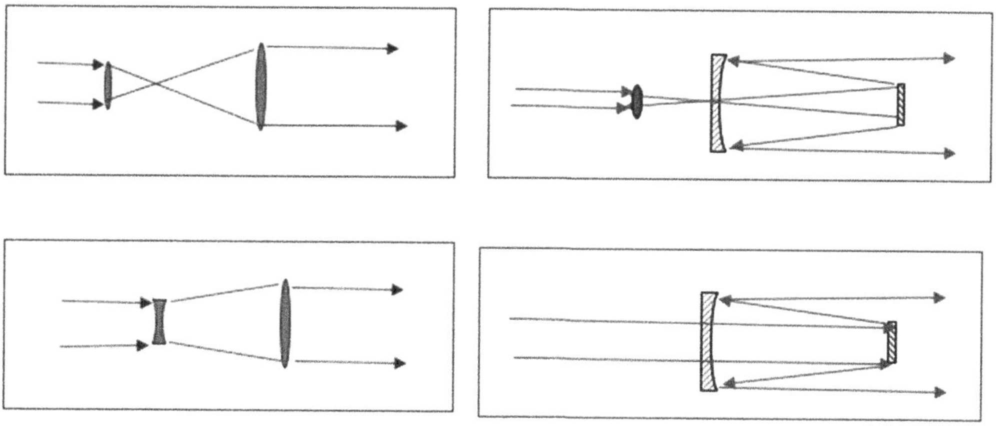

Figure 8.7 Expansion of beam diameter at the source reduces diffraction effects on Rayleigh range and beam divergence.

achieved using a telescope, as illustrated in Figure 8.7. For this reason, a telescope within a laser system is commonly referred to as a beam expander.

A beam expander design that does not include intermediate focal spots in the optical train is referred to as a relay-less beam expander and is better suited for applications requiring thermal stability, such as high-energy laser systems.

When the Gaussian beam is truncated, the total power of the truncated Gaussian beam is reduced by the following equation.

$$P = (1 - e^{-2\gamma})P_0 \tag{8.14}$$

where γ is the Gaussian truncation parameter which is the square of the ratio of the truncated beam radius a and Gaussian beam radius w_a such that $\gamma = (a/w_a)^2$. Figure 8.8 shows the power transmission of a Gaussian beam as a function of beam radius ratio (a/w_a). To transmit 99% of the power, the radius of the aperture needs to be 1.52 times larger than the beam radius such that $a/w_a = 1.52$.

Unlike the irradiance profile of an ideal Gaussian beam that is represented by a Gaussian function, the irradiance profile of the truncated Gaussian beam is computed using a Fourier optics based on the Huygen–Fresnel principle where the propagating wavefronts are represented as a superposition of many secondary wavelets. To provide more practical solutions, approximations are often made depending on the beam travel distance z from the aperture. The Fresnel number which is a dimensionless quantity that characterizes the diffraction pattern of a wave passing through an aperture or around an obstacle, is given by

$$F = \frac{r_a}{\lambda z} \tag{8.15}$$

where r_a is the radius of the truncated beam and z is the distance from the aperture or obstacle to the observation point. For the Fresnel diffraction (near-field diffraction) region characterized by $F > 1$, the irradiance profile exhibits ripples as shown in Figure 8.9. For the Fraunhofer diffraction (far-field diffraction) region characterized by $F \ll 1$, the irradiance profile of a truncated Gaussian beam shapes to an airy disk as shown in Figure 8.9.

By exploiting the circular symmetry of a laser beam, the general scalar diffraction expression discussed in Section 2.4.1 can be reformulated in a radial coordinate system using a Bessel function

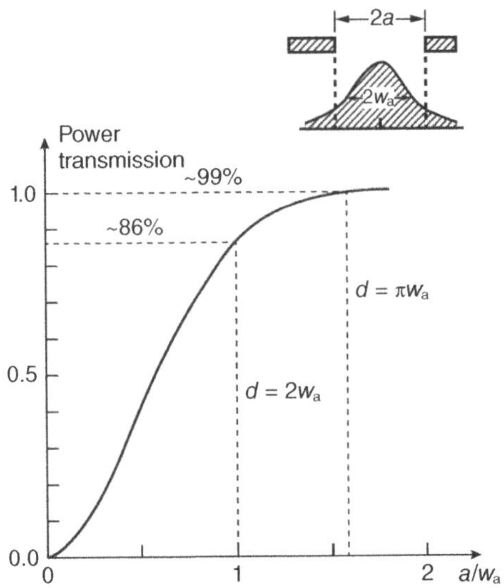

Figure 8.8 Power transmission of truncated Gaussian beam.

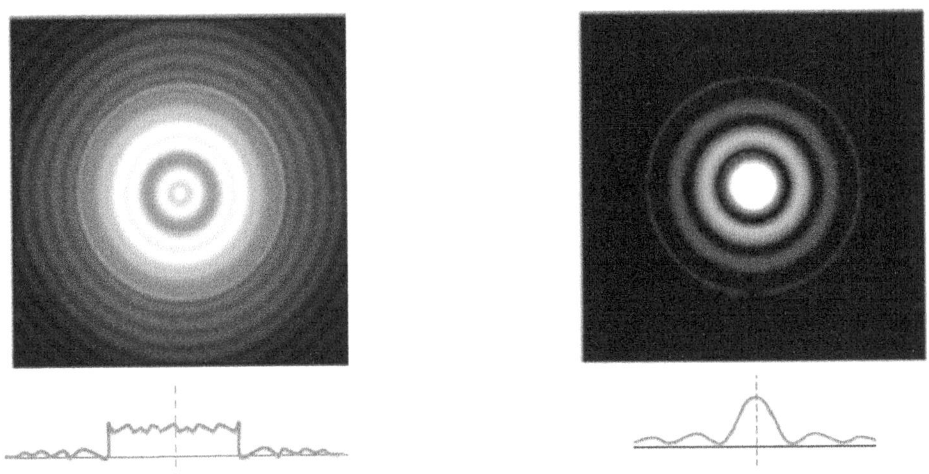

Figure 8.9 Near field and far field diffraction effects on truncated Gaussian beam profile.

representation. According to the Fresnel diffraction model proposed by [1], the transverse irradiance at radial coordinate r of a beam focused at range R, but observed at axial distance z is given by

$$H_T(r) = 4\left(\frac{R}{z}\right)^2 \left| \int_0^1 \sqrt{H(\rho)} \exp\left[i\frac{\pi D^2}{4\lambda}\left(\frac{1}{z} - \frac{1}{R}\right)\rho^2\right] J_0(\pi r\rho)\rho d\rho \right|^2 \tag{8.16}$$

where $\sqrt{H(\rho)}$ is the irradiance profile of a laser beam in the aperture at normalized aperture radius ρ, J_0 is the zero order Bessel function, r is a radial position in units of $R\lambda/D$. Using this equation, on-axis irradiance $H(0, z)$ as a function of propagation distance z for a Gaussian beam focused at range R can be found as

$$H(0, z) = \frac{\pi}{4}\frac{P}{R^2}\left(\frac{D}{\lambda}\right)^2\left(\frac{R}{z}\right)^2 \frac{2\gamma}{(\Phi_0^2 + \gamma^2)} \coth\gamma - \frac{\cos(\Phi_0)}{\sinh(\gamma)} \tag{8.17}$$

and for a top-hat beam

$$H(0,z) = \frac{\pi}{4}\frac{P}{R^2}\left(\frac{D}{\lambda}\right)^2\left(\frac{R}{z}\right)^2 \frac{\sin\left(\frac{\Phi_0}{2}\right)}{\left(\frac{\Phi_0}{2}\right)^2} \tag{8.18}$$

where P is the total power transmitted and

$$\Phi_0 = \left(\frac{\pi D^2}{4\lambda}\right)\left(\frac{1}{z} - \frac{1}{R}\right) \tag{8.19}$$

Figure 8.10 illustrates the on-axis irradiance profiles of a focusing top-hat beam and a focusing Gaussian beam as a function of the observation range z. The parameters used are $\lambda = 1064$ nm, $D = 1.5$ m, $P = 15$ kW, and $R_f = 500$ km. Figure 8.11 presents a comparison between a focused beam and a collimated beam.

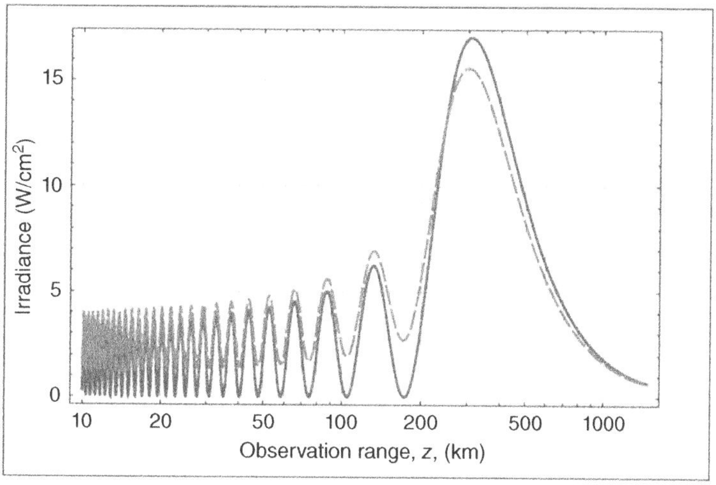

Figure 8.10 On-axis irradiation of a focusing uniform intensity beam in dotted line and Gaussian beam in solid line, $\lambda = 1064$ nm, $D = 1.5$ m, $P = 15$ kW, and $R_f = 500$ km.

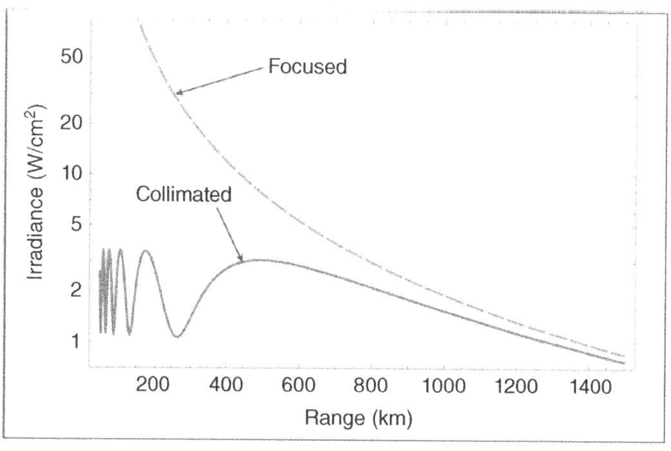

Figure 8.11 On-axis irradiance of truncated Gaussian beam – focused beam in dotted line versus collimated beam shown in solid line, $\lambda = 1064$ nm, $D = 1.5$ m, $P = 15$ kW.

The irradiance profile of a laser beam as a function of a radial position r in the focal plane can also be found. Laser system performance metrics such as Strehl ratio and Power-in-the-bucket are typically based on the irradiance profile of a laser beam in the focal plane. For a truncated Gaussian beam,

$$H(r,R)=\frac{\pi}{4}\frac{P}{R^2}\left(\frac{D}{\lambda}\right)^2 4\frac{2\gamma}{1-e^{-2\gamma}}\left[\int_0^1 \exp(-\gamma\rho^2)J_0(\pi r\rho)\rho d\rho\right]^2 \tag{8.20}$$

where the peak irradiance of the truncated Gaussian beam is at the center in the focal plane given by

$$H_0=\frac{\pi}{4}\frac{P}{R^2}\left(\frac{D}{\lambda}\right)^2 \tag{8.21}$$

And the brightness of the beam at the focus is

$$B=\frac{\pi PD^2}{4\lambda^2} \tag{8.22}$$

8.3 Laser Beam Aberrations

Up to this point, it has been assumed that the laser beam propagates in a vacuum without any aberrations. In addition to the diffraction that influences the beam's irradiance profile, the irradiance of the laser beam can be further reduced by various aberrations within the laser system. This section begins by defining several key metrics commonly used to quantify the effects of laser beam aberrations. Subsequently, the impact of these aberrations is analyzed using these performance metrics.

8.3.1 Laser Beam Aberration Metrics

The Strehl ratio is commonly used in laser systems to quantify the effect of various laser beam aberrations. A simple definition of Strehl ratio is given as

$$SR=\frac{H_{\text{aberrated peak}}}{H_{\text{diffraction limited peak}}}, \tag{8.23}$$

which is the ratio of the peak irradiance of an aberrated optical system to that of the diffraction-limited (unaberrated) system. When the Strehl ratio is determined at the focal point (beam waist) of a laser system, the Strehl ratio represents the height of the measured point spread function compared to the height of the diffraction limited point spread function. When the Strehl ratio of the entire laser system is known, the irradiance profile of the laser beam can be written as

$$H_{\text{target}}(r,R)=SR_{\text{sys}}H_0(r,R) \tag{8.24}$$

where $H_{\text{target}}(r,R)$ is the irradiance profile at the target as a function of the location in the beam r and the range to the target R, SR_{sys} is the Strehl ratio of the entire laser system, $H_0(r,R)$ is the irradiance profile in vacuum for a diffraction limited system. Strehl ratio of the entire laser system can be further written as

$$SR_{sys}=SR_1\cdot SR_2\cdot SR_3\cdots\cdot SR_n \tag{8.25}$$

where $SR_1,\cdots,SR_n$ refers to the Strehl ratio of individual aberration sources in the laser system discussed in the subsequent sections.

For ideal Gaussian and top hat laser beams, the peak irradiance in the far field occurs at the center of the beam irradiation profile. However, the irradiance profile of the real-world laser beam often deviates from the ideal far field diffraction patterns due to various laser beam aberrations. Although the Strehl ratio provides a convenient measure that quantifies the effect of laser beam aberration on the laser system's performance, the peak irradiance in the laser beam may not accurately represent system performance, especially when there are significant phase and amplitude fluctuations in the beam.

Instead of the Strehl ratio that uses the peak irradiance in the beam profile, power in the bucket (PIB) metric uses the irradiance profile of the entire beam to measure the total power of the laser beam within a certain radius (bucket size). For a diffraction-limited Gaussian beam, PIB as a function of a bucket radius of r_{bucket} is given as

$$P_{\text{bucket}}\left(r_{\text{bucket}}\right) = P\left[1 - e^{-\left(\frac{r_{\text{bucket}}}{w(z)}\right)^2}\right] \tag{8.26}$$

where P is the total power in the Gaussian beam and $w(z)$ is the beam radius at distance z from the beam waist. While the Strehl ratio and power-in-the-bucket are useful performance indicators, no single metric can comprehensively quantify laser beam aberrations and their effects on system performance. Depending on the laser system application, other system metrics – such as wavefront error, optical transfer function, encircled energy, irradiance (intensity), and scintillation index – also need to be considered. For clarity and consistency, the analysis herein follows the framework established in [2], with subsequent sections focusing on quantifying the influence of specific laser beam aberrations primarily through Strehl ratio evaluations.

8.3.2 Laser Beam Quality

A real-world laser device has deviations from the ideal laser beam characteristics, which must be considered when analyzing laser beam propagation performance. The M^2 factor, or beam quality factor, is a metric that quantifies the deviation of a real laser beam by comparing its beam waist to that of a diffraction-limited Gaussian beam.

$$M^2 = \frac{w(z)\theta(z)}{w_0(z)\theta_0(z)} \tag{8.27}$$

where w(z) is the real beam diameter at location z, $w_0(z)$ is the beam diameter of a diffraction-limited Gaussian beam, θ is a real beam divergence angle, and θ_0 is an ideal beam divergence angle. For a focusing beam, $\theta/\theta_0 = 1$, and the beam quality factor become the ratio of beam diameters at the beam waist such that

$$M^2 = \frac{w(0)}{w_0(0)} \tag{8.28}$$

For an ideal Gaussian beam, the beam quality factor M^2 is equal to 1 and higher M^2 value indicates a lower beam quality. The beam quality factor M^2 is typically provided in the laser specifications. The Rayleigh range and divergence angle of a laser beam with a beam quality factor M^2 becomes

$$z_0 = \frac{\pi w_0^2}{M^2\lambda} \tag{8.29}$$

$$\theta = \frac{M^2\lambda}{\pi w_0} \tag{8.30}$$

As a result of a lower beam quality, the Strehl ratio which is the ratio of the aberrated peak irradiance to ideal peak irradiance becomes

$$SR_{BQ} = \frac{2P/\pi w^2}{2P/\pi w_0^2} = \frac{1}{(M^2)^2} \tag{8.31}$$

8.3.3 Optical Component

Optical components such as mirrors used in the telescope have less than perfect reflectivity and surface accuracies from its ideal shape. The Strehl ratio of the optical component is

$$SR_{OCO} = SR_{REF} SR_{WFE} \tag{8.32}$$

where SR_{REF} is the Strehl ratio of the reflectivity and SR_{WFE} is the Strehl ratio of reflected wavefront error

$$SR_{WFE} = e^{-(2\pi WFE)} \tag{8.33}$$

8.3.4 Telescope Central Obscuration

Modern telescopes, such as Cassegrain-type telescopes, have central obscurations as depicted in Fig. 8.12, where the beam center is blocked by the secondary mirror when the beam is expanded for propagation. Telescope design with central obscuration allows for a reduction in the size and weight of the telescope while maintaining the high optical resolution. To describe the central obscuration, a rectilinear obscuration ratio is defined as

$$\epsilon = \frac{D_o}{D} \tag{8.34}$$

The irradiance profile of a truncated Gaussian beam with central obscuration observed in the focal plane is given by

$$H_{co}(r,\ R) = \frac{\pi}{4}\frac{P}{R^2}\left(\frac{D}{\lambda}\right)^2 4\frac{2\gamma}{e^{-2\gamma\epsilon^2} - e^{-2\gamma}}\left[\int_{\epsilon}^{1} \exp(-\gamma\rho^2)J_0(\pi r\rho)\rho d\rho\right]^2 \tag{8.35}$$

When $\epsilon = 0$, the above equation reduces to Eq. 8.20. For a top hat beam, the irradiance profile with central obscuration observed in the focal plane is

$$H_{TC}(r) = \frac{\pi}{4}\frac{P}{R^2}\left(\frac{D}{\lambda}\right)^2 \frac{1}{1-\epsilon^2}\left[\sqrt{H_A(r)} - \epsilon^2\sqrt{H_A(r)}\right]^2 \tag{8.36}$$

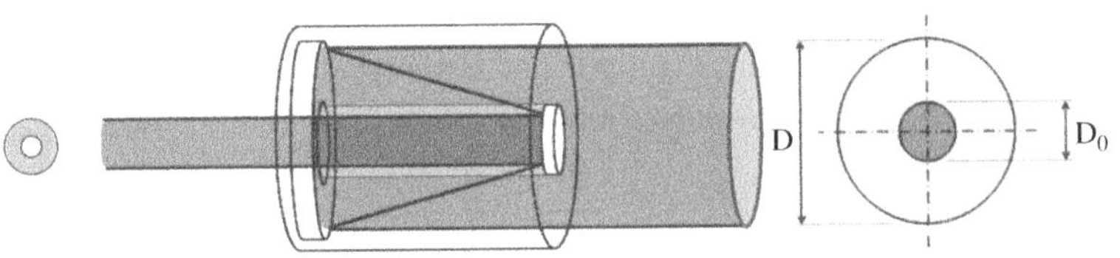

Figure 8.12 Telescope central obscuration.

with

$$H_A(r) = \left[\frac{2J_1(\pi r)}{\pi r}\right]^2 \tag{8.37}$$

An approximation for the far-field peak irradiance of a beam with a central obscuration is given by

$$H_{co} = \frac{\pi}{4}\frac{P}{R^2}\left(\frac{D}{\lambda}\right)^2\left(1 - \epsilon^2\right) \tag{8.38}$$

Corresponding Strehl ratio due to the central obscuration is

$$SR_{co} = 1 - \epsilon^2 \tag{8.39}$$

8.3.5 Optical Jitter

Optical jitter is a ubiquitous source of aberrations in laser systems referring to the overall tip and tilt of the wavefront. Optical jitter can arise from various sources, including platform jitter, tracker system-induced jitter, and atmospheric jitter. As a result of optical jitter, the centroid of the beam's intensity shifts around the aimpoint, causing a blurring of the diffraction-limited size of the beam and reducing its average intensity. The irradiance profile, when accounting for optical jitter, is given by

$$H_J = \frac{\pi}{4}\frac{P}{R^2}\left(\frac{D}{\lambda}\right)^2 \frac{1}{1 + \frac{\pi^2}{2}\left(\frac{\sigma_\theta}{\lambda/D}\right)^2} \tag{8.40}$$

where σ_θ is an RMS motion of the laser beam tilt. The irradiance profile including the central obscuration is given as

$$H_{J\text{-}CO} = \frac{\pi}{4}\frac{P}{R^2}\left(\frac{D}{\lambda}\right)^2 \frac{1 - \epsilon^2}{\left(1 + \frac{\pi^2}{2}\left(\frac{\sigma_\theta}{\lambda/D}\right)^2 \frac{1}{f(\epsilon)}\right)} \tag{8.41}$$

with

$$f\left(\epsilon\right) \approx \frac{1}{1 + 1.7\epsilon^{1.8}} \tag{8.42}$$

The Strehl ratio of the full aperture tilt without central obscuration is

$$SR_{tilt} = \frac{1}{1 + \frac{\pi^2}{2}\left(\frac{\sigma_\theta}{\lambda/D}\right)^2} \tag{8.43}$$

The Strehl ratio of the full aperture tilt with central obscuration is given as

$$SR_{TILT\_CO} = \frac{1 - \epsilon^2}{\left(1 + \frac{\pi^2}{2}\left(\frac{\sigma_\theta}{\lambda/D}\right)^2 \frac{1}{f(\epsilon)}\right)} \tag{8.44}$$

8.3.6 Higher Order Aberrations

Higher-order aberrations refer to wavefront distortions beyond the overall tip and tilt of the wavefront. These aberrations are typically caused by atmospheric turbulence. When higher-order aberrations are present, the laser beam diverges beyond the diffraction-limited diameter and may eventually break up as the ratio of D/r_0 increases, leading to a reduction in both the Strehl ratio and the peak irradiance of the beam.

In laser systems employing AO, higher-order aberrations are often the result of residual wavefront errors, which arise from limitations in the AO system. These limitations can include wavefront sensor (WFS) noise, deformable mirror fitting errors, servo lag, focus anisoplanatism, and other factors.

For a rms phase error (σ_ϕ) of less than 2 radians, the Strehl ratio due to the high order beam aberrations is represented by the formula

$$SR_{HO} = e^{-\sigma_\phi^2} \tag{8.45}$$

The Strehl ratio due to the uncompensated high order aberrations is written in terms of the Fried's coherence length r_0

$$SR_{HO} = \frac{16}{\pi}\int_0^1 u\left[\cos^{-1}u - u\sqrt{1-u^2}\right]\exp\left[-3.44\left(\frac{D}{r_0}\right)^{\frac{5}{3}} u^{5/3}\right]du \tag{8.46}$$

which are often approximated with the following equation.

$$SR_{HO} = \frac{1}{\left[1 + \left(\frac{D}{r_0}\right)^{\frac{5}{3}}\right]^{\frac{6}{5}}} \tag{8.47}$$

A simpler equation is also used for the Strehl ratio

$$SR_{HO-TILT} = 1.02\left(\frac{D}{r_0}\right)^{\frac{5}{3}} \tag{8.48}$$

$$SR_{HO-TR} = 0.134\left(\frac{D}{r_0}\right)^{\frac{5}{3}} \tag{8.49}$$

where $SR_{HO-TILT}$ includes both tilt and higher order aberrations and SR_{HO} only includes the higher order aberrations.

8.3.7 Atmospheric Transmission

Atmospheric transmission is another major factor that degrades the laser beam propagation performance. Strehl due to transmission is

$$SR_{\text{TRANS}} = e^{-\sigma x} \tag{8.50}$$

where σ is the atmospheric extinction coefficient including both absorption and scattering of a laser beam and x is the distance of transmission.

8.3.8 Scintillation

As the laser beam experiences diffraction and scattering due to atmospheric turbulence, different sections of the beam can interfere with one another. This interference results in bright spots, caused by constructive interference, and dark spots, caused by destructive interference, in the far-field pattern. This phenomenon, known as scintillation, is illustrated in Figure 8.13. Scintillation is a characteristic effect specific to laser systems, arising from the monochromaticity and coherence of the laser beam, and is often a limiting factor in laser system performance.

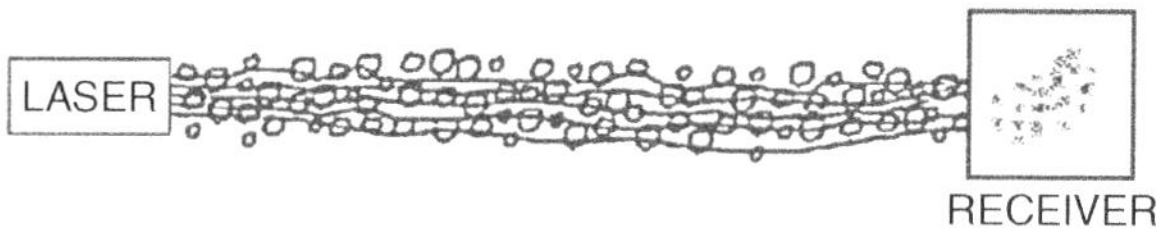

Figure 8.13 Illustration of laser beam scintillation caused by atmospheric turbulence.

In the case of a laser beam, log-amplitude variance of a laser beam (Rytov variance) at a propagating distance R for a plane wave and a spherical wave is given by

$$\sigma_\chi^2 = 0.56k^{7/6}\int_0^L C_n^2\left(z\right)\left(R - z\right)^{\frac{5}{6}}dz \text{ for } a \text{ plane wave} \tag{8.51}$$

$$\sigma_\chi^2 = 0.56k^{7/6}\int_0^L C_n^2\left(z\right)\left(z/L\right)^{5/6}\left(R - z\right)^{5/6}dz \text{ for a spherical wave} \tag{8.52}$$

where σ_χ^2 is the variance of intensity. For a constant C_n^2, such as the horizontal path, the log-amplitude variance becomes

$$\sigma_\chi^2 = 0.307C_n^2k^{7/6}R^{11/6} \text{ for a plane wave} \tag{8.53}$$

$$\sigma_\chi^2 = 0.124C_n^2k^{7/6}R^{11/6} \text{ for a spherical wave} \tag{8.54}$$

The corresponding variance of irradiance fluctuations scaled by the square of the mean irradiance leads to the scintillation index σ_I^2

$$\sigma_I^2 = \frac{\langle I^2\rangle}{\langle I\rangle^2} - 1 = \frac{\sigma_H^2}{\langle I\rangle^2} \tag{8.55}$$

where $\langle I^2\rangle$ is the variance of the intensity, $\langle I\rangle^2$ is the mean intensity, and σ_H^2 is the variance of irradiance. The scintillation index and log-amplitude variance are related to the following equation

$$\sigma_I^2 = A\left[e^{4\sigma_\chi^2} - 1\right] \tag{8.56}$$

where A is the aperture averaging factor

$$A = \left[1 + 1.07\left(\frac{kD^2}{4R}\right)^{\frac{7}{6}}\right]^{-1} \tag{8.57}$$

Then, the Strehl ratio due to scintillation is given as

$$SR_{SCI} = e^{-\sigma_\chi^2} \tag{8.58}$$

It is noted that the log amplitude variations shown in Eqs. 8.53 and 8.54 are only valid for $\sigma_\chi^2 < 0.3$ and the intensity fluctuations no longer increase indefinitely as the turbulence strengthens.

8.3.9 Additional Sources of Laser Beam Aberrations

There are other laser beam aberration sources specific to the laser system applications. For a high-energy laser system, thermal blooming caused by the heating of air can degrade the laser beam propagation performance. Assuming that the absorption coefficient is constant along the beam propagation path, Strehl for thermal blooming is given as

$$SR_{BLO} = \frac{1}{K' N_B^m} \tag{8.59}$$

where $K' = 0.0625$ and $m = 2$ for an infinite Gaussian beam and $K' = 0.01$ and $m = 1,2$ for a top-hat beam. N_B denotes the thermal blooming number defined as

$$N_B = \frac{P\alpha}{\kappa\lambda k} \tag{8.60}$$

where P is a laser power, α is an absorption coefficient of the medium, κ is a thermal conductivity, and $k = 2\pi/\lambda$ is a wavenumber.

In addition to the previously mentioned aberrations, there are application-specific aberrations – such as aero-optic aberrations – caused by rapid airflow near the aperture of high-speed platforms, such as aircraft-mounted telescopes. The impact of aero-optic aberration is strongly influenced by the airflow patterns and the geometry of the system. When a simpler representation of the aberration effects is not feasible, the propagation of laser beams through turbulent airflow and intense turbulence conditions is typically modeled using computer simulations of wave optics. These simulations are employed to accurately determine the electromagnetic field at various distances.

8.4 Laser Beam Control Components

This section provides an overview of the key components used in laser beam control systems. It specifically addresses beam control elements that have not been previously covered, including the beam director, fast steering mirrors (FSM), optical inertial reference unit (OIRU), and light sensors.

8.4.1 Beam Director

A beam director refers to a system or a device that is responsible for the large angle steering and pointing of a beam to a specific area or a target object. It ensures that the beam stays focused on a particular point, adjusts the angular direction, and controls the overall path of the beam. A beam director also defines the field of regard of the beam control system, which is the total area that can be observed and targeted at any given time. The major component of a beam director is a telescope and a gimbal system, which are configured to provide desired beam control characteristics depending on applications. In general, the beam directors are configured as a gimbaled telescope as in Figure 8.14(a–c) and gimbaled flat mirrors as in Figure 8.14(d–f).

The azimuth-elevation (AZ-EL) gimbaled telescope shown in Figure 8.14a is the most common beam director system that can provide good hemispherical coverage. The AZ-EL gimbaled telescope often employs a Coude path such that the optical beam transits from one gimbal to another along the trunnion axes to eliminate the beam walk during gimbal rotations.

Gimbaled flats such as heliostats and coelostats are good for scanning applications since they do not suffer from gimbal lock and do not require a separate Coude path for beam steering. However, the turning flats employed in beam director need to be sized bigger than clear aperture,

$$D_F = \frac{1}{\cos\theta_i} D_A \tag{8.61}$$

where D_F is the diameter of the turning flat mirror, D_A is the diameter of a clear aperture, and θ_i is the incident angle of the beam.

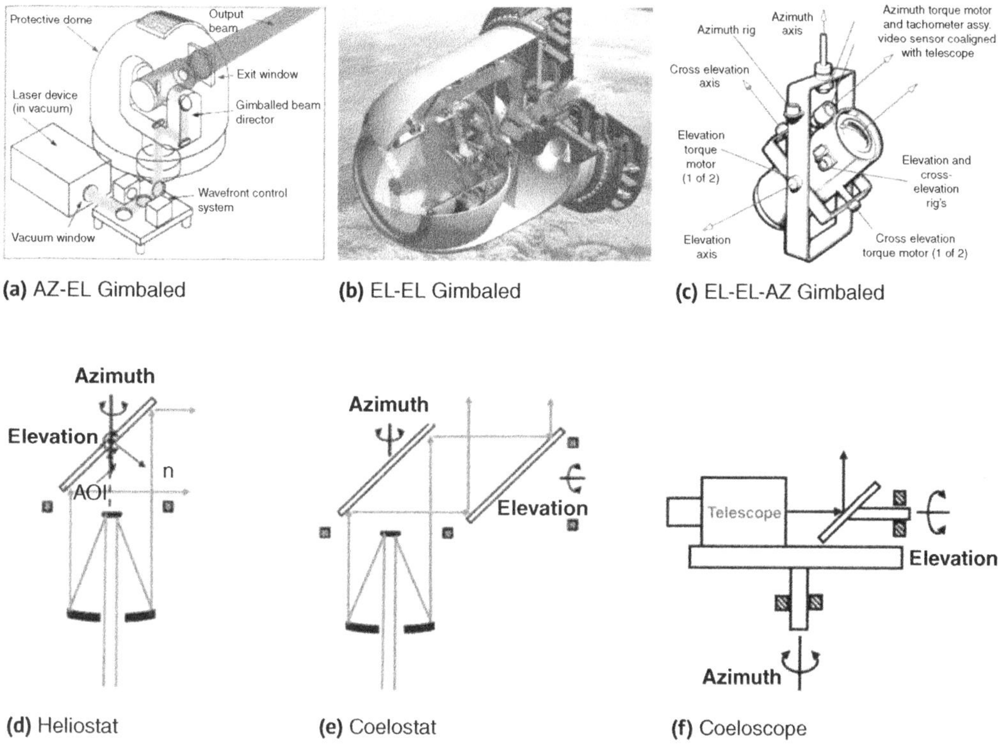

Figure 8.14 Beam director types based on gimbal configurations.

In laser beam control, the telescope in the beam director can be used for both transmitting and receiving optical beams. As a transmitter, the telescope of the beam director expands the beam to reduce diffraction effects and to control the beam shape up to the diffraction limit. As a receiver, the telescope in the beam director provides a bigger bucket to collect more light and increase the resolution of the image of a target up to the diffraction limit. The telescope also allows for the optical beam to be steered within its field of view without moving the gimbals for precision tracking and pointing. Telescope design is typically a trade-off process between desired performances and constraints. Table 8.1 summarizes typical telescope design considerations for a beam director used in laser systems.

8.4.2 Fast Steering Mirrors

FSMs are the backbone of precision beam control functions such as laser beam alignment, LOS stabilization, target tracking, and fine beam pointing. Basics of the fast-steering mirrors are discussed in Section 6.4.1.

In laser beam control systems, FSM are designed with varying aperture sizes, tip/tilt ranges, and operating bandwidths to meet specific beam control requirements. For instance, Figure 8.15 illustrates two types of FSM used in the airborne laser system. One of the FSM have a tilt range of ± 1 mrad, which is responsible for fine beam steering and LOS stabilization. The other mirror, referred to as the beam walk mirror (BWM), has an angular range of ± 8 mrad. In this example, the BWM is used in pairs to relay the laser beam over relatively large distances onboard the platform,

Table 8.1 Telescope design considerations for laser systems.

Design considerations	Impact
Telescope design	The overall design of a telescope such as telescope type, aperture size, refractive, reflective, on-axis, off-axis, focal, afocal, focal length, f number, and number of surfaces affects overall size, weight, optical performance, and cost
Field-of-view	Larger field of view provides larger angular range that the optical beam can be steered without moving gimbals, but it is difficult to maintain optical performances across the entire field of view
Focus control	Focus adjustment integration into the telescope desirable for a focal telescope that simplifies the optical beam control system design
Magnification	Ratio between the expanded beam diameter at the clear aperture and the small beam diameter at the other end of the telescope. The tilt angle of the small beam is related to the tilt angle of the expanded beam by the magnification
Optical coating and reflectivity	Optical coating and reflectivity need to support high transmissivity of the optical beam for designated wavelengths
Reflected wavefront error, surface roughness, scratch/dig	Trade-off between optical performance and the manufacturing complexity and cost
Resonant frequency	Design mechanical resonant frequency of the telescope structures to avoid structure–control interaction limiting the gimbal control bandwidth
Size, weight, inertia	Size, weight, and inertia of the telescope directly affect the gimbal control system performance

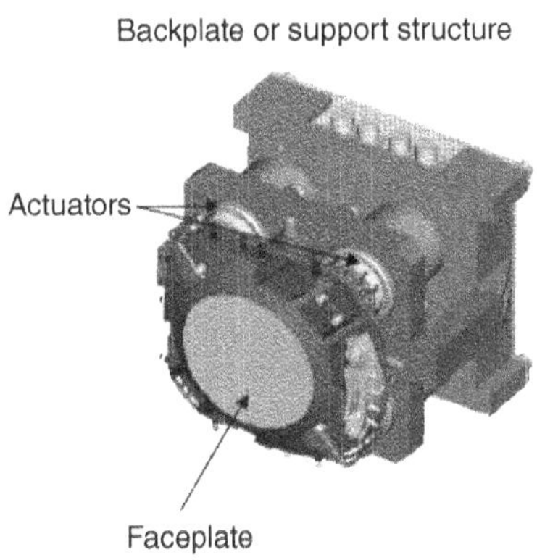

	FSM	BWM	
Weight	<200	<200	lb
Tilt range	1	8	+/– mrad
Angular acceleration	2000–6000	1000–4000	rad/s^2
Quiescent jitter	<0.001	<0.001	mrad
Caged bandwidth	800	200	Hz
Gain margin	10–15	10–15	db
Phase margin	45–55	45–55	degree
Residual reaction	1	1	%
Power consumption	100–150	100–150	w

Figure 8.15 Fast steering mirror and beam walk mirror used for airborne laser system.

providing both tilt and translation. This setup allows for angular motion of the beam while preventing it from walking off the optical path.

For low-power laser systems and imaging applications, pupil relay optics can be employed to position a single FSM in the re-imaged pupil plane, thereby avoiding beam walk-off. However, pupil relay is not feasible for a high-energy laser beam, as focusing a high-energy laser beam along the beam path will present significant laser safety concerns.

8.4.3 Optical Inertial Reference Unit

Compensation of optical jitter is a critical task in many laser systems, as it significantly enhances overall laser system performance. An OIRU is a device designed to provide a stable reference beam, which is essential for measuring and correcting optical jitter for laser systems with significant platform vibrations. In many laser beam control applications, a reliable reference source – such as a bright natural guide star or reference beacon – may not always be available. Therefore, OIRUs are often referred to as pseudo-star references. The inertial pseudo-star reference unit, as illustrated in Figure 8.16, offers a base rejection of over 100 dB and achieves an RMS jitter of the reference beam in the sub-microradian range.

OIRUs are divided into platform-based and strapdown-based units. Platform based unit shown in Figure 8.17a employs a reference laser source mounted on an actively controlled stabilization platform using a feedback control system using sensors and actuators. Strapdown-type unit (Figure 8.17b) employs a beam steering mirror that stabilizes the reference laser using open-loop control system.

Since platform-type units have an advantage of using a closed-loop control system, it will be more robust against disturbances and uncertainty in the system parameters. The platform-based OIRU also provides filtering of high frequency optical jitter from the mechanical flexure design of the platform as shown in Figure 8.18. Strapdown-type, however, is smaller, simpler, and cheaper than platform-type units.

To provide an optimal reference beam, OIRU are typically mounted on a telescope as shown in Figure 8.19 to create a reference beam path comparable to a natural guide star reference. The reference beam could be injected from the side or the back of the primary mirror using a retroreflector to sample all the optical surfaces that could contribute to optical jitter of the laser beam.

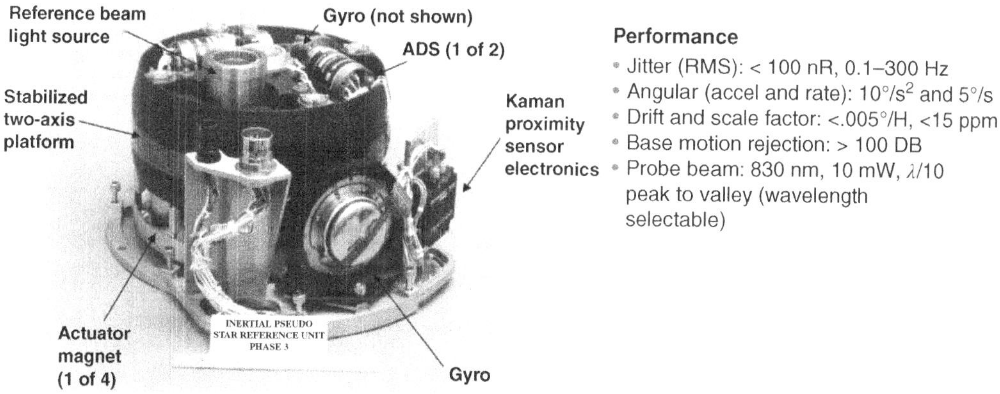

Figure 8.16 Inertial pseudo-star reference unit (IPSRU).

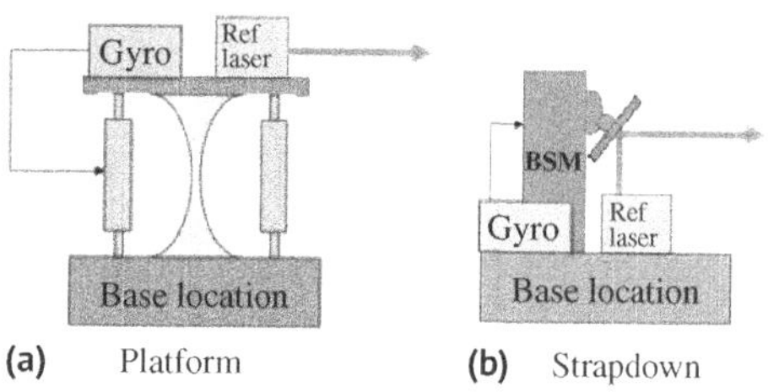

Figure 8.17 (a) Platform based OIRU and (b) strapdown OIRU. *Source:* [2].

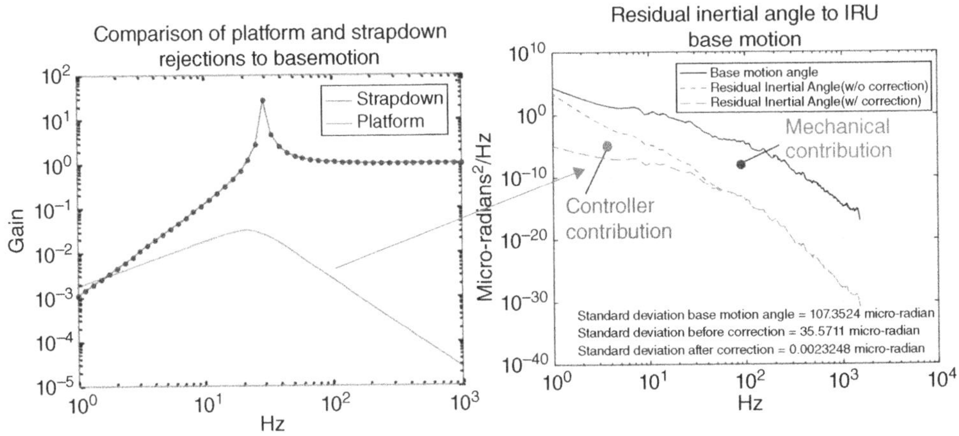

Figure 8.18 (a) Base motion rejection comparison between platform based OIRU versus strapdown OIRU, (b) platform based OIRU base motion rejection showing both controller and mechanical contributions.

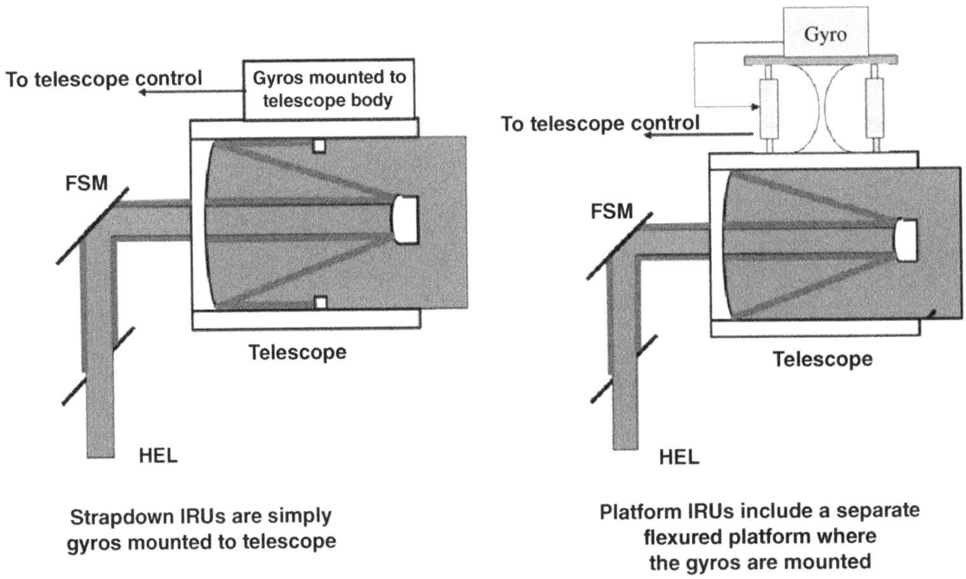

Figure 8.19 OIRU integration on the telescope.

8.4.4 Light Sensors

Light sensors detect and measure the intensity, presence, or changes in light levels. As shown in Figure 8.20, light sensors can function as a non-imaging light detector or as an imaging sensor with an array of detectors.

The materials used in light sensors differ based on the specific wavelength range of light they are designed to detect. Each material exhibits distinct responses to various wavelengths due to its unique optical and electronic properties. Figure 8.21 illustrates the range of materials utilized as light sensors across the optical wavelength spectrum.

Figure 8.20 Non-imaging light sensor and imaging light sensor.

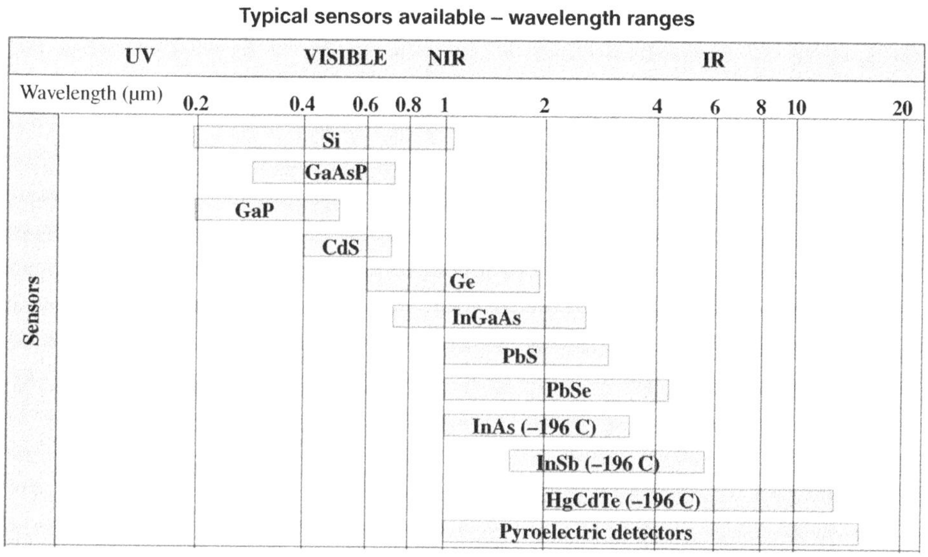

Figure 8.21 Light sensor materials for different wavelengths.

8.4.4.1 Photodiodes

Photodiodes are semiconductor devices that generate a current when exposed to light. As depicted in Figure 8.22, photodiodes are widely used in radiometric and position sensing applications such as LIDAR detection and ranging, laser rangefinders, spectroscopy, optical communication receivers, quad cells, and lateral effect sensors.

A quad cell provides a linear response to the location of the focused beam given by

$$p_X = \frac{A + C - B - D}{A + B + C + D}$$

$$p_Y = \frac{A + B - C - D}{A + B + C + D} \tag{8.62}$$

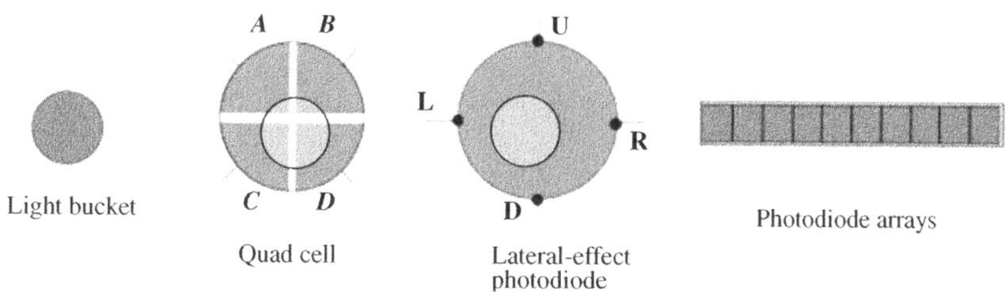

Figure 8.22 Photodiodes for laser beam control.

where p_x and p_y are the positions of the focused beam in x and y directions, and A, B, C, and D are current values of each cell. Since quad cell consists of four individual photodiode cells, there are discontinuity gaps between cells. Lateral effect sensor, on the other hand, consists of a single photodiode with multiple electrode contacts arranged on opposite sides to provide location of the focused beam, with position in x and y direction is given by

$$p_x = \frac{L - R}{L + R}$$

$$p_y = \frac{U - D}{U + D} \tag{8.63}$$

8.4.4.2 Focal Plane Arrays

A focal plane array (FPA) is a light sensor array primarily used in imaging systems, including video tracking sensors and WFSs in laser systems. However, FPAs can still be used for precise focal beam spot tracking and laser beam alignment. The FPA consists of a two-dimensional array of photosensitive elements – such as photodiodes, phototransistors, or infrared detectors – that capture light or electromagnetic radiation and convert it into an electrical signal. Common sensor technologies used in FPAs for laser systems include charge-coupled devices (CCDs), complementary metal-oxide-semiconductor (CMOS), and infrared detectors.

CCDs are used for capturing light and converting it into electrical signals, especially in digital cameras and imaging systems. They are composed of an array of semiconductor photodetectors that capture and hold the light signal temporarily before the readout to form images.

CCDs are distinguished by two illumination types, frontside illuminated CCD and backside illuminated CCD as shown in Figure 8.23. Backside illuminated usually has higher quantum efficiency. CCDs have various ways to read out and transfer the pixel data as shown in Figure 8.24.

Figure 8.23 (a) Frontside illuminated CCD and (b) backside illuminated CCD.

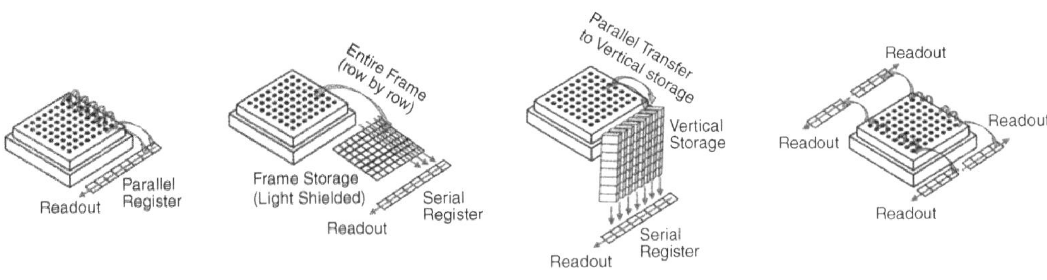

Figure 8.24 CCD image sensor readout mode, (a) parallel shifting readout, (b) frame transfer, (c) interlaced, (d) multiple readout.

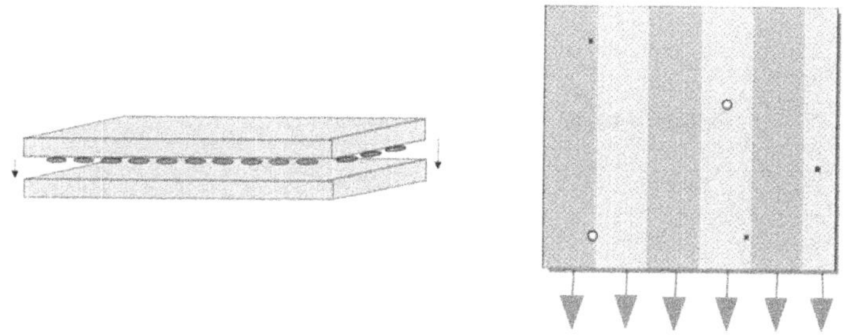

Figure 8.25 IR mosaic detector and non-uniformity due to multiple readouts.

CMOS sensors are similar to CCDs but integrate an amplifier and other circuitry (like transistors) directly at each pixel, allowing independent charge-to-voltage conversion. CMOS sensors are typically smaller, faster, more energy efficient, and less expensive than CCD sensors.

Infrared sensors detect light in the infrared spectrum, typically outside the visible range. Infrared sensors use a mosaic detector consisting of an array that is arranged in a matrix pattern. Infrared sensors often use multiple readouts, as illustrated in Figure 8.25, and require careful calibration to compensate for non-uniformity in sensor output. Infrared sensors are usually shielded from the surroundings and actively cooled to minimize the noise.

The signal-to-noise ratio (SNR) of a focal plane array is given by the following equation,

$$SNR = \frac{s_{\text{signal}}}{s_{\text{noise}}} = \frac{s_{\text{signal}}}{\sqrt{N_r^2 + N_s^2}} \tag{8.64}$$

Here, s_{signal} refers to the mean signal when calculating SNR per pixel, and to the total signal when calculating SNR over an area or time. s_{noise} represents the noise standard deviation per pixel for SNR per pixel and total noise standard deviation for SNR over area/time. Noise is further divided into readout noise N_r and shot noise N_s. When an unresolvable object is illuminated by a monochromatic light source and measured using a light sensor, a point-spread function will form in a focal plane array where the signal power can be written as

$$s_{\text{signal}} = H \cdot A \cdot T \cdot QE \cdot \Delta t \cdot \frac{\lambda}{hc} \tag{8.65}$$

where s_{signal} is the signal in photoelectrons, H is the spectral irradiance at entrance pupil, A is the collection area of the entrance pupil ($A = \pi D^2/4$ for circular aperture), T is a net system transmission, QE is a quantum efficiency, Δt is integration time, λ/hc is photons per Joule. In case the object radiant intensity is available instead of irradiance H, s_{signal} of an unresolvable object is

$$s_{\text{signal}} = \frac{I}{R^2} \cdot A \cdot T \cdot QE \cdot \Delta t \cdot \frac{\lambda}{hc} \tag{8.66}$$

where, I is the object's radiant intensity, and R is the range to the object. For a resolvable extended object, collected signal is

$$s_{\text{signal}} = L \cdot A \cdot d\Omega \cdot T \cdot QE \cdot \Delta t \cdot \frac{\lambda}{hc} \tag{8.67}$$

where L is the spectral radiance of the extended object, A is the emitting area of the object which is $(\text{FOV} \times R)^2$, $d\Omega$ is a solid angle of collection ($\pi D^2/4R^2$ for circular aperture). When a broadband light source is used, the signal equation must be integrated over the spectral band.

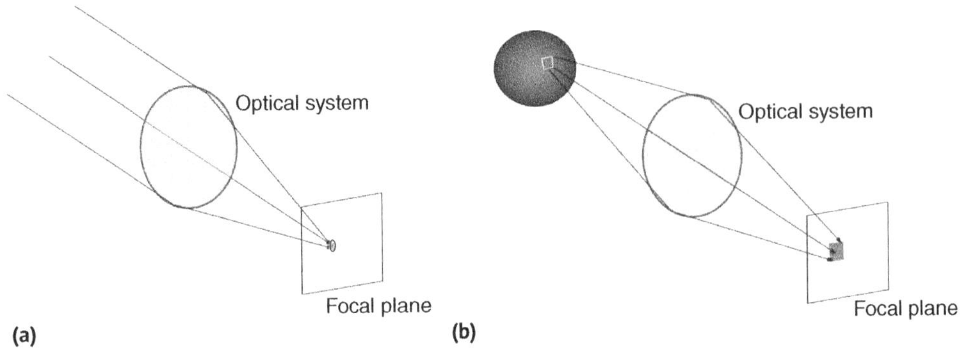

Figure 8.26 Visual object tracking, (a) unresolvable object (b) extended object.

8.5 Visual Object Tracking

Visual object tracking involves detecting and tracking a specific object across a sequence of images over time to determine its position as it moves through successive frames. This task is challenging due to various factors, including background clutter, varying illumination conditions, occlusion, changes in appearance, and differing motion patterns. In laser systems, visual object tracking is crucial for accurately determining the location of a target.

Figure 8.26 illustrates two common visual object tracking scenarios in laser systems: tracking of an unresolvable object and tracking of an extended object. For instance, optical jitter control using an OIRU reference beam is shown in Figure 8.26a, where the task can be effectively performed using quad cells and lateral effect sensors. For extended objects or those with a cluttered background that are difficult to identify and segment within a scene, focal plane arrays are more appropriate. When using a focal plane array, centroid tracking and correlation tracking algorithms are commonly used in laser systems.

8.5.1 Centroid Algorithm

The centroid tracking algorithm generally starts by segmenting an object and placing a gate around the object. The gate can be fixed or dynamic in size depending on the target object.

To provide better performance in the presence of sensor noise, centroid algorithm often employs a thresholding process shown in Figure 8.27. In the binary thresholding, the pixels above the threshold value *TH* become 1s and below 0s. Due to the binary conversion, it is less sensitive to intensity variations on target, which is more suitable for extended target object. In type I thresholding, signal above *TH* retains the value and below set to zero. In type II thresholding, *TH* value is subtracted from the signal when the signal is above *TH* and set to zero when the signal is below *TH*. The centroid of the target object within the gate after the thresholding process is given as

$$
\begin{aligned}
\varepsilon_j &= \frac{\sum_i \sum_j iS(i,j)}{I_T} = \frac{E_i}{I_T} \\
\varepsilon_j &= \frac{\sum_i \sum_j jS(i,j)}{I_T} = \frac{E_j}{I_T} \\
I_T &= \sum_i \sum_j S(i,j)
\end{aligned}
\tag{8.68}
$$

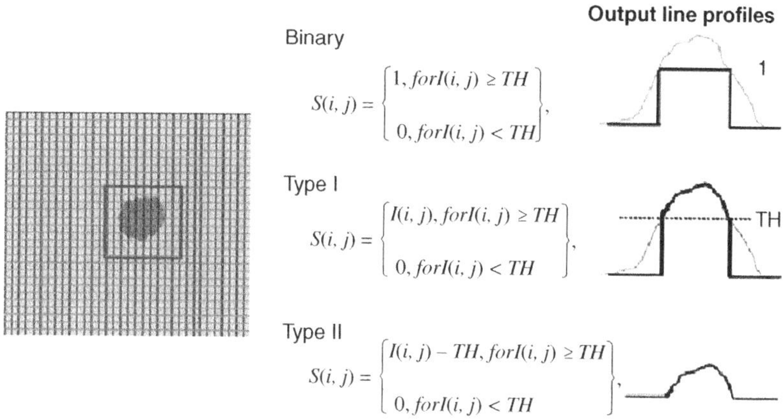

Figure 8.27 Thresholding for centroid algorithm.

Centroid tracker is better suited for a target object that can be segmented from the background and best suited for objects with parabolic intensity distribution. The choice of threshold properties depends on the problem and is essential for the best performance.

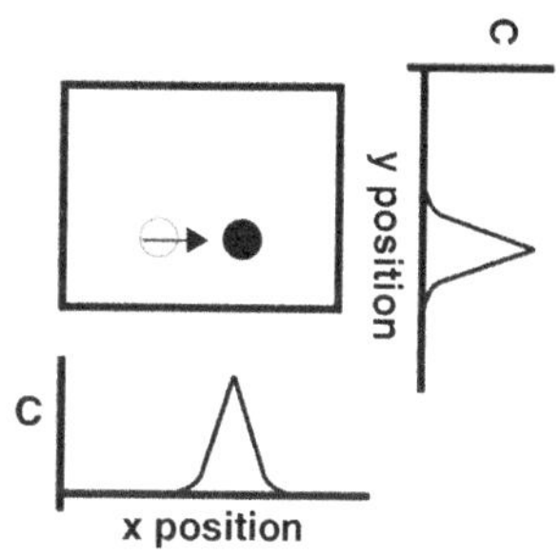

Figure 8.28 Correlation tracker concept.

8.5.2 Correlation Tracker

Applications of centroid trackers are often limited to determining the overall location of a target object. Correlation tracker is better suited for locating a target object that is hard to identify/segment within a scene. For correlation tracker, the target is detected in the image by correlating a template of the target (typically derived from the most recent scene) with the image. Fig. 8.28 illustrates how a circular object template is moved in x and y directions to compute correlations and determine the location with the highest correlation. Correlation tracker can be numerically efficient by leveraging the optimized fast Fourier transformation since convolution in time domain is multiplication in Fourier domain.

The correlation tracker algorithm is given as

$$C\left(x,y\right)=\frac{\iint S(u,v)M(u+x,v+y)dudv}{\left\{\iint[S(u,v)]^2dudv\iint[M(u,v)]^2dudv\right\}^{1/2}} \tag{8.69}$$

where C is the cross-correlation function, S is the latest scene, and M is the map reference.

8.6 Beam Control for Laser Systems

8.6.1 Acquisition, Tracking, and Pointing of Laser Systems

Acquisition, tracking, and pointing (ATP) refers to a set of integrated techniques used to detect, track, and direct the optical beam towards a specific target. ATP functions are typically divided into acquisition and coarse tracking, fine tracking and pointing, and other application specific functions. In this section, ATP functions on acquisition and coarse tracking and fine tracking and pointing functions are discussed.

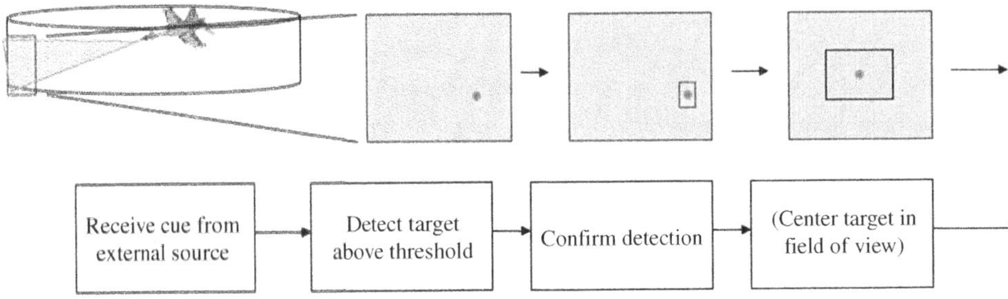

Figure 8.29 Target acquisition process.

8.6.1.1 Acquisition and Coarse Tracking

In laser beam control systems, target acquisition refers to the process of establishing the optical LOS and aligning the target at the center of the acquisition tracking sensor's field of view. During the optical acquisition of a target, the initial location is typically provided by an external source, as the detection range and field of view of the acquisition sensor are generally limited by the optical resolution required to detect distant target objects.

Figure 8.29 illustrates the target acquisition process, where the target may appear as a point source. Detection algorithms in this phase are driven by the radiometry of the target signal, background, and noise. Given that the field of view of the acquisition sensor is broader than that of the fine tracking sensor, the tracking system based on the acquisition sensor is often referred to interchangeably as an acquisition tracker, coarse tracker, or wide field-of-view (WFOV) tracker. Given the wide range of angular corrections required for coarse tracking, coarse tracking is typically controlled by gimbal actuators of the beam director.

8.6.1.2 Fine Tracking and Pointing

Fine tracking and pointing systems typically rely on a fine tracker sensor and an FSM to direct the laser beam at a precision aimpoint on a target. Fine tracker is also referred to as a narrow field-of-view (NFOV) tracker since the field of the view of the fine tracker sensor is narrower than the WFOV tracker.

Figure 8.30 shows an example laser beam control testbed at the Naval Postgraduate School. The testbed employs a WFOV track loop, an NFOV track loop, align and interface optics, and laser source are major components of the beam control system, the schematic of which is shown in Figure 8.31.

WFOV tracking involves two feedback control loops: the inner control loop for inertial stabilization and the outer control loop for tracking and pointing, as depicted in Figure 8.32. The inner loop is a rate control loop, consisting of components such as the gimbal, power amplifier, controller, rate gyros, rotational stage motors, and encoders. This loop ensures stabilization of the LOS in inertial space, compensating for external disturbances and following the input rate command generated by the WFOV tracker. The outer loop is an angular position control loop, which incorporates the WFOV camera and video tracking algorithm. It calculates the error between the LOS and the target's center, then transmits the error signal to the rate command of the inner control loop.

The NFOV tracking loop is an angular position control loop that consists of the NFOV camera, FSM, and video tracking algorithm, also shown in Figure 8.32. This tracking loop identifies errors that could not be compensated by the WFOV tracking loop and adjusts LOS to minimize the pointing error between the target and the LOS. Various optical components, such as mirrors,

Figure 8.30 NPS laser beam control testbed [3].

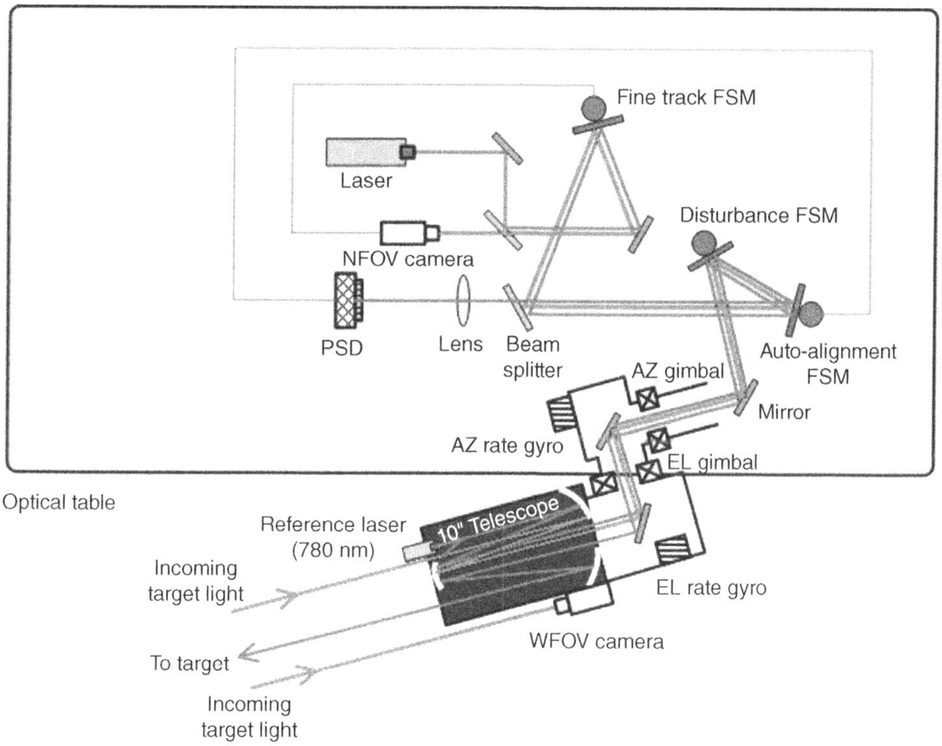

Figure 8.31 Schematic of NPS laser beam control testbed.

lenses, and beam splitters, are mounted on the optical table to establish the optical path between the laser source, target, and sensors. The auto-alignment control loop continuously detects any optical misalignment between the reference laser and the position-sensing device sensor, realigning the optical path as needed.

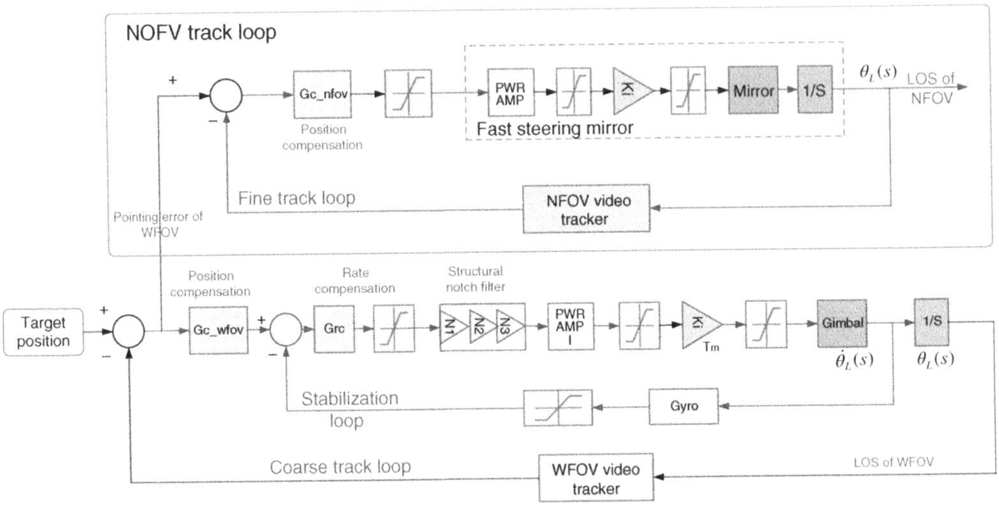

Figure 8.32 Control block diagram of WFOV and NFOV tracking and pointing control.

8.6.2 Optical Jitter Control for LOS Stabilization

In laser systems, mitigating optical jitter is a critical task typically addressed through additional beam control components and strategies. Effective optical jitter mitigation enhances the overall performance of the optical beam control system and is also a prerequisite for wavefront compensation using AO. The primary sources of optical jitter are platform jitter and atmospheric jitter. Since laser system platforms can range from stationary to highly dynamic, strategies for mitigating platform jitter may vary. For stationary platforms, jitter control may rely on a fine tracker based on target imaging. In contrast, for highly dynamic platforms, active jitter mitigation, typically using an inertial reference unit, is required. Atmospheric jitter mitigation, on the other hand, depends entirely on precise tracking of the aimpoint on the target, as laser beacons cannot be used to mitigate atmospheric jitter.

8.6.3 Adaptive Optics for Laser Systems

Adaptive optics plays an increasingly crucial role in many laser beam control systems. The fundamentals of AO are covered in Chapter 6. Adaptive optics systems for laser applications differ from imaging systems in that the laser beam is pre-compensated, based on the principle of reciprocity. A comparison of AO systems is shown in Figure 8.33.

8.7 Free-space Laser Communication System

Free-space laser communication systems for satellite-to-satellite, satellite-to-ground, and deep space communications are becoming increasingly important due to the growing data rate requirements of modern space-based sensors. Figure 8.34 illustrates optical communication experiments conducted by NASA. In October 2013, the Lunar Laser Communication Demonstration (LLCD) aboard the Lunar Atmosphere and Dust Environment Explorer (LADEE) satellite successfully demonstrated communications over a distance of more than 239 000 miles between the Moon and Earth. This demonstration achieved a data download rate of 622 Mbps and an error-free data

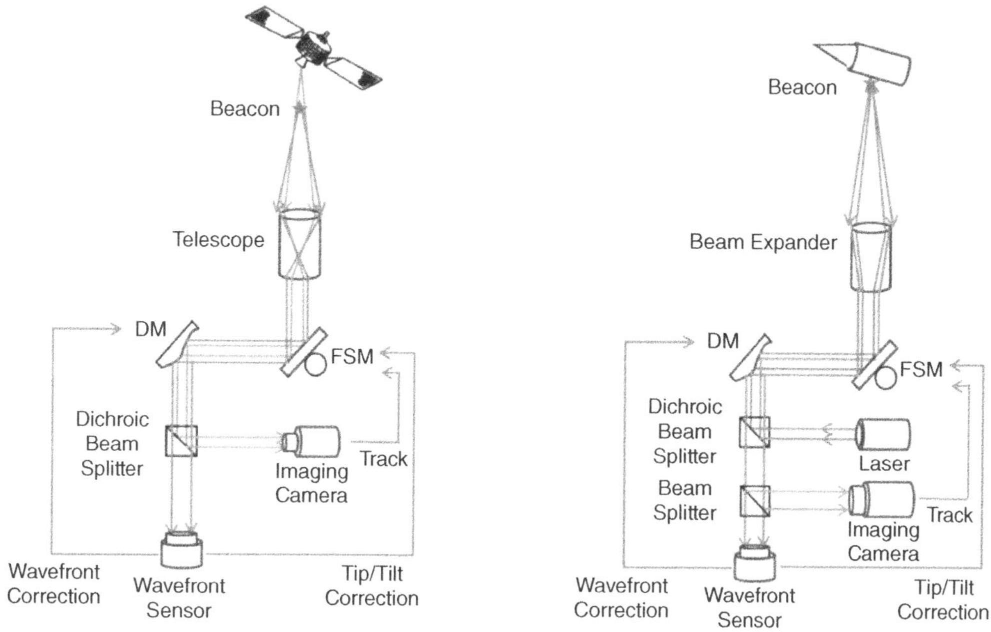

Figure 8.33 Comparison of AO system, (a) imaging system and (b) laser system.

Figure 8.34 Illustrations of NASA's Lunar Laser Communication Demonstration (LLCD) and laser communications relay demonstration (LCRD). *Source*: NASA/ http://esc.gsfc.nasa.gov/267/271.html/ last accessed on March 15, 2025.

upload rate of 20 Mbps. Additionally, the Laser Communications Relay Demonstration (LCRD), launched in December 2021, is NASA's first bi-directional, end-to-end laser communications relay, aiming a data rate of 1.244 Gbps – twice the data rate of the LLCD.

The primary advantage of using a laser beam over a microwave beam lies in the significant increase in data rate for a given aperture size and the enhanced power efficiency and security provided by the narrowband optical beam. The main disadvantage of the laser beam, compared to the microwave beam, is its susceptibility to weather conditions and atmospheric turbulence. Additionally, the effect of optical jitter is more pronounced in laser beams due to their narrow beam width.

In satellite-to-ground laser communications, as depicted in Figure 8.35, distinct differences exist between the uplink and downlink laser beams due to atmospheric turbulence, which is most concentrated near the Earth's surface. The ground-based uplink laser beam experiences significant wavefront distortion during the initial stages of propagation, followed by a turbulence-free path for the remainder of its journey. This results in considerable beam divergence at the beginning, causing a signal power disadvantage of up to 30 dB compared to the downlink laser beam. However, by the time the uplink laser beam reaches the satellite, its wavefront remains relatively flat, as the beam propagates without turbulence over the long distance.

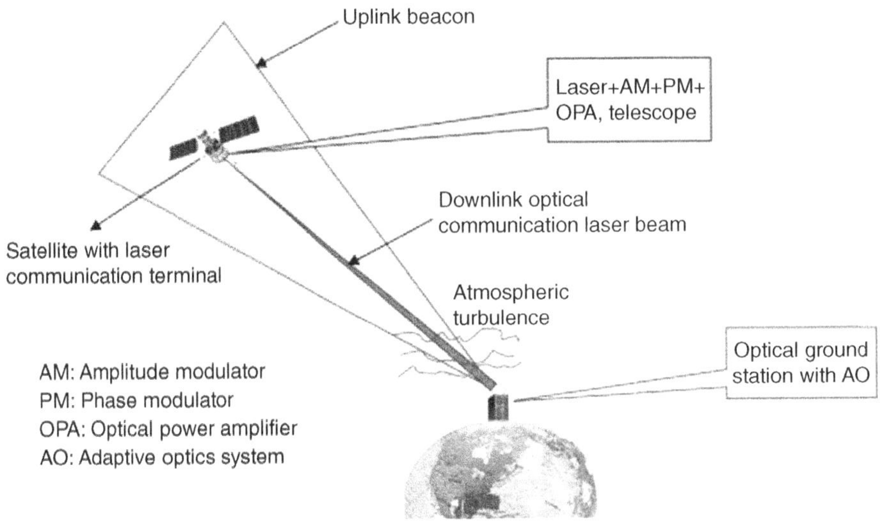

Figure 8.35 Overview of satellite-to-ground laser communication.

The downlink laser beam experiences much less initial beam divergence but exhibits significantly larger wavefront errors at the optical ground station compared to the uplink laser beam. Therefore, the use of a larger receiving telescope and an AO system can greatly enhance system performance at the optical ground station. In satellite-to-satellite laser communications, where atmospheric turbulence is absent, the optical link becomes a fully symmetric duplex connection.

8.7.1 Space Laser Communication Terminal

The laser communication terminal in the satellite payload typically consists of an optical modem, an optical module (OM), and controller electronics. The OM employs a gimballed telescope and other beam control components that are interfaced with controller electronics and optical modem to transmit and receive modulated optical beam.

To analyze the beam control system for a space laser communication terminal, the optical layout of the terminal designed for the Mars Laser Communication Demonstration (MLCD) is shown in Figure 8.36. The key beam control components include a gimbaled telescope, a MIRU, a FSM, a point head mirror (PAM), a quad cell, and a beacon/transmit beam sensor. The subsequent sections provide a detailed discussion of the laser beam control system based on this optical layout.

8.7.1.1 Telescope

The space laser communication terminal typically does not require a very large telescope aperture since the receiving laser beam is already significantly diverged due to the atmospheric turbulence concentrated near Earth's surface. The size of the telescope aperture, however, affects the diffraction-limited beam width θ_{BW} of the transmitting laser beam, which is twice the beam divergence

$$\theta_{BW} = 2\theta_D = \frac{2\lambda}{\pi w_0} \tag{8.70}$$

where θ_D is the beam divergence and w_0 is the beam waist. The actual beam width of the transmitted downlink beam is larger than the diffraction-limited beam width due to the atmospheric turbulence. For example, a space laser communication terminal with a 30 cm aperture at a

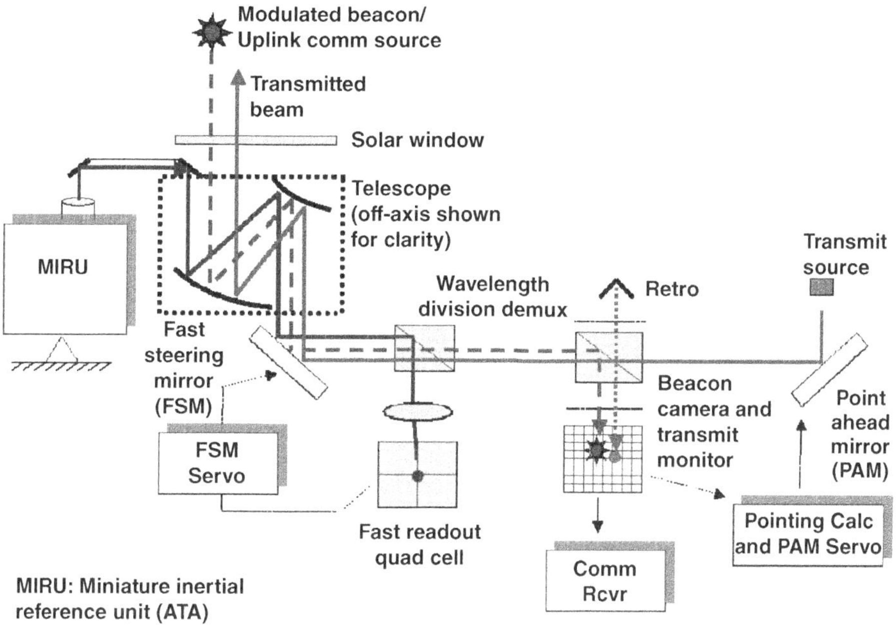

Figure 8.36 Optical layout of laser communication terminal for Mars Laser Communication Demonstration (MLCD) concept. *Source*: NASA/ https://www.nasa.gov/general/nasas-first-two-way-end-to-end-laser-communications-system/ last accessed on March 15, 2025.

wavelength of 1060 nm typically has 3–5 μrad of beam width. For the case of LCRD shown in Figure 8.36, a 10 cm reflective telescope produces a ~15 μrad downlink beam width.

8.7.1.2 Optical Beam Jitter Control for LOS Stabilization

For the space laser communication terminal, optical beam jitter control is the most significant factor in operating the laser communication system effectively. There are various disturbance sources in the satellites including low frequency thermal distortions and gross platform motion and high frequency platform vibrations The magnitude of these disturbances is typically orders of magnitude larger than the optical beam jitter requirements, and both passive isolation and active stabilization are required to reject the angular disturbances. For a space laser communications terminal, the typical RMS beam jitter requirement is less than 30% of the beam width. In Figure 8.36, a OIRU serves as a reference optical beam and the optical jitter is measured by a fast read-out quad cell and controlled by an FSM. In the case of LCRD shown in Figure 8.36, the fiber coupled optical telescope is directly mounted on the voice-coil actuated, inertially stabilized Magnetohydrodynamic Inertial Reference Unit (MIRU) platform that uses magnetohydrodynamic angular rate sensors to detect angular disturbances.

8.7.1.3 Acquisition, Tracking, and Pointing

To establish and maintain laser communication with an optical ground station (OGS), an uplink beacon from the ground station is typically used. Use of an uplink beacon is, however, limited for deep space communications where the link range is very large. In that case, other tracking references such as Earth, Sun, star tracker, and other inertial measurement and navigation sensors need to be used. In the case of LCRD, a simple quadrant detector is used as a spatial acquisition detector with a field of view of approximately 2 μrad. It is used both for detection of a scanned uplink beacon signal, and as a tracking sensor for initial pull-in of the signal.

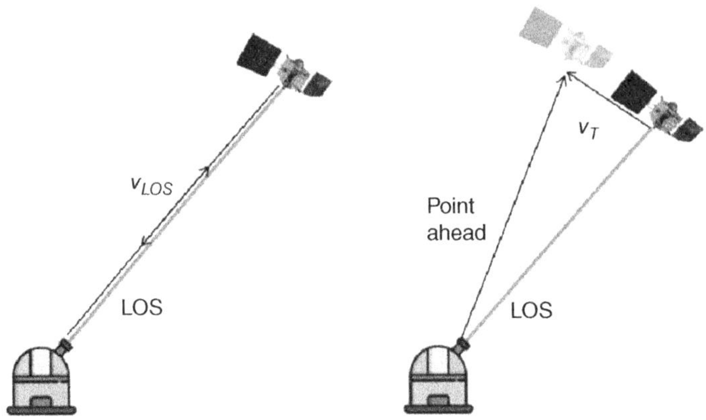

Figure 8.37 Doppler and point ahead for laser communication.

For fine tracking and pointing shown in Figure 8.36, a modulated uplink beacon from the ground station is used as a reference beam, and a single focal plane array is used for both tracking and pointing of a receiving beacon beam, as well as the alignment of the transmitting laser beam. One important consideration in pointing the transmitting laser beam is the point ahead bias. For satellites operating in the lower Earth orbit, the point-ahead bias is typically much larger than the beam divergence and needs to be pre-compensated using a steering mirror such as the point ahead mirror (PAM) shown in Figure 8.36. PAM $\Delta\theta$ is proportional to the velocity perpendicular to the line of sight v_T shown in Figure 8.37b such that

$$\Delta\theta = \frac{v_T}{c} \tag{8.71}$$

Another consideration in the laser communication terminal is the Doppler effect. Doppler is proportional to the LOS velocity between terminals v_{LOS} shown in Figure 8.37a, such that

$$\Delta f \approx \frac{v_{LOS}}{c} f \tag{8.72}$$

where Δf is the frequency shift due to Doppler effect and f is the frequency of the optical beam. The Doppler effect can be either pre-compensated at the transmitter or tracked-out with narrow-band spectral filtering at the receiver.

8.7.2 Optical Ground Stations

An OGS is a ground-based facility designed to establish and maintain an optical link with the space laser communication terminal. For the OGS, atmospheric turbulence is the most significant factor in operating the laser communication system effectively. For a downlink laser beam to ground, the use of larger aperture increases the acquired signal power; however, the effects of turbulence need to be mitigated when the aperture size is greater than the Fried's coherence length r_0. Therefore, OGS typically employs an AO system and/or multiple smaller receive apertures.

For LLCD, an array of four 40 cm receive reflective telescopes and four 15 cm refractive transceiver telescopes was used in the OGS. The OGS1 for LCRD shown in Figure 8.38a uses a 1-m aperture telescope for both receiving and transmitting a laser beam. The optical link for OGS1 is required to operate at as low as 5° solar angle, 20° elevation, Fried coherence length (r_0) of 5.2 cm, and wind speeds of 2.3 m/s. Coupling the downlink signal into the fiber interferometer under these atmospheric conditions requires an AO system to correct wavefront aberrations

Figure 8.38 (a) LCRD optical ground station 1. *Source*: NASA/Public Domain. (b) LCRD optical ground station 2. *Source*: NASA/[7]/last accessed on March 15, 2025.

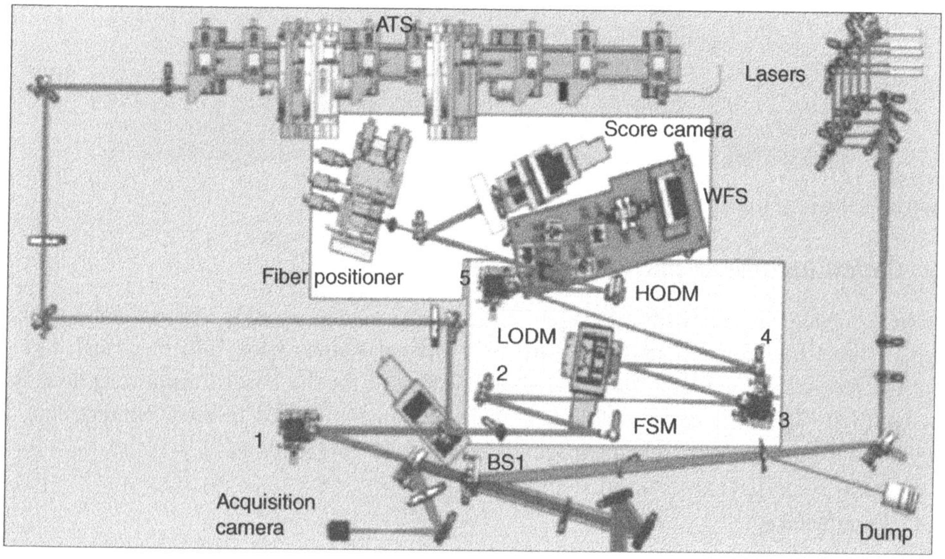

Figure 8.39 The IOS opto-mechanical layout. *Source*: [8].

caused by atmospheric turbulence. The OGS1 for the LCRD also employs communication and beacon lasers, receiver and transmitter modems, the monitor and control software, and a laser safety system.

To explore the beam control system of an OGS, the optical layout of the Integrated Optical System (IOS) designed for the LCRD is shown as an example in Figure 8.39. The portion of the downlink beam (~1%) from the telescope is acquired by an InGaAs NIR acquisition camera that has a 280 μrad field of view, which is large enough to acquire a laser beam. The acquired downlink beam further forms pupils on each of three subsequent active mirrors: a FSM that compensates for the bulk tip/tilt motion of the beam, the high-order deformable mirror that corrects for the high spatial frequency aberrations, and the low-order deformable mirror that corrects for the lower order spatial frequencies as well as residual tip/tilt error. The portion of the downlink beam (~20%) is used by a wavefront sensor for AO corrections and another portion of the light (~10%) is sent to the scoring camera to measure the quality of the post-AO corrected beam. A six-axis fiber positioner holds the receiver fiber and compensates for slow thermal or mechanical drifts. The transmit laser beam originates from fiber laser sources is injected using a dichroic beam splitter (BS1). For diagnostic testing of an AO system, an optically aberrated beam from the atmospheric turbulence simulator can be injected into the AO system instead of the received downlink beam.

Figure 8.40 Photo of the flight modem 1 for LCRD.

8.7.3 Optical Transmitter and Receiver Modem

An optical modem is responsible for converting electrical data signals into optical signals for free-space laser transmission and for converting optical signals back into electrical data signals at the receiver. Figure 8.40 illustrates Flight Modem 1 for the Laser Communications Relay Demonstration (LCRD), which integrates both laser transmitter and receiver components. Key components of the optical modem are discussed in the following sections.

8.7.3.1 Laser Source

The laser source is chosen to meet specific requirements such as wavelength, power output, beam quality, and efficiency to ensure efficient data transmission through the atmosphere. For the LCRD, a low-power (<10 mW) fiber-coupled laser diode is used.

8.7.3.2 Coding and Modulation

Coding is used to improve the reliability and efficiency of the communication link by adding redundancy and protecting against errors that might occur during laser beam transmission. The LLCD uses, in both the uplink and downlink, an error-correction code paired with a long (up to 1 second) data interleaver, which acts to mitigate the turbulence-induced signal fading. This construct allows reliable error-free communications even during stressful atmospheric turbulence.

Once the data is error-protected through coding, it proceeds to the modulation stage. The LCRD modems support two modulation formats: differential phase shift keying (DPSK) and pulse-position modulation (PPM). DPSK is a phase modulation technique that encodes data based on the phase difference between consecutive symbols. This modulation scheme is particularly beneficial in environments where maintaining phase coherence is challenging, such as in turbulent atmospheric conditions. By detecting phase changes between symbols rather than relying on absolute phase, DPSK helps mitigate the effects of environmental disturbances that could otherwise degrade the signal. Due to its superior noise tolerance, DPSK is well-suited for high data rate operations, particularly in near-space applications. Additionally, it enables reliable communication even when the Sun is within the field of view.

PPM was the primary modulation technique used for the LLCD and used in LCRD. In PPM, data is encoded by varying the position of a pulse in time. The position of each pulse in a sequence corresponds to the data value being transmitted. Since PPM is highly photon efficient (few signal photons required per bit at the receiver), it is suited for deep space operations, where data rate is reduced as efficiency is increased.

8.7.3.3 Fiber Amplifier

A fiber amplifier is an optical amplifier that uses an optical fiber doped with rare-earth elements, such as erbium, ytterbium, or dysprosium, to amplify optical signals directly, making it highly efficient and useful for long-distance communication systems. Erbium-doped fiber amplifier is the most common type of fiber amplifier that is also used in LCRD. In LCRD, the fiber amplifier can produce up to 10 W of optical power but is limited by software to a maximum power of 7.3 W during operation. After accounting for transmission loss of the transmit telescope and the window in the dome, the maximum power emitted into free space outside the dome is 5.4 W.

8.7.3.4 Demodulation

Demodulation is the process of extracting the original information from the modulated signal. In optical DPSK, a Mach–Zehnder interferometer or Michelson interferometer can be used to split the incoming signal into two beams, one of which is delayed by one bit duration, and then the beams are interfered to extract the data.

8.7.3.5 Detector

The downlink receiver typically employs single-photon detector technology in conjunction with error-correction codes, enabling power-efficient reception of wideband downlink signals. Multi-element arrays of these detectors provide the required count rates to support high-speed downlink reception. Each array is generally sized to facilitate simple and efficient coupling to the multimode fiber from its telescope.

8.7.4 Optical Link Analysis

The optical link equation relates the transmitted and received optical power to predict the SNR and the bit error rate. The optical link equation includes the effect of free space propagation losses, antenna gains, and other optical aberrations discussed in Section 8.3. The simplified optical link equation is given by

$$P_R = P_T\left(\frac{\pi D_T}{\lambda}\right)^2\left(\frac{\lambda}{4\pi R}\right)^2\left(\frac{\pi D_R}{\lambda}\right)^2 \eta_T e^{-\left(\frac{4\pi\sigma}{\lambda}\right)^2} e^{-8\left(\frac{\theta}{\theta_{div}}\right)} \eta_R L_{PR} L_{MM} L_{atm} \tag{8.73}$$

where

P_R: received power; P_T: transmitted power; $\left(\frac{\pi D_T}{\lambda}\right)^2$: transmitter optical antenna gain; $\left(\frac{\lambda}{4\pi R}\right)^2$: receiver optical antenna gain; $\left(\frac{\pi D_R}{\lambda}\right)^2$: free space propagation loss; η_T: transmitter optical efficiency; $e^{-\left(\frac{4\pi\sigma}{\lambda}\right)^2}$: Strehl ratio loss; $e^{-8\left(\frac{\theta}{\theta_{div}}\right)}$: transmitter pointing error loss for Gaussian beam η_R: receiver optical efficiency; L_{PR}: receiver pointing error loss; L_{MM}: mode matching loss L_{Atm}: atmospheric loss.

8.8 High-energy Laser Systems

A high-energy laser system uses focused laser beams to deliver concentrated energy to a target, causing damage through intense heat by melting, burning, or damaging its structure. High-energy laser systems are increasingly considered for security and defense applications aiming to disable sensors and electronics or destroy missiles, drones, aircraft, space debris, and other incoming threats. The major benefits of high-energy laser systems include speed of light engagement and reduced operational cost since the only resource needed for the laser is the electric power supply. Figure 8.41 shows examples of high-energy laser system experiments recently conducted by the US Navy, including the Optical Dazzling Interdictor Navy and the High Energy Laser with Integrated Optical-dazzler and Surveillance.

8.8.1 Overview of High-energy Laser Beam Control

The engineering of a high-energy laser system requires a thorough investigation of various system components and their impact on overall performance. Table 8.2 categorizes the high-energy laser system into four key areas – lasers, beam control, propagation, and target – and summarizes the components under each category. It is evident that beam control plays a critical role as an enabling technology, closely interconnected with other aspects of the high-energy laser system.

Similar to other laser systems, high-energy laser beam control is responsible for several critical functions, including target acquisition, tracking, pointing, LOS stabilization, aimpoint selection

Figure 8.41 (a) Optical Dazzling Interdictor Navy. *Source*: USN/MC Elisha Smith/Public domain. (b) High Energy Laser with Integrated Optical-dazzler and Surveillance. *Source*: NASA/ https://www.eoportal.org/satellite-missions/stpsat6-lcrd#optical-module/ last accessed on March 15, 2025.

Table 8.2 High-energy laser system engineering.

Lasers	Beam control	Propagation	Target
Solid-state	Target type	Beam Properties	At target irradiance
Fiber	Beam Director	Diffraction	Vulnerability
Continuous wave	Acquisition	Scattering	Protection
Pulsed	Tracking	Absorption	
Wavelength	Pointing	Turbulence	
Power	Optical jitter Control	Aero-optics	
Energy	Adaptive Optics	Thermal Blooming	
Beam quality	Aimpoint selection	Scintillation	
Run time	Damage assessment		

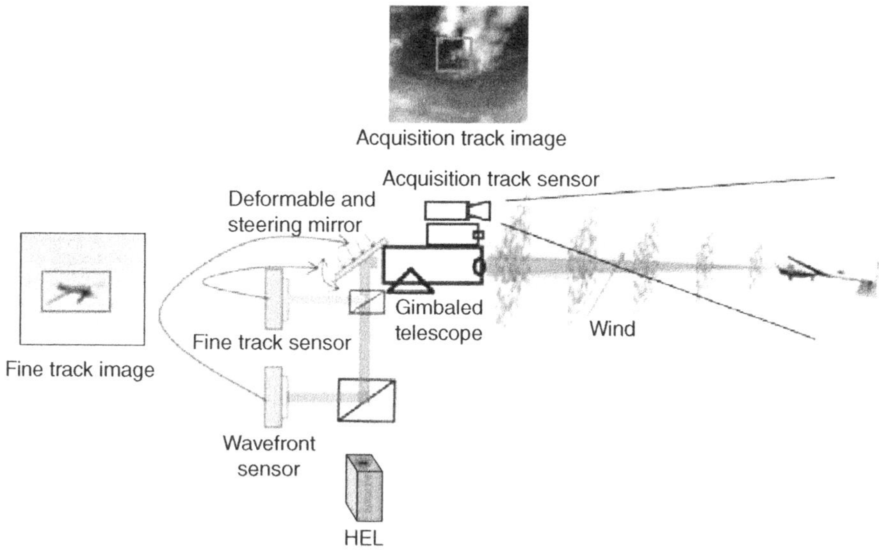

Figure 8.42 Overview of HEL beam control system. *Source*: USN/MC Elisha Smith / Public domain.

and maintenance, as well as compensation for atmospheric turbulence and other optical aberrations in the laser beam. Using the layout of the high-energy laser beam control system shown in Figure 8.42, an overview of each beam control function is provided in the following discussion.

8.8.1.1 High-energy Laser Source

High-energy laser systems can use various laser sources, such as solid-state lasers, free-electron lasers, chemical lasers, and fiber lasers. Fiber lasers are among the most widely used high-energy laser sources due to their ability to deliver high beam quality and high power in a compact form factor. Fiber lasers typically use doped fibers to generate amplified high-energy laser beams. Commercial high power fiber lasers are widely available as shown in Figure 8.43 with up to 10 kW of laser power from a single fiber.

To further scale up in laser power further, multiple fiber laser modules can be combined. Spectral beam combination (SBC) and coherent beam combination (CBC) are two different techniques used to combine multiple laser beams into a single high-power beam. Both methods aim to increase the overall power while maintaining beam quality, but they differ significantly in their principles and applications.

As shown in Figure 8.44, SBC combines laser beams with different wavelengths into a single output. Each individual beam has slightly different wavelength, and they are combined using optical elements like diffraction gratings or prisms that merge the beams spatially while keeping their spectral properties. CBC, on the other hand, combines multiple beams that are coherent, meaning they have the same wavelength, phase, and frequency. The beams are aligned in such a way that their phases are controlled using additional phase modulator for each fiber, allowing them to interfere constructively to generate a single, high-power beam. This process requires precise control of the phase of each beam to ensure that the beams remain coherent and combine effectively. It often involves phase-locking techniques, where the relative phase of the beams is synchronized.

High-power pulsed lasers are also considered for certain high-energy laser system applications. Current ultrashort pulse laser technologies include Ti-sapphire lasers and dye lasers. High output peak power usually requires chirped pulse amplification of a seed pulse from a mode-locked laser. Dealing with high optical powers also needs nonlinear optical phenomena to be considered.

	 **High-power laser cabinets** YLS series	 **High-power fiber laser racks** YLR series
Wavelength	1070 nm	1007, 1010, 1030, 1070 nm
Single-mode power	up to 10 kW	up to 3 kW
Multi-mode power	1 to 125+ kW	up to 8 kW
Multi-mode fiber diameter	50, 100, 150, 200 μm (Feed fiber output) up to 1000 μm (Process fiber)	50, 100, 200 μm
Power tunability	10–100%	10–100%
Modulation frequency	up to 10 kHz	up to 50 kHz
Cooling	Water or air	Air cooling up to 1 kW

Figure 8.43 Commercial high-power fiber laser source from IPG photonics with up to 10 kW of single-mode power output. *Source*: IPG Photonics Corporation/[10]/last accessed on March 15, 2025.

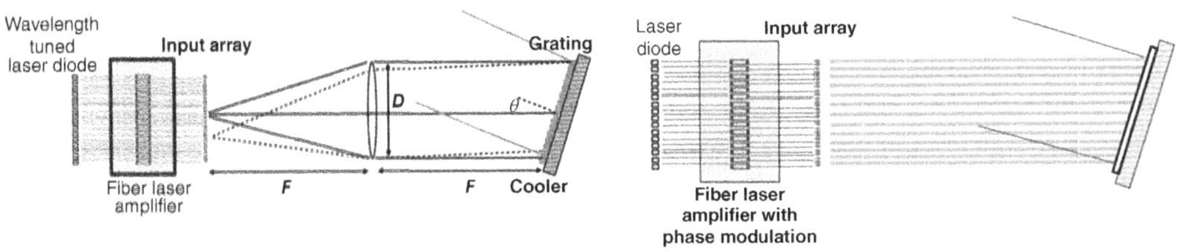

(a) Spectral beam combination (SBC) **(b)** Coherent beam combination (CBC)

Figure 8.44 Beam combination methods (a) spectral beam combination (b) coherent beam combination. *Source*: IPG Photonics Corporation/ https://www.ipgphotonics.com/products/lasers/industrial-cw-fiber-lasers/high-power-fiber-lasers#ftrresources/ last accessed on March 15, 2025

8.8.1.2 Acquisition, Tracking, and Pointing

High-energy laser systems are typically employed to engage targets under various environmental conditions, making ATP a complex challenge that often requires additional considerations. For typical high-energy laser systems, ATP operations rely on visual object tracking of an uncooperative target. Target acquisition and visual object tracking can be particularly difficult due to cluttered backgrounds, random target motion, dynamic and low-light conditions, as well as the increased atmospheric turbulence effects associated with low-elevation targets.

In Figure 8.42, a target illuminating laser is utilized to enhance the SNR of the target image, thereby improving tracking and pointing performance. While active tracking with a laser illuminator can enhance tracking performance, particularly against dim targets, its effectiveness may be limited by changes in target appearance induced by the illuminator pattern. Additionally, scintillation due to atmospheric turbulence and speckle pattern caused by the light reflections from a rough target surface can further constrain tracking performance.

Figure 8.45 NPS high-energy laser beam control research testbed.

Figure 8.45 shows a picture of the High Energy Laser Beam Control Research Testbed (HBCRT), developed for high-energy laser beam control research at NPS. Figure 8.46 depicts the operations of the acquisition tracker and fine tracker for a stationary target, showing a typical 150 μrad pointing error when using the acquisition tracker alone, and a typical 5 μrad pointing error when the fine tracker is used for precise aimpoint tracking.

8.8.1.3 Optical Jitter Control for LOS Stabilization

In high-energy laser systems, the primary sources of optical jitter are platform jitter and atmospheric jitter. Since the platforms used for high-energy laser systems range from stationary to highly dynamic, the strategies for controlling platform jitter can vary. For stationary platforms, jitter control may depend on a fine tracker based on target imaging. In contrast, for highly dynamic platforms, active jitter mitigation utilizing an inertial reference unit is necessary. Atmospheric jitter mitigation, on the other hand, relies entirely on precise tracking of the aimpoint on the target, as laser beacons cannot be used to mitigate atmospheric jitter.

8.8.1.4 Atmospheric Turbulence Compensation

Atmospheric turbulence-induced laser aberrations represent the most significant limiting factor in the operation of high-energy lasers. The laser beam typically propagates horizontally at low elevations, where atmospheric turbulence has the greatest effect. Compensating for deep atmospheric turbulence, characterized by strong scintillation and substantial spatial and temporal wavefront aberrations, remains an active area of research. Figure 8.47 shows the example AO experiments performed using the HBCRT for enhanced imaging of a scaled aircraft target.

8.8.2 Airborne Laser System

The US Air Force airborne laser (ABL), shown in Figure 8.48, was developed for missile defense, with the laser mounted on a specially modified Boeing 747 to engage and disable ballistic missiles during their boost phase. The ABL was a collaborative effort between the Air Force Research Laboratory, Boeing, TRW (now Northrop Grumman), and Lockheed Martin. The system utilizes a megawatt-class COIL, which generates the primary beam at a wavelength of 1.315 μ, with a power output of approximately 2.5 MW.

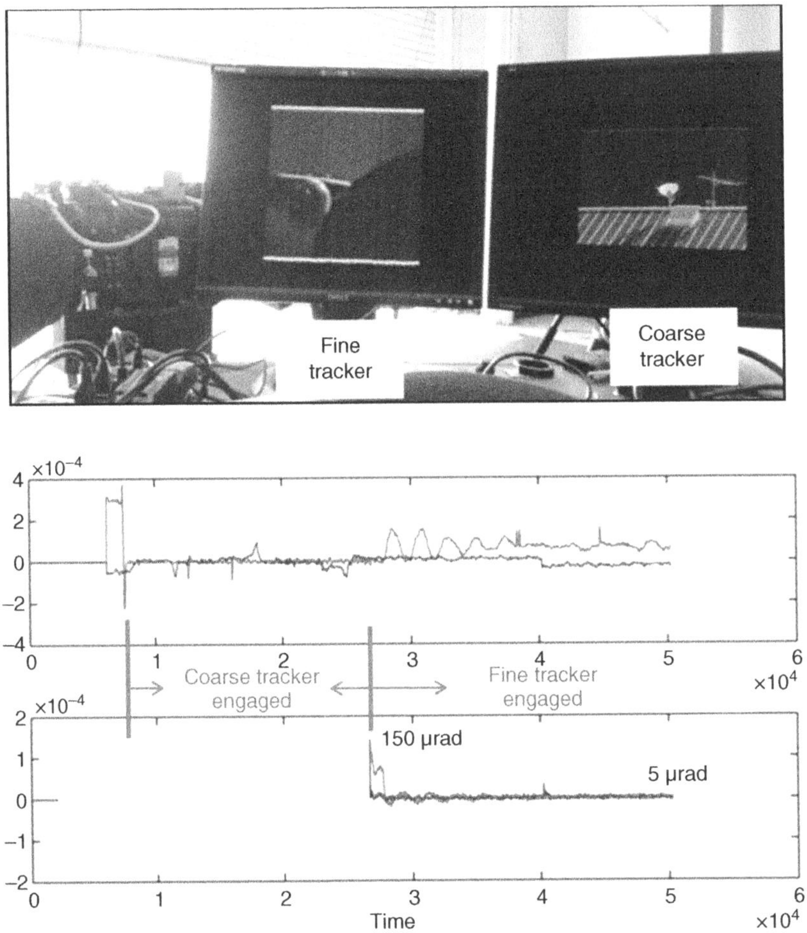

Figure 8.46 Acquisition tracker error and fine tracker error from NPS HBCRT.

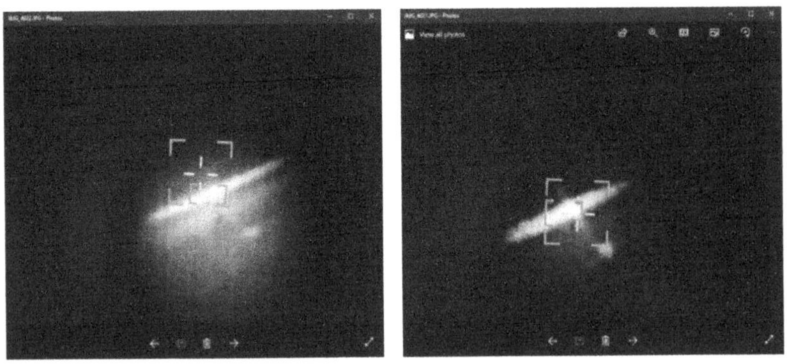

Figure 8.47 Laboratory AO testing of the HBCRT testbed demonstrates enhanced imaging of a scaled aircraft target.

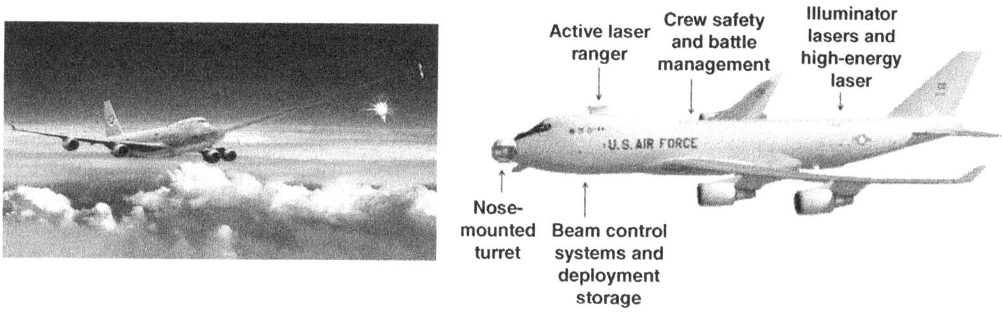

Figure 8.48 Airborne laser system.

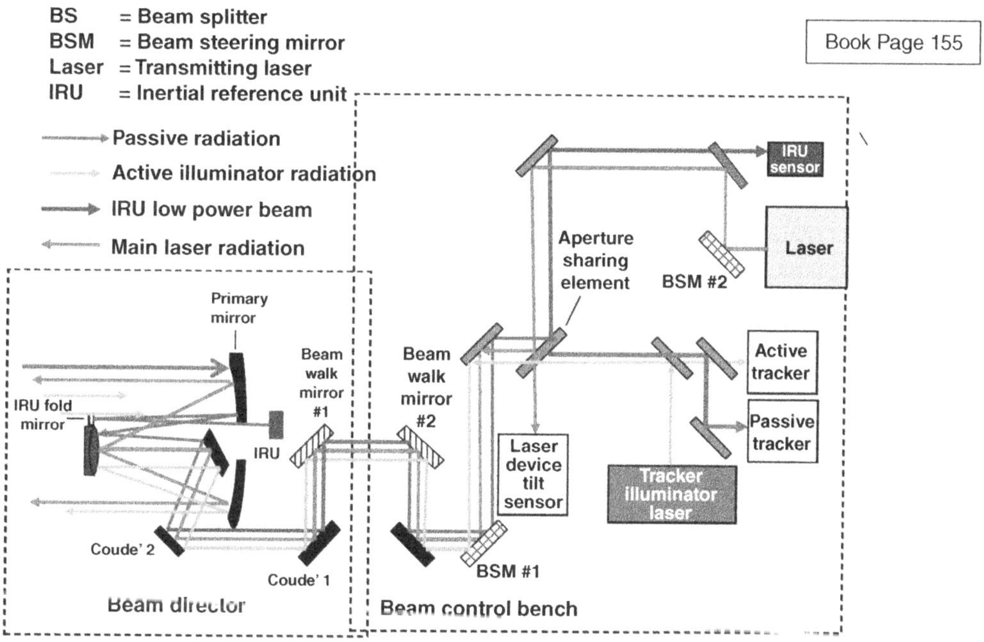

Figure 8.49 Representative optical layout of airborne laser system. *Source*: Verdict Media Limited/ https://www.airforce-technology.com/projects/abl/ /last accessed on March 15, 2025.

The ABL serves as a prime example of a complex high-energy laser beam control system, incorporating various functions essential for effective beam control. Using the representative layout of the ABL system shown in Figure 8.49, the key beam control functions are discussed in the following sections.

8.8.2.1 Acquisition, Tracking, and Pointing of ABL System

Acquisition of a target is achieved by pointing the beam director to a commanded location provided by the acquisition sensor, such as a wide-angle radar. Passive wide field of view trackers establish the coarse tracking of a target. A tracker illuminating laser, shown in Figure 8.49, is turned on and irradiates the target. An active tracker with a smaller field of view performs the fine tracking of a target and adjusts the beam steering mirror (BSM1) to the selected aimpoint.

8.8.2.2 LOS Stabilization of ABL System

Alignment and optical jitter control of the optical train for LOS stabilization need to be continuously operated before and during ATP system operations. For ABL, an IRU mounted on the back of the telescope primary mirror is used as a reference beam for both alignment of the optical train and optical beam jitter control. The system also employs a laser device tilt sensor to measure the optical jitter of the laser source and other non-common beam path errors between the high-energy laser beam and the ATP system beam, which are compensated by a beam steering mirror (BSM2), in Figure 8.49.

References

1 Virendra N. Mahajan, Uniform versus Gaussian beams: a comparison of the effects of diffraction, obscuration, and aberrations, J. Opt. Soc. Am. A 3, 470–485, 1986.
2 Paul Merritt and Mark Spencer, Beam Control for Laser Systems, Second Edition, DEPS, 2018.
3 Dojong Kim, Duane Frist, Jae Jun Kim, Brij Agrawal, A HEL Testbed for High Accuracy Beam Pointing and Control, NPS Technical Report, NPS-MAE-09-001, 2009.
4 http://esc.gsfc.nasa.gov/267/271.html.
5 https://www.nasa.gov/general/nasas-first-two-way-end-to-end-laser-communications-system.
6 Don M. Boroson, Abhijit Biswas, Bernard L. Edwards, "MLCD: overview of NASA's Mars laser communications demonstration system", Proceedings Volume 5338, Free-Space Laser Communication Technologies XVI, 2004.
7 https://www.nasa.gov/missions/tech-demonstration/new-ground-station-brings-laser-communications-closer-to-reality/.
8 Keith E. Wilson and Lewis C. Roberts Jr., "Recent developments in adaptive optics for the LCRD optical ground station at Table Mountain", Proc. International Conference on Space Optical Systems and Applications (ICSOS), 2014.
9 https://www.eoportal.org/satellite-missions/stpsat6-lcrd#optical-module.
10 https://www.ipgphotonics.com/products/lasers/industrial-cw-fiber-lasers/high-power-fiber-lasers#ftrresources.
11 https://www.airforce-technology.com/projects/abl/.

Index

Optical Beam Control: Imaging Satellites and Laser Systems, First Edition.
Brij N. Agrawal, Jae Jun Kim and Sachin Agrawal.

Printed and bound by CPI Group (UK) Ltd, Croydon, CR0 4YY
17/09/2026
14964257-0002